OUR BATTLE FOR THE HUMAN SPIRIT

Scientific Knowing, Technical Doing, and Daily Living

Contemporary society has become permeated by scientific and technological modes of thinking and doing that have had an enormous impact on our lives and our relationships. Yet – despite the seemingly endless possibilities of technical knowing and doing – this development, which began to gain momentum in the closing decades of the twentieth century, has led directly or indirectly to a raft of environmental, social, and economic ills such as global warming, the rise in social inequality, and the erosion of community life.

This book, the fifth volume in Willem H. Vanderburg's ambitious study of the relationship between technique and culture, explores our secular myths and the way we live the contradictions in what is increasingly a global civilization. Vanderburg argues that our lives have become desymbolized and impoverished by our reliance on discipline-based approaches and our disregard of the interrelatedness of all human activity within the biosphere. In successive chapters, he systematically analyses how scientific knowing and technical doing are reduced to cults of fact and efficiency, how economic growth gives rise to the "anti-economy," and how disembodied communal and personal life gives rise to the "anti-society" and epidemics of anxiety and depression. On a scale that is at the same time boldly sweeping and intimately personal, *The Battle for the Human Spirit* examines the consequences of these developments and goes on to show the possibility and necessity of reordering our thinking to achieve a more liveable and sustainable world.

WILLEM H. VANDERBURG is the founding director of the Centre for Technology and Social Development and is now professor emeritus at the University of Toronto.

WILLEM H. VANDERBURG

Our Battle for the Human Spirit

Scientific Knowing, Technical Doing, and Daily Living

UNIVERSITY OF TORONTO PRESS
Toronto Buffalo London

Toronto Buffalo London
www.utppublishing.com
Printed in the U.S.A.

ISBN 978-1-4875-0047-4 (cloth) ISBN 978-1-4875-2035-9 (paper)

Printed on acid-free, 100% post-consumer recycled paper with vegetable-based inks.

Library and Archives Canada Cataloguing in Publication

Vanderburg, Willem H., author
Our battle for the human spirit : scientific knowing, technical doing, and daily living / Willem H. Vanderburg.

Includes bibliographical references and index.
ISBN 978-1-4875-0047-4 (cloth). – ISBN 978-1-4875-2035-9 (paper)

1. Science and civilization. 2. Technology and civilization.
3. Civilization, Modern – 21st century. I. Title.

CB478.V389 2016 303.48′3 C2016-901494-0

This book has been published with the help of a grant from the Federation for the Humanities and Social Sciences, through the Awards to Scholarly Publications Program, using funds provided by the Social Sciences and Humanities Research Council of Canada.

University of Toronto Press acknowledges the financial assistance to its publishing program of the Canada Council for the Arts and the Ontario Arts Council, an agency of the Government of Ontario.

Canada Council for the Arts Conseil des Arts du Canada

ONTARIO ARTS COUNCIL
CONSEIL DES ARTS DE L'ONTARIO
an Ontario government agency
un organisme du gouvernement de l'Ontario

Funded by the Government of Canada Financé par le gouvernement du Canada Canada

Contents

Preface

As members of a civilization permeated by scientific knowing and technical doing, we all live a terrible contradiction. We individually and collectively act as if this knowing and doing have no limitations. If they did, would our civilization not be engaged in a frantic search for alternatives that could take us beyond these limitations, realizing that our contemporary ways of life fundamentally depend on this scientific knowing and technical doing? Since there is no such all-out effort, the implication is either that there are no limits, or that these limits are of no consequence. In that case, we live as if the gods of the past have returned in secular garb. We are not aware of any entities in our universe, and certainly not of any human activities in all of human history, that are without limits, other than the gods every society has created for itself. At the same time, we pride ourselves on being secular, and if we admit at all to any religiosity it is deemed of a spiritual kind – although the differences between these two are somewhat unclear.

If scientific knowing is omnipotent in relation to all other human knowing, and if technical doing is all-powerful among all other doing, it follows that both this knowing and this doing would become universal in character. This has led to a global civilization during the last fifty years. Humanity has put itself into a position where "one size fits all" when it comes to knowing and doing. At the same time, we see this as entirely unrelated to an endless succession of economic, social, and environmental crises, which surely are symptoms of local inadaptation.

This work will explore the way we live the above contradictions in what is increasingly a global civilization. We will examine scientific knowing as a cult of the fact, technical doing as a cult of efficiency, economic growing as a cult of the anti-economy, anti-societies with the cult of disembodied life, and self-actualization in the face of anxiety and depression. To put it in another way: knowing, doing, economic housekeeping, living together in societies, and the living of our daily lives have become so desymbolized by our disregard of how everything is related to everything else and thus evolves and adapts in relation to everything else that we have turned them into their opposites. We can then "improve" them on their own terms with a total disregard of how they are integral to a humanity living in a biosphere. Each chapter will first examine the consequences of these developments, which began to gain momentum in the closing decades of the twentieth century. They eventually established an absolute reign of discipline-based approaches, which implies an intellectual division of labour that assigns one category of phenomena to a discipline distinct and autonomous from all the others. Following this diagnosis, we will show that a kind of resymbolization is possible in order to create a very different synergy between knowing and anti-knowing. In subsequent chapters, we will follow the same approach to doing and anti-doing, economies and anti-economies, societies and anti-societies, and life and non-life.

This volume is the fifth and perhaps the last of my study of the relationship between technique and culture, as approaches based on desymbolization and symbolization respectively. It began with the study of the *Growth of Minds and Cultures* as it occurred before industrialization began. It was followed by an examination of how we are trapped in *The Labyrinth of Technology* by means of discipline-based approaches, which make it impossible for engineers, managers, and regulators of technology to understand how technology influences human life, society, and the biosphere and (even more important) to make use of this understanding to adjust design and decision making to ensure the desired result but at the same time minimize undesired harmful effects. The statistics are mind-numbing, but they form the diagnosis on which a treatment based on preventive approaches is developed in great detail. The third volume examined in some detail how Western civilization is *Living in the Labyrinth of Technology*, thus backing us into an impossible corner characterized by the multiplication of debt, a diminishing capability of sustaining human life, and a growing impossibility that all these efforts could be sustained by the biosphere. Under the circumstances, it is easy to understand why we took the decisions and adopted the approaches we did, but when regarded from a broader perspective, it should have been evident that the results in the long term would be disastrous. Finally, all this was examined from the perspective of desymbolization to develop the beginning of a project of resymbolization, beginning with university reforms, in order to diminish and eventually end *Our War on Ourselves*. The present volume builds on the previous four in such a way that what the reader needs to know about them is provided in the Introduction. For a more comprehensive treatment of the issues raised in this Introduction, the reader is referred to the previous works as well as to more extensive entries into the literature, which will not be repeated in this volume. It should be noted that the first volume was entirely based on what today is referred to as brain plasticity, but I am far from convinced that we are anywhere near recognizing the implications of this discovery for what has been going on for nearly three centuries in Western Europe and which subsequently transformed the entire world.

In sum, there is no prerequisite the reader must have for the reading of this volume other than being a concerned member of humanity open to the argument that our current economic, social, and environmental crises are the product of the way we have structured our knowing, doing, and growing. The basic argument is that, when we wire the components of an electronic circuit to perform the function of an amplifier, there is no way of changing this function other than by reorganizing the circuit. Similarly, we cannot dissociate our incredible successes at improving the performance of everything in isolation from everything else (to which discipline-based approaches are ideally suited) from the accompanying disordering of human lives, societies, and the biosphere, which is directly related to dealing with these entities as collections of distinct phenomena. Nevertheless, this chaos is completely dissociated from the way we have "wired" our affairs during the last century, as if this "wiring" had little or nothing to do with our crises. We have been behaving as if the output of a circuit can be dissociated from its organization. We invoke all kinds of political solutions to what is actually a structural and organizational problem of our knowing, doing, growing, and living. It has trapped us into the illusions of the great secular political religions of the twentieth century: communism, national socialism, and hard-line democracy (of the kind we find in the United States and increasingly in western Canada). If my reader believes in these political illusions, there is no point in reading this book. We must be willing to re-examine each and all presuppositions in the face of the possibility that the current system could destroy human life and the planet as we know them, because we have created a system of knowing and doing that disorders the fabric of all life. I hope that therefore everyone will have the courage to take another hard look at much of what we take for granted, thanks to our secular myths that orient and guide our civilization in an ultimately unknowable universe.

Whenever feasible, any theory should be tested over time. Since these kinds of studies are rarely undertaken, I can only furnish the following evidence. A French journalist, Jean-Luc Parquet, who discovered the work of Jacques Ellul in the late twentieth century, published a book, *L'homme qui avait presque tout prévu*, whose literally translated title is "The Man Who Foresaw Almost Everything." The same kinds of sentiments were expressed by several people whom we interviewed for a five-part CBC Ideas program in 1979. I recall very well when I first read Ellul. After only a chapter and a half of *The Technological Society*, I was astounded by how well his description of technique represented the way my engineering mindset worked, and I went to France to spend four and a half years with him, beginning as a NATO post-doctoral fellow. While there, I wrote my first book, *The Growth of Minds and Cultures*. It began with an examination of the role culture played in human lives and societies prior to the emergence of technique. After he had read it, just before my return to Canada, his response was that he was fairly sure he had dealt with much of it in some of his publications. I asked him for the references, and a few days later, he reported that he had not dealt with it after all. I assured him that this was not surprising because I believed that the theory expressed in my book was very much implicit in all his work. It became the basis for my examination of technology and later of technique. This confirmed that, decades after he first published his books on technique, the theory still accurately described what was happening. Since the emergence of the internet and the social media, my only surprise has been the speed at which the interaction referred to as "technique changing people" has transformed individual and collective human life. Finally, based on the theory of technique, I was able to develop (through the Centre for Technology and Social Development at the University of Toronto) an undergraduate and graduate program in preventive approaches for the engineering and management of technology. This approach was adopted by the former Premier's Council of Ontario, which asked me to co-chair a round table examining how these approaches could be used to restructure the Ontario economy by introducing a more preventive orientation in higher education. Years later, the Natural Science and Engineering Research Council and the Social Sciences and Humanities Research Council of Canada utilized this same approach as one of the models used for the STS21 program to achieve similar objectives on the federal level. Both initiatives became the victims of Conservative governments, who preferred to rely on "natural" markets. A preventive approach to technical and economic development was also recognized as one of twenty-five recent leading Canadian innovations by the Canada Foundation of Innovation in 2002 and by the Canadian Academy of Engineering, which inducted me as a fellow in 2009. Nevertheless, these developments provide anecdotal confirmation of the ongoing relevance of the theory of technique.

I wish to thank many generations of students at the University of Toronto, who came from backgrounds as diverse as engineering, sociology, environmental studies, political science, and religious studies, and whose questions and encouragement kept me going in academic surroundings that are not very hospitable to iconoclastic challenges. They particularly made their sentiments known when they refused to have me retire. No teacher could wish for a greater compliment, and I am delighted to make my contribution to what I hope will be a more liveable and sustainable future by continuing my teaching on a part-time basis.

This work is the most ambitious interdisciplinary synthesis I have tackled thus far, but without it, it would be impossible to show how our structural problems in the organization of science, technology, the economy, society, and our personal lives are deeply intertwined from both a diagnostic and prescriptive perspective. Hence, I was very pleased with the excellent suggestions from my anonymous referees for somewhat lightening the task of the reader. I am grateful for their valuable contributions. I wish to thank my wife, Rita, for helping copy-edit the manuscript, and also Esther Vanderburg and Justin Wong for word processing my dictated text. Thanks also to Hannah Wong for preparing the index, to Judith Williams for adding the final literary touches to the manuscript, and to Doug Hildebrand and Anne Laughlin for guiding the overall process. May all our work help sow the seeds of change towards a more liveable and sustainable world.

Willem H. Vanderburg
Peterborough, 2015

OUR BATTLE FOR THE HUMAN SPIRIT

Scientific Knowing, Technical Doing, and Daily Living

Introduction: Our New Spiritual Masters

Limits to Knowing and Doing

We live as if science can know everything and as if material, social, and spiritual technologies can do everything. This conclusion can be reached by anyone who takes a good look at our civilization – there is not a single institution (including universities, government ministries, corporations, consulting firms, and non-governmental organizations) that is scrambling to find alternatives to our science and technology because it has encountered an inherent limit to the scientific approach to knowing or the technical approach to doing.

A simple example will illustrate how astonishing this is. Our science is organized by means of disciplines, each specializing in one category of phenomena. Physicists examine physical phenomena, chemists chemical phenomena, biologists biological phenomena, economists economic phenomena, sociologists social phenomena, political scientists political phenomena, and so on. This intellectual strategy of examining human life and the world one category of phenomena at a time is well suited to the study of situations where the influence of one category dominates the influences of all others so that the latter can be neglected. This strategy is also well suited for situations in which the only changes occur in the category of phenomena under investigation, with the result that the influences of all the other categories remain the same. It is clearly ill suited to those situations where multiple categories of phenomena make non-trivial contributions. The first two kinds of situations are most common in non-living entities, while the third tends to be found in living wholes. Consequently, discipline-based science will encounter significant limitations when examining human life, society, and the biosphere because it will tend to treat them as if they have the characteristics of non-living entities. For centuries we have looked for the living among the dead: first by the mechanistic world view, which depicted life in terms of classical machines, and later by the information-based world view, which depicts life in terms of computers. Both traditional and information machines are built up from separate and distinct domains. Each domain receives one or more inputs from other domains, transforms these inputs into a desired output by endlessly repeating a process based on a single category of phenomena, and transfers this desired output to another domain until the overall desired result is achieved. These machines would not work very well if their constituent domains were unable to keep at bay the categories of phenomena dominating neighbouring domains, since this would result in an interference with the one category of phenomena on which that domain depends. As a result, the "world" built up in this manner is well suited to discipline-based approaches, and it is in this context where these approaches have begun to proliferate on a historically unprecedented scale.

In contrast, in living wholes, we encounter very few domains because everything is related to everything else and everything evolves in relation to everything else. Categories of phenomena whose influences are negligible in the short term or long term are almost non-existent. As a result, the probability of any situation repeating itself or reoccurring later is extremely low, in contrast to the "world" of machines, whose domains are based on endless repetition. For nearly two millennia, the division and separation of human life into "parts" has been the goal of a great deal of research and thought, beginning with Socrates, Plato, and Aristotle searching for the rules "underneath" human skills, and then with Western philosophy pursuing the goal further. It has culminated in decades of artificial intelligence research still being unable to solve the so-called "frame" problem, that of dividing human life into the equivalent of domains (also referred to as scripts and micro-worlds).[1] Rules may be regarded as micro-domains which repetitively turn any input into an output. Algorithms can thus be built up from these micro-domains. If the organizations of human brain-minds were constituted from such domains, they could be simulated by logic, in which case they could not evolve and neural plasticity would be impossible. In sum, there is no evidence that I am aware of that human life is divisible and separable into domains. The intention is not to define human life but to differentiate it from non-human life, a distinction that is essential in the sociocultural context of our civilization.

The reasons why discipline-based approaches to knowing and doing burst on the scene on a historically unprecedented scale during the later phases of industrialization are now beginning to come into focus. What is not so clear is why we are not scrambling for alternatives, since most of the challenges faced by our civilization result from the collisions between the "world" of machines on the one hand and human life, society, and the biosphere on the other. This is the conclusion reached in an earlier study.[2] Even if we found exceptions or were able to come up with counter-arguments to this finding, the bottom line is that we live as if discipline-based science has no limits we need to be concerned about, with the result that all our institutions resist any other approaches to knowing. This puts our civilization in a situation analogous to that of a contractor coming to our home to make some repairs but bringing only a hammer because he is so lacking in experience that he has never encountered a situation in which the limitations of hammers have become obvious. Is this not the equivalent of entrusting all our knowing to discipline-based science? Could this be at the heart of our economic, social, and environmental crises? In this respect, are we any different from earlier civilizations that treated any entity in their experience which appeared to have no limits as a god? Is our attitude to science really as rational and secular as we think?

From the above argument, it follows that our discipline-based approaches to doing take on the characteristics of technologies, and thus have the same limitations. The organization of our universities, including their professional faculties, suggests that our civilization has not yet encountered, or fails to recognize, any significant limitations to our discipline-based approaches to doing. Nevertheless, as a detailed study has shown, there are many such limitations, with far-reaching consequences for our civilization.[3] Activities such as engineering design, business management, and medical practice depend on discipline-based knowledge, but they cannot be reduced to its application. By proceeding as if this is more or less possible, our civilization is denying their status as arts that cannot be built up from domains the way applied sciences can.[4] Any art fundamentally depends on experience.

It may be argued that, just as physics is the model for discipline-based science, engineering disciplines such as operations research, fluid mechanics, and stress analysis are models for technical doing. The discipline-based approaches to doing have taken on all the characteristics of technologies regardless of their areas of application, thereby constituting the phenomenon that Max Weber[5] described as rationality around the beginning of the twentieth century and which Jacques Ellul[6] referred to as technique some fifty years later. Simply put: technique attempts to improve human life and the world as if they were built up from separate and distinct domains dominated by a single category of phenomena.

The issues raised above have their counterpart in all human experience. The "world" we see has all of the characteristics of the "world" of classical and information machines.[7] Anything we see exists in its own space separate from everything else. When we focus on a particular entity in our field of vision, it can be only one thing, thus excluding the possibility of its being inseparable from something else without which it cannot be understood. As a result, this "world" is built up from separate and distinct entities according to the principles of non-contradiction and separability. To this we can add the principle that each entity can be defined in terms of its own characteristics, thus making closed definitions possible. The principles of non-contradiction, separability, and closed definitions imply that everything within this "world" is open to measurement and quantification by itself, independent from everything else. Doing so opens the door to mathematical representation and applied logic. Such mathematical representations will be logically consistent internally and also consistent externally with what they represent, with the result that these representations are of an applied, as opposed to a pure, character. Finally, because this "world" apprehended by seeing conforms to the organizational principles of non-contradiction and separability, it can be built up one constituent at a time. Each additional constituent will not affect any of the previously added ones. It is a domain-like world, which we will refer to as *reality*, to distinguish it from the kinds of worlds in which individual and collective human life was lived prior to industrialization.

The Other World

In contrast, the world that babies and children enter into by symbolizing their experiences, acquiring the language of their community, and adopting its way of life and culture is a world with diametrically opposite characteristics. These are the cause and effect of their being members of a symbolic species. Their lives in the world begin with their development as embryos, which creates everything required for a human body by means of the differentiation of stem cells into all the kinds of different cells necessary for building every organ and tissue. It is as if the "parts" are grown within the developing body, which is made possible by each stem cell containing the entire "biological blueprint" in the form of the DNA. As a result, every cell in the body – a brain cell, liver cell, heart cell, etc. – is a particular expression of the whole. This evolving internal differentiation of stem cells through which the bodies of pre-born babies begin to take shape and grow is grafted onto all prior biological developments summed up in the DNA. Somewhere in this process, humanity became a symbolic species – a process that oriented the development of the brain in a unique direction, thus enabling us to relate anything detected by the senses and nerves to anything detected in the body at the same time, as well as to anything lived previously.[8] Whatever is thus detected is symbolically represented by neural and synaptic changes to the organization of the brain. In this way, the brain as an organ of the body is gradually transformed into a brain-mind unique to a symbolic species. Because of this orientation, a symbolic differentiation of experience is grafted onto the biological differentiation of cells that produces the body. Jointly, they sustain human lives in the world.

The complexities of the brains and symbolic worlds of babies and children are similarly built up by progressive internal differentiation on the symbolic level. As noted, every cell in a human body is a local expression of the biological whole – a kind of living analogue of how each point in a hologram enfolds the whole. As a result, each cell is both internally and externally connected to all the others: internally by means of the DNA, and externally by the obvious ways. It is the diametrically opposite architecture to those of classical and information machines. Although we still understand very little about the symbolic functions of our brain-minds, a great deal of human life and behaviour can be explained as resulting from a similar kind of progressive differentiation and integration – that of the experiences of babies and children as cause and effect of a parallel differentiation of the organizations of their brain-minds. They gradually learn to live symbolic lives in which each experience enfolds that life, thus becoming a moment of their lives and the life of their community. It will gradually become evident that it is the culture of this community that now supplies the equivalent of the DNA and makes this symbolic differentiation possible.

We will be building on an earlier comprehensive study of the role of culture in individual and collective human life whose general orientation may be summed up as follows.[9] *Homo sapiens* first lived in groups and later in societies, all characterized by unique symbolic cultures that are distinguished from the cultures of animals. These symbolic cultures are learned by the members of each new generation, thus making them relatively distinct from what is genetically transmitted. Such cultures depend on what has become known as brain plasticity: each and every lived experience of *homo sapiens* modifies the organizations of their brain-minds by means of neural and synaptic changes that symbolize these experiences, thus permitting them to be remembered and lived as moments of the people's lives. This symbolization of all experiences suspends the members of a community sharing a language into a symbolic universe that makes life liveable in an ultimately unknowable world by generally preventing relativism, nihilism, and anomie. Consequently, these symbolic universes depend on collective beliefs that are ultimately sustained by what (in cultural anthropology, archaeology, the sociology of religion, and depth psychology) have been referred to as myths. For example, when people lose their trust in any of society's entities (such as a currency, a bank, a corporation, or a government), it will no longer be able to function and could provoke a crisis in a symbolic universe. Gods can cease to exist in our universe, laws may become inapplicable, and moralities can be discarded. There is little, if anything, that ultimately does not depend on implicit or explicit beliefs, in turn sustained by myths. The rise and fall of a civilization is ultimately a question of the viability of the symbolic cultures of its member societies.

Symbolization involves three somewhat distinct and yet interdependent levels, which we will refer to as symbolization, experience, and culture. Symbolization begins by what we can detect from our surroundings by means of our senses, and from our bodies and minds by means of our nervous systems. The five senses correspond to five dimensions of experience of the external world, and these are complemented by those of our "inner worlds." This first level of symbolization discriminates between what can be lived by taking on a meaning for our lives relative to everything else and what remains "noise" or random stimuli as yet devoid of any possible significance. Something can be internalized as part of an experience only if that experience can be related to previously symbolized experiences, thereby taking its place relative to them in our lives. Only then are we able to live it. The lives of babies and children are thus protected from meaninglessness. They detect only what they are able to symbolically give a place in their lives, by relating it to everything else through subsequent levels of symbolization.

Experience, the second level of symbolization, becomes possible by the way the organizations of our brain-minds differentiate all experiences from one another and integrate them into our lives. The organizations of our brain-minds thus symbolize the body of experiences that constitute our lives. Despite current beliefs, there is not a shred of evidence to suggest that we are able to live two things at the same time as a kind of multitasking. Everything that is detected as potentially having a meaning for our lives will contribute towards an experience. Whatever we direct our attention to or whatever attracts it will be lived as the foreground, and everything else will be lived as the background. A new experience will be directly differentiated from the previously symbolized ones that most resemble it, to symbolically take its place as a unique and distinct moment of our lives. Each new experience confirms all previously symbolized ones in either a cumulative or a non-cumulative manner, thus evolving our lives as sustained by the organizations of our brain-minds. Since each experience is differentiated from all the others, its meaning and value are symbolically established relative to all others, and subsequent experiences will constantly confirm or modify the meanings and values of all previous ones. Since our lives are lived as members of a community characterized by a symbolic culture, such meanings and values are widely shared. Consequently, the meanings and values of a great many experiences are established within the context of the collective life of a community. All this is possible because the organizations of our brain-minds are the equivalent of the "interpolation" and "extrapolation" of all our experiences symbolized as neural and synaptic modifications of these organizations. This "interpolation" and "extrapolation" accumulate a great deal of metaconscious knowledge within the body of experience, which, according to the evidence of the social sciences, we use in our daily lives, even though we have neither directly learned it nor received it from genetic processes. In this way, we acquire cultural elements such as eye etiquette, conversation distance, a way of arranging our lives in time and space, our social selves, the social selves of significant others, the social roles of others, the way of life of our community, and a great deal more. This metaconscious knowledge implied in the organizations of our brain-minds is used by our metaconscious social selves to live each experience against the background of our entire lives, thereby symbolizing it as a moment of our lives. These lives can work in the background as a mental map, gyroscope (or compass), or radar, depending on the kind of society in which we live. Hence, the body of experience from which our lives are built up permits the symbolization begun on the first level to deepen significantly.

Culture, the third level of symbolization, deepens it further in order to jointly sustain the lives of the members of a community in time and in an ultimately unknowable universe, by dealing with the threat of the unknown. In our daily lives we never encounter the kinds of situations created by philosophers when they question whether the unknown is able to radically call into question what we know and live. We live as if what we know and live is entirely sufficient and trustworthy for making sense of and living in the world. As members of a community, we are sustained in doing so by the organizations of our brain-minds, which create the deepest metaconscious knowledge by "interpolating" and "extrapolating" all our experiences of our lives in the world, with the result that the unknown is symbolized as more of what we already know and live. Culture complements and completes what happens on the other two levels of symbolization: anything that is radically other cannot take a place in our lives by taking on a meaning and a value and thus cannot be lived. In this way, we collectively interpose a symbolic universe between our community and an ultimately unknowable universe evolving in time. All our community's elements and institutions (including its science, technology, economy, social structure, political organization, legal framework, morality, religion, and aesthetic expressions) are dimensions of the way it culturally mediates all our relationships with everyone and everything. As I will soon discuss in more detail, this metaconscious transcending of the experiences of all our lives that have not been suppressed or lost will treat any future experiences as "gaps" between our experiences, and thus as extensions of what is known and lived towards which we are evolving. With the unknown thus symbolized as more of what is known and lived, its threat is eliminated and we can confidently proceed with our lives in the world. We know where we are going. We are in touch with everything around us and with what is yet to come. As a result, the door to relativism, nihilism, and anomie is closed most of the time. Metaconsciously, we have created a firm ground to stand on in time and in an ultimate unknowable universe. There is no need for any kind of ontology. This symbolization of the unknown as more of what is known and lived corresponds to our current understanding of the role of myths in a symbolic culture. Their functions are somewhat analogous to the DNA for biological life. These myths are the ultimate identity of all symbolic cultures.

Today most symbolic universes have been so highly desymbolized that they have been largely replaced by what we have referred to as reality. Nevertheless, for most of human history, symbolic cultures worked so effortlessly in the background of individual and collective human life that they remained largely unnoticed until industrialization, urbanization, secularization, and desymbolization began to weaken them and a lack of symbolic support became increasingly evident. As reality invaded symbolic universes, some observers eventually were led to believe that symbolization could be reduced to symbol processing in information machines. How this became thinkable has been made clear in our previous studies. From a historical perspective, such a belief amounts to an expression of what is thinkable within our current secular myths – something that would have been unthinkable in all earlier symbolic cultures.

This brief overview of symbolization, experience, and culture will be illustrated with a number of examples. Babies appear to be born with very few innate structures of the brain that allow them to make sense of what is happening. For example, they are able to follow movement, but they must learn to focus on what is moving. They can suck or move their limbs in limited ways, but they have no idea how this relates to their lives, and thus how to live these experiences. The organization of their brain-minds works in the background to protect them from being overwhelmed by the blurs of stimuli detected by their senses and nerves. These cannot be added to their lives until they can be symbolized, that is, related to everything else in their lives, and this usually begins by living each new situation as more or less similar to one they have already lived before.

By means of symbolization, babies and children build up a relative knowledge of their lives in the world. The feelings that they will later name as being hungry or fed, being wet and dirty or comfortable, being lonely or loved, and a great deal else, come and go; this coming and going becomes associated with visual blurs, aural blurs (voices and other sounds), skin pressures, mouth movements, and odours. By associating these with each other, babies can derive an experience with a rudimentary meaning. In this way, a number of these rudimentary experiences become differentiated from each other, and the organizations of the brain-minds of babies begin to imply a great deal of metaconscious knowledge as a result of these organizations "interpolating" and "extrapolating" these experiences.[10] It is somewhat analogous to fitting a curve through the experimental data plotted on a graph in order to discover the "law" governing the relationship between the two variables being examined. In the same way, the organizations of the brain-minds of babies imply all the relationships they have built into their lives, and later these will imply the cultural approaches of their communities. For example, this metaconscious knowledge begins to imply a distinction between the physical self and a separate world because babies' experiences of the self involve two or more tactile sensations when touching themselves, while their experiences of the world often involve only one, as when they touch their crib or a toy. This metaconscious awareness in turn changes their embodiment in their experiences by becoming more intentional and social. For example, they learn to use some sounds as vocal signs. It now becomes much easier to track the way they differentiate their lives in the world. Having learned to associate their experiences with a cat at home with the vocal sign of "cat," toddlers may initially use the same vocal sign for their experiences with a small dog at a friend's home. Similarly, having learned the word "bird," they may excitedly point to a small plane overhead and exclaim: "Big bird!" When corrected, the toddler is encouraged to learn something about dogs relative to cats, and about cats relative to dogs. Similarly, he or she begins to learn something about birds relative to small planes and about small planes relative to birds. By means of a detailed analysis of this kind, it becomes apparent that babies' evolving lives in the world move towards entering the symbolic universe of their community. Simply put: this universe is an enfolded whole in which everything in life is differentiated from everything else, with the result that what something is, according to the process of differentiation, is what everything else is not, and new relationships, situations, and entities emerge by initially being lived like an earlier experience. In our examples, dogs emerged from cats and small planes from birds. Hence, the meaning and value of everything is dialectically enfolded into the meanings and values of everything, as is reflected in the dictionary of a language. Everything is defined in terms of other entities, which in turn are defined in terms of still others, and so on, thereby revealing how everything is dialectically related to everything else.

The dialectically enfolded character of symbolic universes can be further illustrated by means of the following examples. Individual human beings are not the "parts" of a society. We are all both individuals and society because, when we grow up in the same community, the organizations of our brain-minds metaconsciously enfold culturally unique ways of skilfully coping with life in the world, including how to live in space and time, how to relate to others using eye etiquette and conversation distance, how to cope with alternatives by means of values, how to fit into the way of life and history of our community, and a great deal more. In this way we have all become individually unique expressions of the cultural whole of our community. We skilfully cope with the world in a way characteristic of our time, place, and culture. To use a biological analogy: growing up in the social womb of our community has enfolded its "cultural DNA" into each experience and thus into the organizations of our brain-minds.

In the same way, all social relationships are dialectical in character because they represent a dialectical tension between our similarities and our differences. When the differences overwhelm the similarities, we grow apart and enter into different worlds, making it difficult to understand each other. If we have shared our life with someone else for a long time, the growing similarities may overwhelm the differences, making it more and more difficult to enrich each other's lives with something new and interesting, and we may get bored with each other. The art of maintaining a healthy relationship comes down to constantly recreating a dialectical tension between being similar and different. In contemporary mass societies, this is exceedingly difficult.

In the same way, healthy groups and societies thrive when their members enrich each other's lives, which, once again, requires a dialectical tension between individual differences and cultural similarities in the group, and between individual diversity and a cultural unity in a society. If the individual diversity overwhelms a cultural unity, a society will disintegrate into fragments. Unfortunately, "us versus them" sentiments proliferate along with these divisions, and if eventually everyone is individually up against the system, the society may collapse.

The extent to which the members of a symbolic species depend on the culture of their community in order to skilfully live in the world becomes obvious when we examine the cases of children brought up by animals or who have lived in isolation for a long time,[11] and children whose entry into the symbolic universe was greatly delayed by a combination of blindness and deafness.[12] It is also apparent when we examine the lives of people suffering from the loss of short-term memory or who are in the early stages of Alzheimer's disease. Their lives disintegrate into a sequence of unrelated moments of existence, which disconnects them from space, time, and the social world. For example, when they enter a building for the first time, they are immediately lost because they cannot remember the last turn they took. They are unable to carry out any conversation because they cannot remember one or two sentences back. They encounter the same problems when they read a book or watch a movie. Similarly, they are unable to remember whether they ate a meal five minutes or three hours ago. Nevertheless, their lives are not completely a sequence of separate moments lived in micro-worlds because the organizations of their brain-minds that developed up to the onset of the disease still continue to work in the background.

In the absence of these kinds of diseases, our lives as symbolized by the organizations of our brain-minds work in the background to help us skilfully cope with the world in a way that is individually unique and culturally typical. What we see is symbolized like everything else and thus integrated into our lives on its cultural terms. We may regard symbolization as a strategy for living our lives in the world based on the indivisibility and inseparability of our lives and the symbolic universes in which they are lived. Divisibility and separability occur in people's lives in the world as a result of mental illnesses affecting the brain (as an organ in our body) or the mind (as a consequence of tragic experiences or an unhealthy culture). In either case, the ability of the organization of the brain-mind to sustain someone's life is immediately affected. In contrast, the living of a healthy life in the world means that all of our life's constituent activities are unique expressions of our personality, upbringing, education, practical experience, values, beliefs, and culture – in short, expressions of our entire lives working in the background, with the result that each and every experience is symbolically transformed into a moment of our lives.

Living our lives in the world depends on a great deal of metaconscious knowledge.[13] We have noted that the organizations of our brain-minds in effect "interpolate" and "extrapolate" all our symbolized experiences into a life, because each and every experience is symbolized by neural and synaptic changes to these organizations, thus symbolically relating it to all the other experiences at the same time. In other words, without these organizations there would be no life as we know it. Because our lives are lived with others, the shared experiences enfold the lives of the members of a society into one another in a manner that is deeply affected by its social structure and institutions. By including all this in the way we differentiate all our life's experiences from each other, we metaconsciously know which aspects are subjective and which are objective (as part of a working culture).

Toddlers and children metaconsciously learn, from a great many face-to-face conversations and interactions, the eye etiquette of their culture, the appropriate conversation distance, and the body rhythms for signalling deeper connectedness. All these are unique to a culture and never explicitly learned. Similarly, the "interpolation" and "extrapolation" of a great many eating experiences (differentiated by the organization of the brain-mind) may suggest to a toddler that neither parent appreciates it if he messes around with food. Similar experiences with others will imply much the same thing. Eventually, the symbolic equivalent of the cluster of differentiated eating experiences will imply, first, the metaconscious knowledge to the effect that no one appreciates messing with food, and later, the metaconscious knowledge of the objective value of the culture that food is not to be spilled. In this way, all the values associated with particular activities are metaconsciously acquired.

Among the most important forms of metaconscious knowledge are the mental "images" we derive from the "interpolation" and "extrapolation" of our experiences with significant others. It permits us to say to a friend: "I never thought you would do this. I am simply shocked." We did not have to analyse all the experiences we shared with that friend, deduce their personality type, explore what a person with that personality type would do in those circumstances, and, having done so, arrive at the conclusion that we are shocked. On the contrary, our reaction to our friend's behaviour is instant, and requires no analysis of any kind. Since others are the mirror in which we see ourselves, in addition to forming metaconscious mental images of significant others, we also form a mental image of ourselves. This metaconscious knowledge is significant because it means that symbolization is unlike signal processing with an output fed to our consciousness. Instead, the organizations of our brain-minds enfold our social selves, with the result that no distinct and separate self is required to receive the output of our brain-minds. In sum, a great deal of human behaviour can be understood by hypothesizing that the brain-mind grows by means of two processes: differentiation and the integration of our symbolized experiences.[14] Simply put: the process of differentiation distinguishes between situations experienced as fundamentally different, while the process of integration interrelates a person's experiences into a life. The metaconscious knowledge gained from the latter process is the result of each experience being enfolded into our life while that life works in the background of each experience.

Creating a Human World

As a consequence of our being a symbolic species, relationships with our surroundings are always reciprocal in character. As we change our surroundings by externalizing something of ourselves in our behaviour, these surroundings simultaneously change us as we symbolize our experiences of our surroundings as neural and synaptic changes to the organizations of our brain-minds. Moreover, the symbolization of our surroundings goes beyond immediate experience by interpreting everything in terms of actual and potential meaning and value for human life. As a result, a sphere of new possibilities opens up, which interposes itself between ourselves and our surroundings. The significance of this approach will be illustrated in terms of three mutations in human life corresponding to three life-milieus that have occurred until now.[15]

The first occurred when groups of people living by food gathering and hunting began to systematically symbolize everything in their natural surroundings in terms of its potential meaning and value for their lives and the group. It was the dawn of humanity becoming a symbolic species. The second mutation was characterized by the emergence of societies, which interposed themselves between the group and its natural surroundings. A third mutation occurred when techniques of all kinds began to interpose themselves between individuals, groups, and societies.

The first mutation may be illustrated with the following examples. Before a fallen branch on the forest floor can become a human implement, it must first be symbolized as one. For a dead tree trunk floating down a river to become a canoe, it had to be symbolized as such. For death to be anything else but the end of a human animal, it had to be symbolized as such. By systematically symbolizing everything in terms of its potential meaning and value for human life in the world, people opened up a sphere of human possibilities that interposed itself between the group and nature: the possibilities of hunting weapons, canoe-based fishing and travel, and the consolation of an afterlife or reincarnation according to merit.

Similarly, there is not a single human institution that could have been invented without symbolization. For example, a committed sexual liaison became symbolized as the institution of marriage. Symbolization thus led to a sphere of possibilities that interposed themselves between the human group and nature, thereby creating a symbolic universe and also transforming what it was to be human away from the natural and towards the cultural. This self-reinforcing development has made us a symbolic species distinct from all animals.[16]

Suspended in Language and Myths

Along with this newly created sphere of possibilities came the necessity of ordering human life within it. All possibilities could not be equally meaningful and valuable if human life was not to descend into relativism, nihilism, and anomie. The danger of plunging the human group into chaos was avoided by symbolizing the life-milieu (which we today symbolize as nature) as an entity without which it was impossible for the members of the group to understand who they would be, how they would live, and what their world would be like.[17] Consequently, in one way or another, groups directly or indirectly symbolized some elements of that life-milieu as its creators and sustainers having the power of life and death by granting or withholding everything necessary for life. Early on in our development as a symbolic species, the food-gathering and hunting groups did not differentiate themselves from all other life forms as the only symbolic species, with the result that the life-milieu became populated with spirits, powers, and gods who had to be obeyed by means of natural moralities and religions. By symbolizing everything in the natural life-milieu in terms of its potential meaning and value for human life (differentiated from everything else in ways very different from ours), the group created a symbolic universe within which life was oriented and anchored by means of myths. These were the entities in human experience that were symbolized as being all-powerful and without limits in their jurisdictions.

I am using the concept of myths in the sense of contemporary cultural anthropology, the sociology of religion, and depth psychology. Myths cannot be experienced by those who live by them. When they are being displaced by other myths, something of the older myths may be intuited and lead to explicit mythologies. As a result, myths are the infinite contexts that are taken for granted because they are so self-evident that no one during their reign can imagine anything different. They form the horizon of human experience beyond which there is simply more of the same. They also are the ultimate roots and orientation within this horizon, thereby making human life in an ultimately unknowable universe possible by shielding it from relativism, nihilism, and anomie. From a steady stream of new discoveries and possibilities there could emerge some elements that, in one way or another, do not fit into the symbolic universe of the group, with the potential threat of calling it into question. In a sense, myths are to a human group or society what water is to fish.

Elsewhere,[18] I have shown that our contemporary understanding of the role of myths in human life is that these myths represent the deepest metaconscious knowledge implicit in the organizations of our brain-minds. We are thus immersed in the myths of our society just as the members of another society are immersed in theirs. The members of neither society are critically aware of their own myths: it is the other society that appears to be taking certain things for granted, not the members of our own. When we become aware of what another society takes for granted, its culture may appear inferior, thereby justifying the myths of our own society. Myths accomplish this and much more by symbolizing the unknown as more of what is already known and lived, which will be discovered and lived in the future. Accordingly, the myths of a society mediate between its symbolic universe and what lies beyond. Individual and collective human life is thus symbolically suspended in time and in an ultimately unknowable universe. The symbolic universe constitutes the absolute frame for human life.

We experience something similar during each moment of our daily lives. There are no "edges" around our field of vision because our brain-minds have symbolically interpreted them out as having no meaning. The limited portion of our surroundings that we visually experience is thus symbolized as integral to all of our surroundings: there is no meaningful barrier between what we see and what lies beyond. In an analogous manner, the cocoon of myths by which individual and collective human life is symbolically suspended in time and in an ultimately unknowable universe symbolizes what the members of a group or a society know and live as being integral to a much larger reality, thereby removing the threat of the unknown.

In other words, is impossible to pictorially represent what we have just described in terms of a set of concentric circles: the symbolic universe as the totality of the symbolized experiences of human life in the world as the inner circle, the cocoon of myths as the next circle, and the unknown as the third circle. The cocoon of myths represents the symbolic elimination of the interface between what is known and lived by a group or society and what lies beyond.

When myths perform their indispensable roles, they essentially enfold the unknown as the "interpolation" and "extrapolation" of all human experiences. Myths thus cover over a void with all possible future experiences that fit into and are compatible with what is known and lived, thereby excluding anything radically other. The boundary between the symbolic universe of a society and the unknown beyond is thus eliminated, as is any threat posed by the unknown.

In this way, all human groups and societies have been protected from radically doubting what they knew and lived. They developed an unshakeable confidence and an ultimate religious faith in what their members knew and lived as being entirely trustworthy and reliable – a vantage point and direction for confidently living in a world of which much was unknown. To be sure, the universe was full of surprises, but none of them could pose a radical threat by being entirely different and thus incompatible with human life. Thanks to the shield of myths, human communities were in touch with their time and their worlds, at least in principle. The ongoing stream of discoveries and new possibilities no longer posed any threat, and the door to relativism, nihilism, and anomie was all but closed. The thought experiments philosophers conducted in certain times in an attempt to cast radical doubt on who we were, what life was really like, and whether the world was really as we experienced it were just that: thought experiments that could never be replicated in our lives. In sum, the world is a dialectically enfolded whole of relative knowing and doing.

Relative Knowing and Doing

It would appear that as a symbolic species, humanity implicitly recognized that it was impossible to know anything *absolutely*, that is, on its own terms independent from anything else. All human knowledge was *relative*. Having no direct access to the unknown beyond what was known and lived, the best humanity could manage was to know something *relatively*, that is, know it in terms of the relations it had with everything else in human life and the world. We may suppose, therefore, that symbolization, experience, and culture were invented and evolved on the basis of an implicit recognition of the impossibility of grasping what is there in its totality. They put humanity in a relation of dependence on its gods, which each and every culture created for itself by articulating the intuitions regarding its deepest metaconscious knowledge. The situation may be likened to that of a toddler who confidently strides into a large box store holding her mother's hand. As the child is distracted by the many things on display and begins to explore the store on her own, she may suddenly look up and find that her mother is nowhere in sight. A frantic search follows, and when she is still not able to find her, the child breaks out in tears. In other words, living in an ultimately unknowable world is not a threat when you can trust a loving parent to sort things out when you get stuck. Eventually, the metaconscious knowledge that develops in the organization of the child's brain-mind reduces her reliance on adults, and thus permits the child to grow up and become an adult herself, able to skilfully make sense of and live in the world.

The Birth of Societies

The ways of life and symbolic universes that sustained the food-gathering and hunting groups in the life-milieu we symbolize as nature began to undergo a gradual mutation as this way of making sense of and living in the world became unsustainable in many areas because of population density. Nature simply did not grow enough food through natural ecosystems. Moreover, the supply of game was limited by the eltonian pyramid, which represents the throughput of matter and energy of the carnivores and herbivores in ecosystems. The limits are imposed by the fact that this pyramid is ultimately supported by a limited biomass derived from photosynthesis.[19] A very long and difficult transition towards agricultural ways of life became necessary. There appears to be little doubt that these emerging agricultural ways of life represented a considerable reduction in the quality of human life from those based on food gathering and hunting. Almost simultaneously, civilizations began to spring up in the swampy deltas of the great rivers on different continents, possibly because the groups that failed or were unwilling to defend their territories found refuge there.[20] These groups had to cooperate in order to survive by creating the embryonic beginnings of irrigation projects. They brought with them an enormous stock of knowledge of the natural life-milieu in terms of the relevance its constituents had for human life, as well as their experiences gained with partial agricultural initiatives that had been the intermediaries between food gathering and hunting and the emerging agricultural ways of life.

These developments led to the birth of societies, which interposed themselves between the group and the natural life-milieu. These societies became the primary life-milieu for the group, thus mediating between them and nature, which became the secondary life-milieu. Civilization, experience, and culture became directed to this life-milieu, with the result that groups (and, much later, individuals) began to distance themselves from this life-milieu the way this had previously happened with regard to nature. These developments took several millennia. It would appear that, during the early stages, many peasants were regarded as essentially part of the local estate, belonging to the land like the local animals.

The Cultural Pillars of Western Civilization

Three significant innovations emerged in the symbolization of individuals, groups, and societies. They are worth mentioning because they became the cultural pillars of Western civilization, which gave birth to discipline-based science and technology.[21] The first pillar was the cultural achievement of the Greeks. Their culture was widely admired throughout the known world at that time for having achieved what may be described as a more liveable dialectical tension between individual diversity and cultural unity in favour of the former. Ironically, much of what was passed on by Socrates, Plato, and Aristotle may be interpreted as being anti-cultural and ahistorical because they sought to arrest their culture's decline by looking for the rules "behind" many human skills, and thus "behind" significant portions of their culture. They conducted the first inquiries equivalent to knowledge engineering, which attempted to discover the rules experts used in their decision making. These approaches set Western philosophy on a course that had to be abandoned nearly two thousand years later.[22]

The second pillar was the cultural achievement of the Romans, who also shifted the dialectical tension between individual diversity and cultural unity in favour of the former, but in a much more practical manner. They developed a legal code that gave citizens political rights with respect to the state. No earlier legal codes had ever granted rights to citizens. The Romans created a workable individual diversity with political rights and a dialectical tension with the cultural unity upheld by the state.[23]

The third pillar was the cultural achievement of the Jewish people, who, in contrast to the Greeks, centred human life in experience and history with a great deal of hindsight. Their attitude towards morality and religion has received far too little attention. For example, the opening chapters of the book of Genesis in the Hebrew Bible informed the Jewish people that, in the beginning, the relationship between God and humanity was one of love and communion.[24] It allowed humanity to live in a creation that only its creator fully understood. Consequently, as long as this communion was not broken, the unknown posed no threat to what was known and lived. As creatures, all knowing and doing was referred to God, and there was no need for morality or religion. The closest we can come to this is by the experience of falling in love with someone. In love, we spontaneously seek to please the other, and they respond to that love by doing the same. No morality is required to spell out the good and the evil for the relationship.

In the same opening chapters of Genesis, Adam names the animals to signify the spiritual dominance humanity was to exercise over creation – a dominance by the word, just as God did.[25] In Hebrew, there is no distinction between the name and what it designates. It was not until the fourteenth century that Western civilization began to make this distinction. Today, we experience a complete dissociation between these two because words are used for many purposes, such as tools by advertising and public relations-speak to create a particular effect. However, the situation described in Genesis is that of a humanity whose word is true, and which defines and gives meaning to reality through the word. All the cultures we are describing could not have existed at that time because there was no need for myths any more than for a morality or a religion. Adam did not name the animals in relation to the meaning and value these had for human life because things did not stop there. Life was lived in communion with the Creator.

This interpretation lends support for what happened following the break between God and humanity. Genesis 3 mounts a deadly attack on morality, religion, and magic.[26] The building of cities begins soon after, which culminates in the tower of Babel: humanity wishes to build a city with a tower reaching to the heavens in order to make a name for itself.[27] God named Adam, Adam named the animals, but now humanity desires to establish a spiritual dominance over itself. Humanity wants to make a fresh start in an urban world of its own making, of which it will be the master by excluding God.

Nothing suggests that this account designates the creation of many languages, as is often believed. However, it has the fullest possible significance: these languages are embedded in cultures, these cultures depend on myths, these myths are externalized by creating religious institutions and gods, and these religions reach towards the heavens. Moreover, the first permission granted in what is commonly referred to as the Ten Commandments (which are in fact promises because the Hebrew is in the future tense) is that one day humanity will live without false gods.[28] From a cultural perspective, the Hebrew Bible's account of the history of the Jewish people may be interpreted as a struggle to live without gods, and thus without myths. Although this account may not have had a direct influence on Western civilization, it has had an enormous indirect influence through Christianity, which was built on it. Much was lost when the Roman state made Christianity the official religion of the Roman empire, and it never recovered so that it could become an anti-morality and an anti-religion.

A distinction may therefore be made between the religion any culture creates for itself and whatever may be received as a communication from a transcendent God, which comes from outside a culture. From a Christian perspective, this distinction between religion and faith was well developed by Karl Barth.[29] Karl Marx also recognized the need for an equivalent distinction – if during the fifth and final stage of human history there would still be people who believed.[30] The point is that Christianity is rarely thought of as including a project of attempting to live without myths (or false gods). It is worth asking which myths we serve as we gamble with our planet and our chances for a liveable, sustainable future. Being a little more realistic by living without false gods may well be in order. This is especially true for our discipline-based approaches to knowing and doing. I will therefore argue that this third pillar of Western civilization needs to be kept in mind as much as the first two.

Towards Absolute Knowing and Doing

Initially, civilization, experience, and culture jointly gained the deepest possible knowledge of everything related to everything else. Although in principle this knowledge was relative in character, it had to be absolutized by means of myths to close the door to relativism, nihilism, and anomie. Doing so is not a matter of absolutizing every element of human knowledge, but of connecting all of them to those elements of human experience that appear to have no limits and which can thus temporarily function as the absolute frame for human life in the world. All this happens very gradually as the organizations of the brain-minds of children and teenagers develop this deep metaconscious knowledge as the "interpolations" and "extrapolations" of all their experiences into a life conducted by the way of life and culture of their community.

A very different way of defining spiritual commitment to what was known and lived first emerged in Western civilization. It was based on attempts to gain *absolute* knowledge. During the Middle Ages, the Roman Catholic church gradually developed the concept of an infallible inspiration of the Christian Bible under the influence of Greek philosophy and Roman law. It was a radical departure from the ways the Jewish people regarded these matters. They never had a conception of their God using his creatures as storage devices, dictaphones, and typewriters in order to pass on an absolute knowledge of himself, his creatures, their actions, and his responses. After liberating them from the "double anguish" of life and death, God could not be imagined as one who turned his people into mechanisms for transmitting a message. There were oral traditions, and at some point groups of rabbis had put these into writing. However, it was the Jewish people who discerned what was and what was not a communication from their God. The practical implications were far-reaching.

The Christians regarded their Bibles more as narratives of absolute facts that could then be used to prove or disprove any claims made during times of disagreement, which was basically all the time. As a result, it was unthinkable that any theological fact could be negated or contradicted by a scientific fact. The violence of the struggle between Catholics and Protestants in the sixteenth century can in part be attributed to both sides claiming to possess absolute knowledge of the Christian Bible, in the name of which torture and murder could be justified and ecclesiastical authority legitimized in an absolute manner.

It is easy to understand why, under these circumstances, some Christians and non-Christians searched for ways in which human beings of all persuasions could intellectually journey together. What they could all share could be taken as the most reliable available knowledge. Discipline-based science offered the most promising model. The hope was that any limits to their sharing could be rolled back by scientifically educating people to act as detached and objective observers. They would use a scientific approach in designing experiments to arrive at "experiences" of human life and the world that might be regarded as having the status of "facts" because they could be replicated by anyone having the appropriate training. These "facts" could then be assembled into scientific theories. Physics, chemistry, and astronomy, the forerunners of contemporary disciplines, were demonstrating how this could be done. As noted, their practitioners succeeded because the situations they studied were dominated by only one category of phenomena, which meant that their approaches had limits.

At the time, people hoped desperately for absolute knowledge capable of determining the facts and mathematically (not culturally) integrating them into a new understanding of human life in the world. It was hoped that, for the very first time, humanity could reach through and beyond what was culturally known and lived to take hold of what lay beyond in manageable "chunks," each of which constituted an element of absolute knowledge or fact. From the perspective we are developing, the idea of a scientific fact amounted to endowing scientists with the ability to bypass their myths and thus create an element of knowledge valid for all times, places, and cultures as well as for different stages in the development of science itself. Facts became valid in themselves, and were not considered as relative to everything else that was known scientifically. Similarly, scientific theories represented knowledge that was valid in itself, which would make it universal. This amounted to science accomplishing what no culture had been able to achieve: knowing something *absolutely*, that is, relative to itself as opposed to relative to everything else that was known. It made discipline-based approaches to studying human life in the world thinkable. The absence of a *science of the sciences* capable of scientifically performing the roles cultures had always played made it impossible to integrate all the facts from all the disciplines into an overall understanding of human life in the world that could replace the equivalent based on symbolization, experience, and culture.

I readily acknowledge that this is an oversimplification of the complex historical developments that occurred over centuries. Nevertheless, while cultures essentially reached for the heavens to obtain absolute knowledge through their religions, science would now accomplish the same by an absolute of our own making. Science was able to pierce through all the myths that had shrouded the symbolic universes of all cultures in superstition in order to take hold of what lay beyond. Some people took this a step further, claiming that as science advanced, humanity's need for religion (and thus for culture) would recede.

What aggravated these issues even further is that, ever since the sixteenth century, theologians and scientists confused truth with reality. Theologians assumed that the truth as revealed was reality. The scientists assumed that the reality they studied was the truth.[31] Under the influence of Greek philosophy in general and the concept of universal knowledge in particular, theologians had lost track of the fact that meaning cannot be transmitted directly. It must be imbedded or clothed in a narrative. The Jewish people had always read their Bibles not as transmitting facts but as narratives transmitting meaning. As a result, in the accounts of historical events, we are told something about how one or more people interpreted their meaning and how they related it to their God, and also how God interpreted these events. For example, the two contradictory creation accounts in the book of Genesis contain complementary meanings, which must have been the reason why the rabbis retained both of them and integrated them into a single account, and why it was discerned as a coherent revelation regarding the most fundamental of relationships, as opposed to an account of the origin of the world. All this becomes incomprehensible when the creation accounts are read as a series of facts true in themselves.

Two earlier developments paved the way for the belief that absolute knowledge was within human reach. The first was the previously mentioned invention of universal knowledge by the Greeks, which re-entered Western civilization in the Middle Ages. The second development was that Christian theology became the systematic study of an infallible Bible, making it the subject of the central faculty in the medieval university on the way to becoming a discipline. These distortions of Christian theology and secular science paved the way for a conflict between infallible biblical knowledge and factual scientific knowledge, each claiming to be absolute in its own way.

These developments prepared the "cultural soil" for a radical new spiritual commitment in the historical journey of Western civilization following the Second World War. It confronted the threat of human life being plunged into chaos as a consequence of new discoveries and possibilities that were diametrically opposite to the ones based on the symbolization, experience, and culture of all earlier civilizations. Western civilization now depended on science for absolute knowing and on technique for absolute doing. The former resulted from the secular cult of the fact and the latter from the secular cult of efficiency. Applied to all situations, including those where multiple categories of phenomena make non-negligible contributions, this spiritual commitment means that what can thus be known and achieved is far less comprehensive because the context taken into account is limited to one category of phenomena at a time. As a result, the scientific approach to knowing and the technical approach to doing are highly desymbolizing, to the point where they could threaten our future as a symbolic species.[32] We must remember that our spectacular successes are indissociably linked to our equally spectacular failures to create ways of life that are economically viable, socially liveable, and environmentally sustainable. Despite these increasingly challenging problems, our spiritual commitment to the cult of facts and efficiency remains unshaken.

The Role of Industrialization

How did Western civilization back itself into the present corner? The short answer is that industrialization brought more than we bargained for. It began with an all-out effort to mechanize human work and industrialize production. This involved reorganizing human life and society into distinct and separate domains, beginning with human work. The introduction of the technical division of labour decomposed this work into a sequence of production steps that are endlessly repeated by a worker or a machine assigned to each step. Each production step represents a domain in which the required inputs are transformed into an intermediary output by endlessly repeating a particular process based on an instance of a single category of phenomena. Nothing else is supposed to affect this. The technical division of labour thus transfers human work from a symbolic universe of what makes sense to a "world" of non-sense built up from these domains.[33] Adam Smith recognized that this would make human workers as stupid as they could possibly become, but somehow the new wealth of nations would bring such progress as to make all this worthwhile.[34]

Elsewhere,[35] I have interpreted mechanization and industrialization in cultural terms. No human activity can create or destroy the matter and energy on which it depends. Hence, all the activities of a way of life are connected by a network of flows of matter and a network of flows of energy, which overlap whenever a flow represents a composite of matter and energy. In the industrializing societies of the nineteenth century, this connectedness evolved into the technology-based connectedness of human life and society. Until then, the interdependence of all human activities by flows of matter and energy had been enfolded into the culture-based connectedness of human life and society established by symbolizing these activities in terms of their place and significance in individual and collective human life. For example, many exchanges were an almost invisible part of patterns of kinship, neighbourliness, friendship, or reciprocal feudal obligations.

As a consequence of industrialization, this technology-based connectedness was increasingly built up from domains. This process can be understood as being entirely governed by the inputs received by a domain, the efficient transformation of these inputs into a intermediary output by the repetition of a particular process, the transfer of this intermediary output to another domain, and the continuation of this pattern until a good or service has been produced. This calculus of inputs and outputs could not be effectively managed by means of symbolization because the influences of the broader context of any or all domains had been deliberately designed out and kept at bay by the way the corresponding activities were organized and managed.

In the same way, the economy (already separated from society) became conceptualized, organized, and managed by means of three kinds of domains. In a first kind, an algorithm referred to as *homo economicus* allocates a wage (the input) to the various needs a person has, which once backed by a portion of this wage are transformed into market demands for the corresponding goods and services (the outputs). *Homo economicus* determines these allocations to gain the greatest possible utility from this wage. A second kind of domain takes the capital of an entrepreneur (the input) and allocates it to those goods and services that will jointly ensure the greatest possible accumulation of capital above the point where it is merely renewed (the output). A different algorithm, also referred to as *homo economicus*, maximizes profits within the economic constraint of at least having to renew the capital stock, which means that the cost of producing a particular good or service must be kept as far as possible below its market price. A third kind of domain is occupied by what I refer to as the mechanisms of supply and demand. These operate on the basis of two mathematical functions, each specific to a particular good or service corresponding to its own market. Functions of supply and demand relate the quantity produced or consumed to the cost of production or the offered price respectively. The market price is established at the point of intersection of the two functions, where the quantity produced equals the quantity sold and consumed. This kind of domain thus mediates between the first two kinds. It should be noted that in all three kinds of domains, the broader context is supposed to have no influence on what happens in a domain. For example, if someone is starving to death for lack of money necessary to turn this need for food into demands for food, the corresponding markets will not be affected. Nor will there be any effect on any market if the consequences of production and consumption on third parties are nontrivial. A market cannot be established for a product or service whose use by one party will not diminish its value for another party. Of course, these are well-known market externalities which demonstrate some of the consequences of organizing economies as if they were constituted of these three kinds of domains.

The above description of how markets for goods and services supposedly functioned in the nineteenth century made it possible for political economy to be turned into a discipline. Its overall domain could now be built up from the above three kinds of domains by taking one at a time and in the process excluding any kind of phenomena not included in them. All non-economic phenomena were thus excluded from the discipline. A century later, economies could be described by means of input-output tables and all manner of mathematical models, which "interpolated" and "extrapolated" the "economic facts" as the statistical data gathered by governments and other institutions. Nevertheless, it was already recognized in the nineteenth century that describing a market economy with this kind of approach was incomplete. Adam Smith showed that the markets for goods and services would jointly operate as a great invisible hand that would create the best possible world for most people.[36] The ideological use that has been made of this concept is enormous, but we forget its theoretical significance, which was to point to a fundamental incompleteness of Smith's description of the role markets played in human life and society, and therefore also in the above kind of description implied in modern economics as a discipline. Beyond its domain, there were real people, societies, and a biosphere whose behaviour could not be divided into distinct and separate domains. Smith limited his attention to one positive outcome, not foreseen or taken into account in individual market transactions. Why restrict our attention to a single positive one? Are there other positive outcomes? If there are positive outcomes there are equally likely to be negative ones. Organizing economies one transaction and one market at a time results in market prices that do not reflect the unforeseen and thus unaccounted-for consequences for human life, society, and the biosphere.

All this becomes more than a little absurd when we recall that there is not a single human activity that can create or destroy the matter and energy on which it depends. Human life and society can therefore not be understood one activity at a time, nor can economic ones be divided from non-economic ones. As a result, the attempt to "internalize" some externalities into market prices has proved a daunting if not impossible task. It is also widely recognized by economists that market prices are a poor indicator of the meaning and value goods and services have for human life and society. As a discipline, modern economics has restricted its attention to economic reality, that is, to what can be divided, measured, quantified, mathematically represented, defined in itself, and contribute to a simple complexity built up one element at a time. Everything in human life and society that is dialectically enfolded, and everything in the biosphere that is biologically enfolded (which is all of it), is excluded from this reality. It is as if only what we can see or visually imagine is real. A symbolic universe, a symbolic species, and everything else that lives cannot be a part of this reality.

The problem does not stop here. Every other discipline in the modern university does the same thing. It should not be imagined that when a university creates a centre, faculty, or institute for the environment, we can somehow integrate the disciplines back around the practical issues such as the environmental crisis. This intellectual division of labour makes it impossible to understand that the environmental crisis is something our civilization produces, in the same way it produces cell phones, computers, or cars. Engineering faculties offer optional minors, or courses on the environmental implications of technology, as if this was external to engineering design and practice. Economics offers environmental economics as a kind of add-on. Environmental economics must not be confused with ecological economics, which attempts to get to the root of the problem. The university is thus engaged in modern secular magic that transforms life into non-life, after which it can be broken down into domains that may be understood one at a time and which we hold to be reality itself.

The economic approach by which the industrializing societies of the nineteenth century conceptualized and managed their economies is thus very far removed from the cultural approach on the basis of which things had been done historically. The former dealt with everything *absolutely* while the latter dealt with everything *relatively*, that is, in relation to everything else in human life, society, and the biosphere. It is impossible to divide human life and society into the technology-based connectedness – organized by means of the technical division of labour and managed by "free" markets (liberated from all contexts including human beings) – and the culture-based connectedness – to be left to symbolization, experience, and culture. It is impossible to build up the former one domain at a time and argue that a divided human being exhibits a mental illness of one kind or another.[37] It is impossible to determine market prices one good or service and one market at a time, when the meaning and value of this good or service are determined by how this good or service is related to everything else in human life, society, and the biosphere. Market prices are a highly desymbolized economic shadow of cultural values.

From a historical perspective, Adam Smith was describing a highly unique development in human history. For example, the social fabric of the industrializing societies had been terribly strained, if not destroyed. The migration of farm labour (displaced by a higher agricultural productivity through mechanization) to the new industrial centres created a situation that could hardly be described as new social communities. The usual cultural guidance for human life had been greatly weakened as the consequence of collapsing traditions, and life in the new industrial centres had to be almost completely reinvented from a social point of view. The poverty of wage earners was not simply financial. As Karl Marx and many others pointed out, even marriage had become the privilege of the highest socio-economic strata for the simple reason that working days were so long that, when people came home from the factory, they were exhausted and barely had the energy to eat, never mind being a supportive spouse or a caring parent.[38] In the same vein, being a close and loyal friend, a good neighbour, a responsible community member, and a good citizen had become the privilege of the wealthy. Compared to farm labour, the poverty of factory workers was not merely financial, it was existential. Moreover, their humanity was being undermined by the technical division of labour, which included them as hands, but excluded them in every other way. In the absence of bonds of kinship, friendship, and neighbourliness, there was little left other than survival by means of self-interested behaviour. Economically (but less so existentially), this was also true for shopkeepers and entrepreneurs, who had to struggle to survive under the rigid discipline of renewing and accumulating their capital under highly turbulent conditions. It is astonishing that Smith believed that self-interested behaviour, which historically had always been a threat to the survival of any community, was now the way to build a new and better tomorrow.

It is true that, under these conditions of social destitution and extreme poverty, *homo economicus* approximates the behaviour of wage earners rather well. If someone needs a new pair of shoes and an overcoat, but is able to afford only one of them, following which it will take a significant length of time before the other can be afforded, careful consideration as to their utility for a person's life will be given. In contrast, in consumer societies with a much higher standard of living, these kinds of decisions amount to whether to buy an additional pair of shoes or another coat without either one being very essential to a person's life. Moreover, as John Kenneth Galbraith[39] and I[40] have examined in great detail, the growing dependence on highly specialized scientific knowing and technical doing has completely changed the relationship between the corporation and the economy. Compelling the former to plan to such an extent is to minimize its reliance on markets. This planning includes exhaustive testing of consumer responses to new products, estimating the size of potential markets, and closing the gap between planned supply and consumer response by managing consumer behaviour through advertising. In other words, in mass societies, *homo economicus* is an unsatisfactory model of consumer behaviour.

In nineteenth-century market economies, the values of goods and services as measured by market prices all but displaced cultural values. Money became the common denominator of all values and thus the "value of values" in all the industrializing societies of Western Europe. It has become clear that, with the growth of the role of the market in any culture, a collision is created between the culture's values and market prices, and Karl Polanyi has shown that the greater the role markets play in a society, the more its culture is weakened.[41] The theory of symbolization, experience, and culture explains why this is the case.

In the reorganization of human work, production, and distribution in terms of distinct and separate domains, the role assigned to human beings as *homo economicus* was an extreme reduction of their being a symbolic species. More and more of life was delegated to domains that maximized utility or profit.

Competition ensured that no human values could play a role in all of this. For example, if an entrepreneur woke up one morning with the conviction that he could no longer continue to exploit his workers, he essentially had two options. First, he could supplement their wages from his own capital, but the latter being finite, it would just be a question of time before he would have to conform to his earlier behaviour or cease to play the role of an entrepreneur. Alternatively, he could charge higher prices for his products or services to pay for the higher wages, in which case he would be competed out of the market, go bankrupt, and have to give up the role of entrepreneur. In other words, the "system" itself functioned as a kind of mechanism regardless of the values and cultures of the people and societies involved. Moreover, organizing more and more of human life and society in terms of separate and distinct domains made it possible to know and deal with whatever was happening in these domains by itself, that is, on the basis of absolute rather than relative terms. As a result, such approaches were the diametrical opposite of those based on symbolization.

All this was made liveable by a new limitless certainty in human life. Every year there were more machines, every year these machines were becoming larger and faster, every year there were more factories full of these machines, and every year these factories collectively produced an ever-growing output of goods. There was nothing on the horizon of human experience that appeared remotely capable of changing these trends, with the result that they were seen as limitless. By means of symbolization, these experiences were "interpolated" and "extrapolated" as leading to a growing ability on the part of society to meet the material needs of its members, and thus to reduce and all but eliminate poverty for those who contributed to this progress by hard work. In turn, these achievements would be accompanied by a significant reduction in the social scourges associated with poverty. Eventually, human energies could be diverted from the struggle for survival towards solving other social problems, and everyone would become happier.

In the nineteenth century, the myths of progress, work, and happiness implied that every previous society and civilization had approached the living of life in the world in the wrong way. They had all sought to improve this life by living their moral and religious values and beliefs, and had dealt with material needs within that context. What these new myths implied was that symbolization, experience, and culture should not be relied on to procure humanity's material needs. The Christian maxim of not living by bread alone was turned into: you shall live by bread alone because everything will be then granted as well. In other words, by organizing more and more of human life and society by means of separate and distinct domains, the door was open to a knowing and doing dealing with what happened in these domains one category at a time. Again, this meant dealing with these domains on their own terms, that is, to do so *absolutely* as opposed to *relatively* (i.e., to everything else). As technology became more advanced and expensive, and as the infrastructure was constantly expanded, everything critically depended on the renewal and accumulation of capital as the life-blood of a system ruled by capital as the value of values. Both entrepreneurs and wage earners totally depended on this capital. As a result, there was nothing capital would not be able to accomplish through progress, work, and happiness. Capital became the central myth or sacred of the system named after it.[42]

Although no concept of culture as it is used here existed in the nineteenth century, Karl Marx summed up the spirit of his age when he relegated everything cultural to the superstructure of society except for the forces and relations of production, which constituted the base.[43] Even science was regarded as a part of this superstructure, which, from a sociological perspective, was more or less accurate in his time. A new defining spiritual commitment was now directed at developing the economic base of society, because everything else would follow.

Metaconsciously, Marx had projected the myth of the nineteenth century back to all human history and forward to the future as a kind of human destiny. Everything could be summed up as a succession of five economic bases, each associated with a certain stage of technological advancement. At his funeral, Engels claimed that Marx had discovered the laws of human history.[44] Again, all of this was thinkable within the myths of the nineteenth century, on the basis of which it was entirely self-evident to Marx and Engels that the trends of the time were limitless. It should have come as no surprise, therefore, that after the success of the Russian Revolution, Communism became the first secular religion of the twentieth century.[45] The Communist Party was the new secular "church" with Marx as its prophet. It and it alone knew the truth of human history: where we came from and where we were going. There was no point in sharing political power with other parties that did not know this truth and would therefore stand in the way of reaching the final stage in human history, when all alienation would cease. Millions of lives were sacrificed on the altar of this secular religion. The new myths of capital, progress, work, and happiness played the same role myths always had, except that they took on a secular form because the gods had now come to earth.

During the closing decades of the nineteenth century and the first half of the twentieth century, the rest of the industrializing world did not escape these myths. Almost without exception, economists accepted the centrality of the economy for human life and society. All political differences on this point vanished. Almost everyone agreed that they had to accept their economic alienation and serve the economy with everything they had. Doing anything else was simply unrealistic. It was all but forgotten that we create economies to serve us, and not the other way around. Almost everyone bowed down to these new economic gods. Some economists went even further by proclaiming that all this was natural, based as it was on phenomena such as natural markets and natural levels of unemployment. They must have heard nature command them to obey it. They had forgotten that the economy is a human creation for which we must take responsibility.

What is surprising is the role that Christians played in all of this, as was described by Max Weber in his influential work on the Protestant ethic and the rise of capitalism.[46] All branches of Christianity, including conservatives, liberals, evangelicals, and fundamentalists, reinterpreted their theologies, either to serve or to escape what was happening. Jesus became the answer to questions no one was asking.

It is important to further interpret the above development from the perspective of symbolization, experience, and culture. The reason why the economies of the industrializing societies of the nineteenth century had become so distinct from the remainder of society, and why this development was unique in human history, is rooted in the reversal in the hierarchy between the culture-based connectedness and the technology-based connectedness of human life and society.[47] This reversal is closely associated with a technical approach to doing replacing the cultural approach in industry, an economic approach replacing the cultural approach in the economy. In the remainder of society, the cultural approach remained dominant, even though it had been decisively weakened by the collapse of all traditions.[48] The technical approach was based on the calculus of improving the internal workings of domains by increasing the outputs obtained from the required inputs based on decision criteria taking the form of input-output ratios as opposed to cultural values. The economic approach was based on the calculus of maximizing profits in order to renew and accumulate capital, the maximization of utility in order to survive on meagre wages, and the mediation between these two by the markets for goods and services. Compared to the cultural approach, both of these deal with the world one domain at a time, with the result that these approaches had a powerful desymbolizing influence on human life and society. Traditions weakened and soon collapsed, and the role of culture in the remainder of society came under heavy pressure as a result of this desymbolization. Moreover, an evolving economy required ongoing modifications to the social structure, legal organization, political institutions, morality, and religion.[49] Artists found it necessary to modify their creative expression in response to what was happening.[50] In many cases, the new situations had no base of previous experience, with the result that people had to think them through. However, this reasoning was initially largely embedded in culture. Hence, society was split into two spheres, one in which the role of symbolization, experience, and culture had all but been eliminated, and the other where it still ruled, albeit in a weakened form. Never before had this occurred in any society or civilization, contrary to the interpretation of human history by Karl Marx and his followers.[51]

From a cultural perspective, the situation was also unique in two important ways. First, the evolution of these industrializing societies was no longer primarily sociocultural (based on the culture-based connectedness dominating the technology-based connectedness), but technological and economic (based on the technology-based connectedness dominating the culture-based connectedness). Second, beginning in industry but rapidly spreading to all society, limits to symbolization, experience, and culture were encountered. These were eventually transcended by switching to a growing dependence on discipline-based approaches to knowing and doing rather than on culture-based approaches. Beginning with the former, all technological and economic changes were increasingly tightly coupled. Any change in the local dynamic equilibrium of the technology-based connectedness of human life and society immediately affected adjacent areas, which then affected others, and so on. For example, an increased rate at which a desired output was produced from the required inputs resulted from changes such as the introduction of the technical division of labour, mechanization, technological advances in machines or processes, improvements in the productivity of labour, or combinations thereof. The upstream transformations which produced the required inputs would be strained, and the downstream transformation receiving the output would be overloaded until its throughput could be increased. These developments in turn put stresses on the processes on which they depended and so on, sending a ripple effect throughout the technology-based connectedness. The only possible alternative was for flows to "leak out" as unused outputs or inputs. Given the enormous pressures to renew and accumulate capital, such leaks could not be tolerated very long. As a result, the process of industrialization can be compared to a chain reaction.[52] Any increase in throughput anywhere in the technology-based connectedness would impose a range of adjustments while at the same time being subject to the effects from adjustments everywhere else. Entrepreneurs and wage earners had to hang on for dear life as everything was constantly changing. It is hardly surprising that this technology-based connectedness responded rather poorly to significant economic and social changes, resulting in one economic crisis after another. On this point, Karl Marx's analysis had put the finger on an almost insoluble problem.[53]

In the course of little more than a century, these developments necessitated a fundamental change in the way the members of a society used the organizations of their brain-minds. Two theories are helpful here. The first is the model of human skill acquisition developed by Stuart Dreyfus.[54] The second is the one developed by David Riesman,[55] associating a dominant personality type with different kinds of societies. When we grow up, we learn to skilfully cope with one another and the world by using our culture as symbolized in the organizations of our brain-minds. There are five stages of human skill acquisition, of which the last one depends entirely on intuition, or what we have here referred to as metaconscious knowledge. In most cases, any new situation can readily be symbolized by making sense of it as being similar to and also different from earlier ones. Occasionally, this may not be possible, which requires our thinking the situation through. We then fall back on the third stage of human skill acquisition, essentially characterized by behaviour resembling problem solving. If successful, a new situation that appeared anomalous with respect to a culture becomes normal when this process succeeds in giving it a meaning and value for individual and collective human life. Given the dialectically enfolded character of the culture-based connectedness of human life and society, it is impossible to encounter "hard" anomalies the way this can occur in the logical domain of a scientific discipline. As a result, life tends to go on, with a feeling that something does not quite make sense. It generally takes a great many kinds of these situations before the effectiveness of a culture becomes diminished. This kind of sociocultural evolution is common in traditional societies whose members can use the organizations of their brain-minds as mental maps. Past experience embodied in their working culture is used to find their way in the present. The ways of life of these societies are thus tradition-based, and Riesman suggests that the dominant personality type found in these societies is tradition-directed.

When traditional societies began to industrialize, they were soon engulfed in new situations that had never existed before. For example, it was impossible to deal with technically divided labour in the way craftwork had been dealt with, as is clearly evident from the harsh factory disciplines that had to be imposed. In a great many areas, guidance by tradition became impossible. More and more situations had to be thought through, and this problem-solving behaviour had to be based on the necessities imposed by the technology-based connectedness and not the culture-based connectedness. As a result, goal-directed behaviour began to displace culture-directed behaviour. Until then, engaging in a particular activity involved making it the focus of attention for your life at that moment. By symbolizing it, its meaning and value for individual and collective human life would have acted as a guide, and no specific goal would have been necessary. People's lives fully worked in the background, with the result that they lived their activities as a moment of their lives. The living of human lives could not continue in this manner once traditions became overwhelmed by situations that could not be symbolized as being similar to prior experience. People could no longer use the organizations of their brain-minds as mental maps. They could only use those elements which had remained relatively stable, and this was limited to the deepest metaconscious knowledge of fundamental values, beliefs, and myths. Together, these gave people an intuitive knowledge of the fundamental orientation of their culture, and this could be used to steer them through very turbulent waters. Riesman has suggested that people began to use the organizations of their brain-minds as a kind of gyroscope or compass, with the result that the dominant personality type became inner-directed. In other words, people's desymbolized lives worked in the background in this way. Although these changes correlated with a growing demographic density in the industrial centres, I believe that a more direct explanation of these changes results when they are related to the different phases of industrialization.

At this point, goal-directed behaviour became commonplace. Traditions exploded into numerous goals, many of which simply reflected the necessities imposed by the need to renew and accumulate capital to permit the system to evolve. For many others, these goals reflected the necessity to survive on inadequate wages. Since the organizations of people's brain-minds were built up from the experiences of their lives in the world, it took only a few generations before the organization of their brain-minds no longer shared any common elements. The present had become so completely different from the past, and the future so different from the present, that the sociocultural evolution had become extremely turbulent. Riesman suggests that people began to use the organizations of their brain-minds (their symbolic lives) as a kind of radar to scan what everyone else was doing in order to follow the crowd. The result was the dominance of the other-directed personality in what now had become mass societies characterized by mass production, mass consumption, mass media, and mass advertising.

These three changes in the dominant personality types and their different uses of a culture were paralleled by equally important cultural changes in industry and the economy. Here the limits of symbolization, experience, and culture became evident in different ways. In some of the pioneering industries (textiles, machine building, and metallurgy), there initially had existed a strong correlation between what was accessible by the senses (and thus to experience) and what was actually happening in the machines or processes. In most cases, this correlation was short-lived. As machines became faster and more complex, the internal workings became less and less accessible to the senses, and in the chemical and electrical industries such correlations did not exist from the very beginning. It necessitated the separation of knowing and doing from experience and culture, beginning in Germany but rapidly spreading to other industrial societies.[56] Germany began to replace England as the leading industrial power in Western Europe, and during the opening decades of the twentieth century, it held most of the patents in the chemical and electrical industries. The Germans had applied discipline-based approaches to knowing and doing in industry with spectacular success because machines and processes are built up from distinct and separate domains in which one category of phenomena dominates all others.

These discipline-based approaches to knowing and doing do not merely restrict the context taken into account to a single domain. They also distort it, and in many cases completely transform it. They do not begin by focusing on the local technology-based connectedness of human life and society and keep the culture-based connectedness in the background. The latter context is entirely eliminated along with the interdependence of the technology-based connectedness and the culture-based connectedness. The local technology-based connectedness is then represented in terms of the domain of a discipline. As we have seen, such domains are populated by only one category of phenomena. For example, solid materials are replaced by a continuum, which is a mathematical model based on the properties of that material related to its strength being uniformly distributed throughout space. In the same way, liquids or gases are replaced by a fluid, which is a mathematical model based on the distribution of the relevant characteristics throughout space. In other cases, what happens is represented by a set of equations that jointly describe what is happening in a domain. Such models are exceedingly useful and powerful, and modern engineering would be unthinkable without it. Nevertheless, these kinds of models essentially impose a goal: a continuum for determining the strength of materials, a fluid for modelling the mechanics of the flow over or inside an object, the equations to make a particular activity more effective, and so on. This leads to the distinction between a design exemplar and an analytical exemplar.[57] The former essentially operates on the level of symbolization, experience, and culture, while the latter operates on the level of the domain of a discipline or technical specialty. A design exemplar thus approaches the design of something on the basis of the way we would encounter it in our daily lives, the functions it would perform in those daily lives, and the contributions it would make to these lives, and via them to society. In contrast, analytical exemplars examine and optimize a small detail coextensive with the domain of a discipline. For example, a beam can be designed for strength, for manufacturability, for reliability, for resistance to hazards, and so on. Each aspect requires a different domain of a discipline according to the category of phenomena it deals with, such as stress analysis, metallurgy, and manufacturing. Contemporary engineering has completely confused design exemplars and analytical exemplars, and consequently has been unable to effectively teach design for half a century.[58] Another way of looking at this is that the goals of design and the goals of the various disciplines are on completely different levels. A design exemplar can be subdivided further and further until some of the details can be analysed and optimized by the appropriate analytical exemplars, but never can analytical exemplars be cumulated into a design exemplar.[59] The result has been a proliferation of technical products ill suited to human life, society, and the biosphere.

Rationality and Technique

As a result of these developments, human life and society became increasingly engulfed in goal-directed as opposed to culture-directed behaviour. The best analysis of the situation came out of Germany, where Max Weber described the phenomenon of rationality resulting from all this goal-directed behaviour.[60] Goal-directed behaviour inevitably rearranges the context taken into account. Whatever is directly relevant to achieving the goal is put centre-stage in people's attention. What is marginally relevant is put on the perimeter, and what is not relevant is left out of consideration. From the perspective of symbolization, if a goal expresses the focus of attention for living a moment of our lives, no desymbolization will occur. If a goal is imposed from the outside as it were, it will distort the context taken into account and may affect a person's embodiment, participation, commitment, and freedom in living that moment.[61] From the perspective of symbolization, its meaning and value will thus not be fully symbolized, with the result that it cannot be lived to the fullest extent possible. If a goal is separated from experience and culture, as is the case for all rational approaches related to the domain of a discipline, the context taken into account will be that of a domain, with the result that a person cannot be physically, socially, or culturally embodied in it. All this completely changes their participation, commitment, and freedom.[62] From the perspective of symbolization, a person's ability to live that moment of their life may be significantly limited, if not absent altogether.

In all these cases, an action may be rational with respect to a goal in the sense that nothing irrelevant to that goal is permitted to interfere, but the action itself is no longer undertaken in accordance with its meaning and value for individual and collective human life as understood from the perspective of symbolization, experience, and culture. To varying degrees, the living of individual and collective human life is externalized in a rational action except to the extent that it is taken into account by the goal.

The goals guiding technical and economic actions are not culturally directed in any way. These actions thus take on a certain distance from the symbolic universe of a society, assume a kind of life of their own, and thereby undermine human life in a symbolic universe. Max Weber summed it up in a prophetic manner when he warned humanity that the phenomenon of rationality would shut it into an iron cage.[63] Human life would be held prisoner by all the rational approaches that jointly served this new master. Rationality demanded that the meaning and value of any human activity be established in terms of rational as opposed to cultural goals. Human life now had to be lived in its service. It constituted a growing threat to our future as a symbolic species.[64] The self-congratulations of humanity having become rational, secular, and of age were short-lived in the face of mounting evidence that the disappearance of the traditional myths and religions had been mistaken for their displacement by secular ones. It unleashed an entirely different kind of alienation – that of being possessed by rationality rather than by culture.

In less than half a century, the phenomenon of rationality mutated into that of technique. Jacques Ellul defined it as "the totality of methods rationally arrived at and having absolute efficiency (for a given stage of development) in every area of human life."[65] The phenomenon of rationality had become coextensive with culture in every area of human life, and all the methods now had absolute efficiency as their goal. In other words, the results of these methods were measured on their own terms, and not relative to anything else. The efficiency of any technical effort was not measured in terms of the desired output obtained from various inputs as though that undertaking was connected to its surroundings, but only by the inputs received and the desired output returned. All other interdependencies were externalized by these methods. Technique had transformed the technology-based connectedness of human life and society into a technique-based connectedness: as a network of flows of commoditized matter, energy, combinations of the two (finished materials, parts, subassemblies, and so on), labour, capital, and discipline-based knowledge. Nothing else mattered to this technique-based connectedness, which now developed on its own terms. In the past, technology was ultimately known and lived through myths, which were the absolute points of reference and orientation of human life in the world. Now technique developed on its own terms, unaffected by anything outside of its technique-based connectedness. Since it is built up one domain at a time, it has the characteristics of separability, non-contradiction, closed definitions, quantification and applied logic, and a simple complexity. It completely shut out any dialectically enfolded complexities of symbolic universes inhabited by the members of a symbolic species.

The early phases of these developments in Germany contributed to preparing the "soil" to grow one of the secular political religions of the twentieth century. As noted, the Germans had derived enormous technical and economic advances from the widespread application of discipline-based approaches, to the point that they could fight the First World War against the other industrial nations. It was the first technological war, which was so terrible that some believed it would end all wars. The considerable war reparation payments imposed on Germany aggravated yet another economic crisis of industrial civilization. Hitler performed the unprecedented economic miracle of bringing rampant inflation and unemployment under control in a very short time. It was widely admired across the political spectrum in many countries, especially by many captains of industry. Already in the mid-1930s some foresaw where this would end: another world war and the sacrifice of millions more on the altars of national socialism. Hitler became the prophet of a glorious millennium that awaited the master race, conducted by the Nazi Party. The Nazi Party would lead the German people to this glorious future. It performed all the sociological functions that traditional religions had in the past.[66]

The developments just described raise historical and sociological issues almost universally misunderstood by the philosophers of technology in the latter half of the twentieth century. How can technique as a human creation shut out human life and go its own way, as it were? How can rationality "shut humanity into an iron cage," and how can technique be an autonomous phenomenon as humanity now watches from another cage equipped with all manner of screens? Such questions do not make any sense philosophically, but they do make sense from the perspective of the social sciences and history.

As noted previously, the relationships people have with their social and natural surroundings are reciprocal in character. As people change these surroundings, the surroundings simultaneously change them. The latter interaction is almost always overlooked in the study of industrialization. Yet it is difficult to find a more comprehensive and massive change in human life and society than the one that accompanied industrialization. The flood of new experiences (symbolized as neural and synaptic changes to the organizations of people's brain-minds) therefore brought about substantial changes to human life, society, and cultures in the course of several generations. These kinds of changes were clearly trivial taken one experience at a time, but they accumulated into substantial changes. Hence, as people contributed to the building of entirely new kinds of societies, they were simultaneously affecting what it is to be human.

In other words, the spectrum of possible relations people can have with their surroundings can range from those where their influence on these surroundings is much greater than the influence these surroundings have on them to situations where the opposite is the case. Industrialization is a case in point. As people changed their technology, this technology also changed them, to the point that eventually a great many characteristics of individual and collective human life no longer correlated with the values and beliefs of society. Instead, they reflected the influences of technology (and everything else that came with it) on people. It is for this reason that it was claimed that capitalism alienated all people, rich and poor alike. It is why rationality could shut humanity into an iron cage and why technique behaved as a largely autonomous force on human life and society. Eventually, it became a system functioning at arm's length from human life, and the desymbolizing effects continue to transform what it is to be human. Politicizing these findings constitutes a clear example of what Georges Devereux has called counter-transference reactions: an unconscious reinterpretation of what social scientists observe so as to reduce the threat to what people believe, thereby lowering their anxiety levels.[67] There is little doubt that many of the major changes that occurred in the industrially advanced world in the second half of the twentieth century were not the result of people's values, beliefs, and ideologies but necessary accommodations to technique and everything else that came with it. The recognition of this reciprocal interaction sweeps away most of the usual interpretations of what has been happening to human life and society in much of the nineteenth and twentieth centuries.[68]

As a result of technique changing people, we speak of the economy, society, democracy, or religion as if our world was made up of all kinds of distinct categories of phenomena which are simply "out there" and existing on their own, as is the case for the phenomenon of gravity. As a consequence of the desymbolizing influences of science and technique, we experience very little of the dialectical enfolding of our symbolic universe. It has led to people believing that many significant phenomena influencing our lives are simply natural. Nevertheless, there is no separate economy "out there" any more than there is a society "out there." With globalization, I can safely say that you are a part of my economy as I am a part of yours. In the same way, there is no separate religion: it is a cultural creation which will collapse along with the myth that gave birth to it. In daily life, we express these kinds of sentiments when we explain something by saying: "This is how the system works." In the case of technique, it has become extremely difficult for us to participate in it on human (i.e., on cultural) terms.

The following five characteristics of technique make this clear.[69] A first characteristic is that of rationality. The rational and technical approaches transfer elements from human life and the world into the domains of disciplines and thus from a dialectically enfolded symbolic universe into a logical domain – an action that externalizes the symbolic character of human life and society. As a result, we cannot participate in technique as a symbolic species even though this is what has made us human until now.

A second characteristic of technique is the necessary linking together of all techniques. The technique-based connectedness of human life and society is not the result of human choices that retain the "good" and reject the "bad" ones. As noted previously, when a new technique increases the efficiency with which a set of inputs is transformed into a desired output, the local dynamic equilibrium of the technique-based connectedness is disturbed. A ripple effect of further adjustments spreads out from it. Techniques are linked in order to maintain the dynamic equilibrium of the technique-based connectedness of human life and society, whether they are beneficial to them or not. It is built up one domain at a time without any reference to sense.

A third characteristic of technique is the absence of values in technical choices. The reasons are clear: an efficiency of 82 per cent is always better than 81 per cent, and 85 per cent would be still better. The choice between technical options is thus virtually automatic, which is inevitable since everything technique touches is measured on its own terms. Besides, technical specialists have no idea of the consequences of the choices they make that fall beyond their areas of specialization, with the result that they have no option but to deal with everything on its own terms. For technique, more is always better, but in human life this is rarely the case. In the case of consumer goods, food is a good example. Having too little food is to starve. Having merely enough is to survive. Having a little more than necessary allows for the enrichment of life by an occasional special meal or a party. Having an abundance can lead to abuse, often accompanied by obesity and ill health.

A fourth characteristic is that the evolution of technique is largely driven by feeding on its own problems and by its internal flow of information. The first three characteristics contribute to the expansion of a technical order that causes disorders in human life, society, and the biosphere. In one way or another, these disorders directly or indirectly undermine the efficiency of technique. When this becomes apparent, creative attention applies the technical approach, and new techniques or further advances follow. The root of the problem can never be dealt with by these approaches, for the simple reason that they do not take note of the influence they have on their surroundings in order to make adjustments to prevent harm. The analogy would be to drive a car by concentrating on its performance as indicated by the gauges on the dashboard and only look out the windows when this performance is interrupted by a loud bang. In this way, technique feeds on the disorder it creates: it has prevented economies from producing wealth, societies from sustaining human life, and the biosphere from sustaining all life. A second major way in which technique evolves is the result of every area of specialization within it functioning as a transmitter and receiver of technical information. For example, when a graduate student finds a way of making some device or process more efficient, it will be written up in a paper, presented at a conference, or published in a journal, with the result that this information flows into the system, where it generally triggers other developments which then trigger still others, and so on. In other words: while, to a limited extent, technique continues to grow in response to human needs and desires, the above components cause it to develop more and more at arm's length from human life. It has made technical development extremely nonlinear and unpredictable, as the failure of technological forecasting has made clear. Information regarding innumerable advances flows throughout the system, triggering further developments with minimal intervention.

Finally, the technique-based connectedness is increasingly universal. Striving for the greatest possible efficiency makes no reference to local conditions, with the result that everywhere technical experts arrive at more or less the same results. Technique is also universal in the sense that it has become co-extensive with culture in nearly every society.

These characteristics are closely related to the fact that any technical specialist is suspended in a triple abstraction. Reaching for absolute efficiency begins with a first abstraction: replacing the surroundings of something to be technically improved with the necessary inputs received, and the desired output to be returned to the surroundings. A second abstraction reduces the involvement of a technical expert to those aspects of the transformation of these inputs into the desired output that are commensurate with the disciplines to be applied by the specialists. A third abstraction reduces the question as to which technical option is best for human life, society, and the biosphere to the question as to which one is absolutely efficient. For example, to apply operations research to the running of any aspect of a modern hospital, the world is replaced by the input of ill people and other supplies, and the output of people (hopefully) on the mend. The process of transforming these inputs into outputs is known differently by doctors, nurses, physiotherapists, lab technicians, dieticians, administrators, information specialists, cleaning staff, security guards, and so on. The third abstraction reduces the choice of possible improvements to the one that is the most efficient. As long as we entrust the running of our hospitals to the application of discipline-based approaches to knowing and doing, the triple abstraction within which the participating specialists are suspended is the mirror image of the above five characteristics of the phenomenon of technique. What is true for the way we run our hospitals is true for everything else.

Our civilization has committed human life, society, and the biosphere to a new master, that of knowing and doing everything by means of absolute (as opposed to relative) approaches to knowing and doing. As we shall see in subsequent chapters, it represents the cult of the fact and the cult of absolute efficiency. It has put our civilization into a dilemma. We can improve the performance of everything beyond our wildest dreams, yet at the same time, this performance comes from putting the fabrics of human life, society, and the biosphere on a rack that slowly but surely pulls them apart towards non-life. In contrast, approaches based on symbolization, experience, and culture balance performance with context compatibility.

By way of an example, consider the environmental crisis. Although humanity has declared our contemporary ways of life to be unsustainable, we have made little or no headway in changing any trends other than in a few limited cases in which society has simply said: "No!" Such situations are extremely rare, because in a great many cases it is impractical to do this, given the way all techniques are linked together in a global technique-based connectedness. We have no idea of how deeply the environmental crisis is rooted in our contemporary ways of life, but we could begin with the recognition that if we believe there is a separate environment "out there" we are lost from the start. There is no separate environment any more than there is a separate economy or society "out there." I am a part of your environment as you are of mine, which means that the roots of the problem can be found in the organizations of our brain-minds. We have symbolized the situation in a completely inadequate manner.

We have written an equation suggesting that the environmental damage caused by technology is the product of population, per capita consumption of matter and energy, the "environmental load" put on ecosystems and the biosphere per unit of throughput of matter and energy, and the ecosensitivity of local ecosystems or the biosphere (which determines the environmental damage done per unit of "environmental load" imposed). Research has shown that despite the baby boom following the Second World War, and the sharply rising standards of living, the primary contributing factor to the environmental crisis in the United States was technology. All technologies were becoming more absolutely efficient and less and less context-compatible. It is no wonder that all the environmental restrictions had so little effect. Technique represents a systematic destruction of the fabric of all life. It is simply built into the way we organize our lives in the world. If we are to confront the unsustainability of our ways of life, there is no other solution but to transform them.

Another example comes from economics. John Maynard Keynes and his followers managed to stabilize the economies of the industrially advanced nations following the Second World War. Their point of departure was the principle that the economy was there to serve people and not the reverse. The stabilization worked very well until technique began to transform the economy on the basis of the cult of efficiency. Costs incurred in the production of wealth skyrocketed, with the result that net wealth began to decline, and budget surpluses, to be spent during lean times, all but vanished.[70] The results were masked for a time by transferring wealth from the lower socio-economic strata of society to the upper ones, causing an unjust income distribution.[71] Rather than going back to the root of the problem and examining why Keynesian economics was failing, conservative economists managed to convince society that there was nothing to be done but to let "free" markets take their "natural" course. Unfortunately, most people never heard nature give this command. It made the situation a great deal worse than if Keynesian economics had been adapted to face the challenge technique posed to human life and the planet in general and to the economy in particular.

Examples of this kind can be multiplied almost without limit. Our new defining spiritual commitment to absolute knowing and doing, by its very organization and structure, will disorder everything it touches; unless we wake up soon, a liveable, sustainable future may be beyond our reach for some time to come. The worst we can do, however, is to politicize our situation. No political beliefs or sentiments, from the extreme right to the extreme left, are going to have any effect whatsoever on the organization of our society. Our entire political spectrum has become irrelevant in the face of technique. A gradual mutation in the organization of industrial society will have to begin, using a non-discipline-based diagnosis of what has been happening to human life and the planet during the last half-century. We have simply grown out of touch with ourselves and the world, as we will seek to demonstrate in the next chapter.

Do We Have a Future?

Technique has become the third life-milieu of humanity. Techniques of all kinds now interpose themselves between human beings and everything else they relate to, thus forming a kind of technical cocoon via which we experience the secondary life-milieu of society and the tertiary life-milieu of nature. Although the life-milieu of technique, like that of society, is of our own making, its influence on human consciousness and cultures is now so great that we have become completely alienated and reified by it. There is nothing new here, but we have surrendered our lives to new myths, which include, as we will see, those of science and technique. Our response should be the same as before. We must symbolize this life-milieu to discover the meaning and value its elements have for human life and society, and thus create a sphere of possibilities between ourselves and it. In this way, the struggle for freedom against the alienation and reification imposed by this life-milieu must recommence.

There is an important difference in symbolizing this life-milieu, in contrast with the two previous ones. Technique has an enormous desymbolizing influence on humanity as a symbolic species, thereby diminishing our capacity to create this distance between ourselves and our life-milieu. All relative approaches to knowing and doing have essentially been relegated to our personal lives, while our public life together is now dominated by discipline-based approaches that are absolute rather than relative in character. Our new spiritual masters have thus come into focus.

Our liberation from myth has always involved being as symbolically realistic as possible about our lives and the world. Symbolization, experience, and culture remain the best approaches available to us, since we know very well that all too quickly we will be trapped in new myths. Nevertheless, we must confront science, technique, and the nation-state in terms of what they can really do for us and rely on other things for what they cannot deliver. This kind of realism is best served by symbolizing these human creations in terms of the meaning and value they have for human life and society. It is our only hope. We must create economies that will not bankrupt us because the costs they impose far exceed the wealth they create, and ways of life that can sustain us so that we do not have to rely on antidepressants and other compensatory means. We must protect a biosphere which, when left to itself, will again sustain all life on the planet.

What I am calling for is for humanity to become genuinely secular by being as realistic as we can be with regard to our most powerful creations, which are closely associated with new secular myths. We must shed all present illusions of being secular, based as they are on mistaking the replacement of traditional myths by our present secular ones for their disappearance altogether. There is a great deal that discipline-based science can never know, and a great deal more that discipline-based technique will never be able to accomplish for us. Moreover, everything is political only if the state is seen as omnipotent and able to do everything for us. These are the false gods which our supposedly secular humanity now serves. It has begun a race to the bottom, as all people and cultures must bow down to the globalization of technique-based connectedness, which is the only thing that is protected by so-called free trade (free from all human values and aspirations). Everything else will be sacrificed on the altar of the third secular political religion, namely democracy. It celebrates our supposed freedom of choosing politicians and governments – which faithfully serve our secular gods regardless of the consequences.

It may be objected that I am a dreamer. How can I propose that humanity attempt to live without myths when it has never succeeded in doing so? Nevertheless, we must engage the struggle, which was originally supposed to have been engaged by the Jewish people who were to live without false gods, and by Christians who served a master who desired to rule by freedom and love. Being realistic is next to impossible for humanity because it brings us face to face with who we really are, and that is so threatening that we have always hidden it behind myths. Now our future as a symbolic species and the future of our planet itself hang in the balance. Are we able to confront our false gods? I know full well that a trip to our local synagogue or church will lead to the conclusion that I am a dangerous idealist. Nevertheless, it is this iconoclasm that sustains the present inquiry and which will lead to the courage and hope necessary to put it into practice. As creatures we cannot know anything absolutely, create anything without limits, or make gods to whom we can assign the responsibility for doing what we cannot accomplish. We must rediscover the Jewish and Christian message, but we will have to do it without their present trappings. We must learn to live, know, and do by a word (symbolization, experience, and culture) inspired by the Word. In so doing, we will create a distance between ourselves and our present life-milieu to create a sphere of new possibilities in which life will become more liveable and sustainable.

How We Will Proceed

Each of the following five chapters examines how the process of desymbolization turns knowing, doing, economic housekeeping, living together, and our personal lives into the opposites of what they were when humanity was more fully a symbolic species. Their relative characters have been turned into absolute ones with corresponding secular religious attitudes creating a cult of the fact, a cult of efficiency, a cult of growth, and a cult of life as non-life. The latter is characterized by low levels of embodiment, participation, commitment, and freedom, creating a kind of anti-life on the proliferating media, especially the web. Following a diagnosis of desymbolization, the first four chapters will present practical steps that could be taken if we were to exercise our freedom and thus break with the secular religious attitudes associated with each of the cults. In other words, resymbolization is ultimately a spiritual battle against our new masters. The final chapter will show what this battle will entail: there will be no secular eternal life on the web as a release from the non-life in anti-societies. Only by exercising our new freedom can a liveable alternative be created.

1 The Cult of the Fact: What Discipline-Based Science Will Never Know

How Well Do We Understand Ourselves and the World?

In relation to this question, we have already noted the following. It is highly probable that the reason for our becoming a symbolic species may have included the recognition that we live in an ultimately unknowable world that cannot be grasped directly. In human life as well as in a living biosphere, nothing can be understood absolutely, that is, in itself and apart from everything else. Everything can only be known relatively, that is, in terms of how anything relates to everything else. Every living entity evolves in relation to everything else, and this becomes a part of its evolving "nature" as a unique summing up of its relatedness. Symbolization was invented and developed for dealing with this relatedness of human life in the world. Every experience in a human life is related to all the others, which includes each element differentiating itself from everything else and at the same time expressing all of this as an integrated whole enfolded into other wholes. For a symbolic species, this enfolding is dialectical in character.

Symbolizing human life in the world includes the unknown, which introduces a dependence on myths. These create a horizon of human experience beyond which there is always more to be discovered and lived. Traditional cultures established through symbolization reached for the heavens by means of their religions, intuited and developed from these myths. As a result, the relative approaches to knowing and doing based on symbolization included absolute points of reference and orientation in an ultimately unknowable world.

It was within the symbolic universes created by symbolization and culture that Western civilization introduced absolute approaches to knowing and doing organized by means of disciplines. Humanity began to live and act as if with these approaches it could reach beyond what it knew and lived. Traditional cultures were regarded as hopelessly bogged down by myths. The new disciplines tried to intellectually grasp "chunks" of what existed beyond the known world, which were assigned the status of facts. These facts were not subject to the upsets resulting from new discoveries and possibilities, thus constituting an absolute as opposed to a relative knowledge. Facts became regarded as objective and universal, and thus valid for all times, places, and cultures. Discipline-based approaches to knowing and doing were considered to have achieved what no culture based on symbolizing individual and collective experience had been able to accomplish: knowing something in itself, thereby creating knowledge valid in itself.

The implicit recognition of all cultures that, in life in general and in human life in particular, everything was related to everything else was swept aside by this understanding of life in terms of non-life. In the case of the former, every situation was an expression of a multiplicity of categories of phenomena all making non-trivial contributions, while in the case of the latter, everything could be separated and divided into non-contradictory elements. These could be defined in themselves on their own terms, measured, quantified, and represented by means of applied logic, and they belonged to a complexity understood and built up one element at a time. The "everything being related to everything else" was thus reduced to considering anything one domain at a time. A domain received certain inputs, repetitively transforming them and passing on the desired output to another domain. Nothing within any domain was affected by any of the others, with the result that whatever happened within it could not evolve and consequently had to be based on repetition, with a regular redesign of this repetition to create the quantum leaps required by the evolving of everything else.

The discipline-based approaches to knowing and doing permitted Western civilization to dramatically improve the performance of everything, and at the same time prevented it from ensuring that the elements which had been made absolutely efficient were compatible with and could evolve in relation to everything else. An economic, social, and environmental crisis burst on the scene, which ongoing developments along the same lines could only deepen. Everything living was thus placed on a rack, yielding facts and absolute efficiency. Efficiency was gained by distorting the internal integrality of everything and severely undermining the external synergies with everything else. In this way, science and technique reached for an absolute reality of their own making by obeying the principles of separability, non-contradiction, closed definitions, applied logic, and a simple non-enfolded complexity. Anything that could not be grasped by means of these principles was not "real" and might as well not exist. Science and technique now defined what was real. It became impossible to argue with facts, which, supported by a new secular myth, were no longer limited to the results of a unique kind of human activity carried out by specialists in the domains of their disciplines according to their education and prior experience in the unique setting of a laboratory, or in the intellectual context of a single category of phenomena. All this was guided by the values, beliefs, and myths built up metaconsciously in the organizations of the specialists' brain-minds. These dependencies imposed limits, but these limits were obscured by new secular myths that converted facts into absolute knowledge. The facts were not questioned except under highly unusual circumstances. The dependencies and thus limitations of everything that had been made absolutely efficient were obscured by other secular myths. Even our economic, social, and environmental crisis did not expose these myths. To put this situation into traditional religious language: the absolute facts and absolutely efficient creations are the new mini-gods of our new secular myths, having lost the relative status of all human knowing and doing.

Although this is but a preliminary summary, it can readily be tested by an attempt to make a list of things science will never be able to know, and another of what technique will never be able to accomplish. I will be the first to admit that it is very difficult to come up with such lists, and yet I have often teased my audiences by asking if this means that science is omnipotent in the domain of human knowing and is thus our secular god of knowing, and whether technique is our secular god of doing. The suggestion makes all of us uncomfortable, but it does not appear to change our difficulties with coming up with these lists.

Nevertheless, we should now be able to generate such a list. A comparison between culture-based knowing and science-based knowing shows that the former furnishes us with knowledge of our lives in the world, while the latter can do no such thing. As noted, there is no science of the sciences capable of scientifically integrating the domains of all the disciplines into an understanding that functions on the level of any culture. In the same vein, consider the economic policies of our present-day governments. All such policies submit themselves to economic facts and the mathematical functions and theories that interpolate and extrapolate them. It is possible to put various spins on all of this, but it remains a submission to the facts nevertheless. Things are as they are, and this gives conservative politicians an important strategic advantage, especially when they appeal to all this as being natural. Other politicians have little choice but to bow down to the same facts and theories, which means that political differences on this point have essentially collapsed.

As a result, our investigation of how good an intellectual grip we have on human life in the world is conducted in a secular sacral world of facts and absolutely efficient creations. We cannot escape it and must confront it head-on, knowing full well that this universe ultimately relies on myths, which will pass like everything else. We can learn to *live* in this sacral universe by intellectually mimicking the metaconscious symbolization of these facts. Doing so resembles putting together a jigsaw puzzle: each fact must be examined in the context of as many other facts as possible. Since these facts are nothing else but a limited expression of a network of interdependencies related to human life in the world, their "shape" is determined by this expression, which introduces a constraint on this intellectual approach: it is the diametrical opposite of the one based on disciplines. As these "shapes" are informed by their dependence on context, they will gradually begin to appear as fitting into a single puzzle, and an image of human life in the world will begin to emerge. It must be emphasized that neither this intellectual strategy nor the ones based on disciplines can be adequate in themselves. Potentially, they complement each other because the one strategy tends towards an intellectual asymptote of knowing less and less about more and more, while the other tends towards an intellectual asymptote of knowing more and more about less and less. Benson Snyder[1] has examined how the professional lives and careers of MIT students depend on the two intellectual modes he refers to as numeracy and literacy and how an underdevelopment of the latter tends to have serious consequences for careers and personal lives. Elsewhere,[2] I have examined the limitations of what I refer to as intellectual map making and discipline-based approaches, and how these limitations can potentially be overcome by a dialectical relationship between the two intellectual approaches. This possibility has been destroyed by the sacral universe of secular myths.

We have thus distanced ourselves from the current practices in the social sciences. Anyone who has lived through the professional developments of any of the major disciplines of the last fifty years or so cannot fail to recognize how one intellectual fashion has succeeded another, to the point that many professional journals become their mouthpieces. Few, if any, look at these intellectual fashions in the context of the development of a discipline as well as in the context of what is being studied of human life or the world, including their constant adaptation and change over time. When this is attempted, it becomes embarrassingly clear that none of these intellectual fashions have had a good intellectual grip on what is being studied, because in a very short time they become hopelessly irrelevant. If these theories had been solid, they ought to have understood the deeper changes that would be felt in the immediate if not long-term future. With hindsight, it is possible to separate the "good" theories as explaining their own time with accuracy, as well as what followed, from the "bad" theories that failed to describe what they studied in a durable manner. In this way, the "good" theories have shown evidence of insight and are likely to become promising puzzle pieces of an intellectual map of human life in our world.

Qualitative social theory thus ought to have a historical dimension. In addition to being tested in this way, it can also support and be challenged by quantitative approaches. The overwhelming majority of contemporary social science research instruments, which take the form of surveys, are not based on a comprehensive understanding of the situation under study. As a result, the target group may not recognize the questions as corresponding to their own experience. Instead of reacting to the questions as implying daily-life situations, they see them as hypothetical ones, which they think through without any guarantee that, if ever they were presented with such a situation, they would respond according to their answer. If the organizations of our brain-minds allow us to skilfully cope with our lives in the world, this coping is based on metaconscious knowledge that cannot be applied to unusual situations. These must be thought through, which pushes the response back to stage three "problem solving" of the human skill acquisition model of Stuart Dreyfus.[3] As a result, the tensions between the practitioners of quantitative approaches and the few who still practise the qualitative ones have created a situation to the detriment of both parties. If a research instrument does not closely correspond to the experiences of the target group, the answers and the theories based on them will have little, if any, scientific value. They will not inform us how people would really respond on the basis of the organizations of their brain-minds in the way they skilfully cope with their life in the world. The responses will not even be based on a careful analysis involving the application of prior experience. There may be further distortions if the members of the target group suspect a hidden agenda because of the gap between the situations as they experience them and as they are portrayed in the research instrument. Still further distortions may result when responses are "corrected" in order to be consistent.[4] All this is especially clear in a great many laboratory experiments where subjects are exposed to situations never encountered in daily life, which will give us no insight into how they use aspects of the organizations of their brain-minds in their daily lives. The value of these research instruments is not primarily scientific but technical – they function as social techniques for managing human life.

There are two more issues that need to be considered before we embark on puzzling together an intellectual map of human life in our civilization. The first is related to all of us having been educated in one or more disciplines and some of us having gone on to become practitioners. In other words, we have learned to think and work in terms of domains separated from experience and culture. Since we are not intellectually schizophrenic, somehow we must have learned to fit together the world of daily life and the "world" made up of the domains of our disciplines. Moreover, the unknown beyond these two "worlds" also has to be dealt with. We can therefore expect to be suspended in a set of secular myths that somehow integrate the two "worlds" and symbolize the unknown as more of the same. There is thus a probable interpenetration of what we know on the basis of symbolization, experience, and culture and what we know on the basis of disciplines. For example, when we use words such as science, technology, the economy, mass society, politics, morality, religion, and art, they bring together the experiences of our daily lives, the images and messages from the mass media, and what we know of some disciplines. This will make our task easier in some ways and much harder in others.

The other issue to be considered is the making of an intellectual map of human life in our civilization. Like all maps, it must serve a particular purpose. For example, geographical maps may be designed for automobile drivers, hikers, pilots of small planes, and so on. It is in relation to each purpose that some features become highly relevant, others marginally relevant, and still others completely irrelevant. The same is true for an intellectual map. There can be no pretence of creating an *objective* map of human life in our civilization. We would be left with no criteria to decide which aspects of human life should be retained and placed on the map and which can be neglected. These kinds of choices require values in order to decide what is relevant and what is not relevant, and within the former, a hierarchy of relevance so that what is of minimum relevance can be neglected if the map becomes too crowded. In this regard, I am going to follow a Western intellectual tradition in which some of our greatest theoreticians have selected the value of human freedom in order to examine which features of our civilization have most threatened that freedom through alienation. In the context of understanding human life and the world, this was, in one way or another, the central concern of Karl Marx, Max Weber, John Maynard Keynes, Jacques Ellul, and others. Alienation is the secular equivalent of what, in the Jewish and Christian traditions, is called sin: to be possessed by someone or something, thus making it impossible for people to be the human beings they were called to be. As a result of its three cultural pillars, Western civilization has spread the value of freedom in relation to alienation across the world, to the point that humanity recognizes that slavery is not an acceptable form of human life. The value of freedom is closely related to the purpose of the intellectual map we are seeking, namely, to evolve towards ways of life free from our secular myths and their false gods, and thus more liveable ways of life that can also be sustained by the biosphere.

We live as if we are surrounded by a world made up from separate entities: a society, an economy, markets, democracy, science, the state, mass media, the internet, religion, and much more. They have all been symbolized and named to be included in the vocabulary of the working culture of our society. What are we to make of entities that are not included in our usual vocabulary, that are not mentioned in the mass media, that are not spoken of by politicians and other officials, and that thus are excluded from the mainstream of daily life? If technique is really the growing alternative to culture-based knowing and doing, then our vocabulary, which may well have been adequate before, may not have sufficiently evolved for us to have a good intellectual grip on where we are taking science, technology, and the economy, and where these creations are taking us through their enormous influence on human life and the planet.

Another way of approaching the above question is to examine some of the important concepts in our working vocabulary such as markets, the economy, free trade, industry, society, and the like. For example, are the economies of the first half of the twentieth century, the second half of the twentieth century, and the opening decades of the twenty-first century similar enough to be referred to by the same concept? Or have the differences overwhelmed the similarities, with the result that we are missing something new and important regarding this part of human life in our civilization? If we find this to be the case for a large number of these concepts, could this vocabulary mask some very significant changes that have taken place? For example, are mass societies dominated by the social media and the internet still essentially similar to mass societies with the more traditional media? Are both sufficiently similar to the industrializing societies of the first half of the twentieth century that they can be referred to by the same name? Are the markets described by Adam Smith, which he encountered in his world of the eighteenth century, essentially similar to the markets we hear so much about today? Is free trade as described by the economists of the late nineteenth and early twentieth centuries the same kind of free trade as we have today? Is the economy of today comparable to the economy of fifty or a hundred years ago? We can multiply these kinds of questions indefinitely, but the point is clear: we cannot automatically assume that we have a good intellectual grip on human life in our civilization.

We encounter much the same kind of question in the university. For example, in the 1960s and 1970s there was a considerable debate about whether we lived in an advanced industrial society, a post-industrial society, a post-capitalist society, a mega-machine society, a consumer society, a mass society, a spectator society, a new industrial state, a technotronic age, or the beginning of an information society.[5] When we speak of industrial societies, what we are implying is that almost every aspect may be understood in terms of the effects industry has had and continues to have on it, thus making industry the intellectual key to coming to grips with human life in these societies. In the same vein, the above theories imply that industry is still a very important phenomenon, but it has changed in non-trivial ways, and what the influence of that is on society is unclear, as if society wasn't always advancing, or as if we were not post–a great many other earlier developments. Other theories of the time clearly went beyond industry by recognizing that new and important phenomena had sprung up, which were now the intellectual key to understanding human life in society. The diversity of theories was never narrowed down by at least a few appearing to gain a decisive explanatory advantage over most of the others. Of course, it must not be expected that a diversity of intellectual maps can be reduced to one, given the different purposes and values that underlie them. However, we did not even come close to some kind of intellectual convergence taking place. As a result, almost all the disciplines of the social sciences had competing schools, but these rapidly lost ground as the quantitative approaches encroached or displaced the qualitative ones. This created an illusion of scientific objectivity.

In daily life, the concepts of a language function like linguistic models related to human life in the world; as members of a vocabulary of a language, they jointly create a kind of linguistic map of that life. By acquiring this vocabulary, toddlers begin to make an unintelligible complexity of "noise" into an intelligible complexity, that of a world which gradually converges with the symbolic universe of their community. At first, the words spoken by adults are merely differentiated from all other sounds that have taken on a meaning in their lives. Gradually, the differentiation of all these sounds begins to imply a metaconscious knowledge to the effect that words are very different from all the other sounds, with the result that some of them can take on the status of vocal signs when they become associated with certain kinds of experiences. Gradually, this metaconscious knowledge will also imply that all these vocal signs have much more in common with each other than they do with all other sounds, leading to children's intuiting something of what they will later learn to call a language. This development is further reinforced when they learn to differentiate experiences with a linguistic communication in the foreground from all others. Additional metaconscious knowledge of the role of language in their lives begins to develop, as implied in their more deliberate use of it. Eventually, they metaconsciously learn that what constitutes a language is what all other experiences are not. A distance is thus created between language and experience and between the corresponding forms of metaconscious knowledge. The metaconscious knowledge related to non-language experiences acts as a metalanguage in the context of which words and phrases take on initial meanings, but these are increasingly supplemented from the meanings derived from directly differentiating the language-related elements from one another. This differentiation entirely on the level of language builds a new kind of metaconscious knowledge that forms a secondary context in which words and phrases can take on a further meaning. Eventually the meanings of words and phrases are dialectically built up, primarily from what all the others do not mean, and secondarily from the meanings derived from the metalanguage created by the differentiated experiences not involving language. A double referencing system thus develops. In this way, children gradually learn to enter the symbolic universe of their community and make sense of it the way the adults around them do. The organization of these symbolic universes is mapped by the vocabulary of their mother tongues. It is in this manner that a semantic memory emerges distinct from the episodic memory of the experiences of their lives.[6]

For example, our linguistic map of the world is full of words such as table, chair, couch, and cupboard which jointly organize the "world" of furniture. There are many other such "worlds." If our linguistic maps are designed to make our lives and the world intelligible, how do we know if they are good maps? These maps cannot be tested directly because we are symbolically suspended in an ultimately unknowable world.

The effectiveness of our linguistic maps depends on striking a balance between overgeneralization and undergeneralization. If these maps overgeneralize, they may omit important details that are crucial to our lives. If the way our attention has directed the metaconscious process of differentiation is too general, what has been experienced as essentially alike may turn out to be fundamentally different. The opposite occurs when our linguistic maps undergeneralize by treating something that is relatively unessential to our lives as a fundamental difference. In either case, our intellectual grip on our lives in the world will not be as good as it might be if the balance between overgeneralization and undergeneralization had been struck differently.

Consider something we all experience in our lives: a table. From the perspective of overgeneralization, it might be argued that this concept covers a rather large range of objects, some of which have four legs, others one or two pedestal legs, while still others use two or more sawhorses, and so on. Some tables have rectangular tops, others square ones, and still others oval or round ones. Some tables are much lower than others, some can be extended, some have drawers, and so on. In daily-life experience, we have never had to reflect on whether this concept overgeneralizes or undergeneralizes. Metaconsciously, we must have learned that the differences between all these kinds of tables are less important than their similarities. This importance is relative to their meaning and value for our lives in the world. From this perspective, the differences between the tables are somewhat insignificant in the context of the activities we carry out at these tables, including eating, working, or playing games. The table permits us to have everything we need for these activities in easy range. In other words, eating a meal at a table is not affected by its legs other than in the way people can be grouped around it. The shape of the top does not greatly affect the activities either, although some shapes may be more convenient for an activity than others. It would appear that the way we have directed our attention at tables and its influence on our metaconscious processes of differentiation and integration has struck a good balance between overgeneralization and undergeneralization.

Ludwig Wittgenstein[7] convincingly showed that what the members of any category (such as games or tables) have in common is a network of crisscrossing and overlapping family resemblances. For our purpose here, we must go further than this. In the case of games, they constitute a unique kind of human activity that comes as close as we can get in real life to a domain, since they are rule-based, which remains unaffected when the telephone rings during a game, or when a player momentarily leaves to get a drink. By differentiating the playing of games from all other human activities in this manner, we have correctly differentiated the differences between the many kinds of games (board games, card games, field games) as different versions of essentially the same kinds of activity. Again, there seems little question of overgeneralization.

We have previously insisted on the fact that humanity has always symbolized everything in terms of the meaning and value for human life in the world. Hence, it is not the factual details of tables or games that matter, but what is true for our lives. As another example, consider the descriptions of different diseases in a medical textbook. These descriptions are designed to help medical students learn to diagnose these diseases in their patients. What matters is not a factual description, but the determination of the significance of different symptoms for the lives of their patients as well as for their careers as doctors. Medical students learn very quickly that they cannot take the descriptions in their textbooks as factual. Only a tiny minority of patients may have all of the described symptoms, the patients who have only some of them can have very different combinations, and some patients may be asymptomatic. It is for this reason that medical diagnosis is an art, which functions on the level of experience, and not an applied science, which functions on the level of the domains of disciplines. Hence, there is no difficulty with the observation that most patients who have a particular disease do not exactly fit its description in a textbook.

There are cases where the vocabulary of a language may appear to be inadequate. For example, in the landscapes of far northern areas, largely constituted of snow, ice, and water, human survival is not dependent on a large vocabulary organizing the many different kinds of snow and ice conditions, even though they are frequently a matter of life and death. Whether the snow is dry and powdery or wet and heavy greatly affects the efforts dogs have to make to pull a sled because of the way their paws slip on or dig into the snow, and the drag force the snow exerts on the sled. There does not need to be a close correspondence between the vocabulary of the language used by northern people and the significance that snow conditions have for their lives. This is because of the double referencing system, one to the metalanguage of differentiated experiences and the other to words and phrases being directly differentiated from each other. As a result, the hunters may get together one morning and, having observed the snow conditions, simply say that it's not a good day to go hunting, and they will all know why this is.

Because human life in the world is always evolving, the vocabulary of a language must constantly adapt and evolve as well. For example, as Inuit people evolved, the ways they harnessed dogs to their sleds, the design of the sleds, their hunting strategies, and the meaning and value of different snow conditions for their lives changed. What snow and ice are to Inuit cultures and camels to desert cultures is matched in importance by what capitalism, markets, free trade, science, and technology are to Western cultures ever since industrialization began. The question is whether these concepts have been adequate during all this time, or whether their use glosses over significant developments that have taken place.

First, let us examine the concept of capitalism. It refers to an economic organization in which the renewal and accumulation of capital takes priority over almost everything else. This was the case in all industrializing societies – democratic, socialist, and communist. Building and evolving an industry involved a vast infrastructure including mines, railroads, canals, and an urban habitat for all the people involved. Nothing could be accomplished without ever larger amounts of capital. Consequently, every industrializing society had to make the renewal and accumulation of capital its top priority, regardless of the political regime. In the nineteenth century, capital was regarded as embodied labour because the value of what factory workers produced was greater than their wages, making the difference available for the renewal and accumulation of capital. The more factories were automated, the less this was the case; with high levels of computerization, capital essentially became embodied technique. Moreover, the emergence of technique in industry, after the separation of technological knowing and doing from experience and culture, necessitated major structural changes in the corporation, the role of the market, the role of capital, and the involvement of the government in the economy.[8] Discipline-based approaches changed everything in the democratic as well as in the socialist societies.[9] Technique became what capital had been in the nineteenth century.[10]

Beyond industry and the economy, a great deal of capital is now produced directly from capital itself without any intervening economic activities.[11] Financial techniques have produced a proliferation of derivatives. It is estimated that around 97 per cent of financial flows in global markets have no relationship whatsoever to economic activities.[12] Speculation on an unprecedented scale by hedge funds, banks, and mutual funds (often backed by computer-driven algorithms for market "trading") has pumped unimaginable sums from lower socio-economic strata, to concentrate more and more wealth at the very top. Central bankers, by means of what is euphemistically referred to as quantitative easing, have printed staggering sums of money that have little or no correspondence to economic activities. The tiny percentage of financial flows corresponding to real economic activity has produced an anti-economy by externalizing costs on a scale that appears to exceed wealth creation.[13] In sum, calling the economies of the nineteenth century, the twentieth century, and the opening decades of the twenty-first century by the same name is to live in the land of economic illusions. Doing so implies a level of overgeneralization that glosses over so many fundamental changes with vast consequences for human life and society that we may have lost our intellectual grip on what we call our economy. We shall return to this subject in a later chapter.

Much the same argument can be made for what we call the market: a kind of organizing principle of the above economies. Adam Smith clearly described the functions of individual markets for specific goods or services as well as their overall role as an "invisible hand."[14] Do these markets have anything in common with what we refer to by the same name today? Without going into details, almost no aspect of his description is applicable to our markets: the size of the producers relative to the size of the market they supply, the number of suppliers, the influence individual producers can exercise over the market they serve, and much more.[15] Today, the great invisible hand is knocking over human lives, communities, and ecosystems – almost everything that really matters to us. John Kenneth Galbraith has argued that when highly specialized scientific and technical knowledge becomes the critically important factor of production, corporations have to carefully plan the entire technological cycle of their products and services: their innovation and development, production, marketing, and obsolescence.[16] All this technical planning now dominates the global economy, and the transnational corporations can no longer rely on markets for supplying many of their "inputs" or for absorbing their "outputs." This analysis was confirmed by examining the changes in technology that occurred as a result of the separation of technological knowing and doing from experience and culture.[17] The resulting planning system made the assumption of consumer sovereignty untenable: economic democracy was undermined to the extent that producers could influence consumers by the mass media and especially by advertising.[18] The significance of this development must be understood in the context of the desymbolization of cultures, which now had to be supplemented by the bath of images portraying everything that used to be provided by a tradition: what we should eat and drink; what we should wear; what we must own and drive; how we should entertain ourselves; what questions we should ask our doctors and what drugs we might have to take; what insurances we should carry; how we should decorate our homes, and what we need to do to sell them; what is happening in our world; what our government is doing and who our friends and foes are; what we should be concerned about, and a great deal more.[19] The development of so-called market economies also fundamentally changed our relations with the biosphere. The markets described by Adam Smith functioned in economies whose scale was small compared to the biosphere. Hence, natural "capital" would not be depleted, and natural resources did not have to be priced. Such an assumption is entirely untenable today.[20] As noted, financial markets have undergone a complete transformation as a consequence of technique, with equally vast implications. In sum, to make economic policy as if we continue to live in a market economy may well help to explain our current structural economic problems.

Much the same can be said about the concept of free trade. When it was first proposed over a hundred years ago, the mobility of investment capital was almost always national and not global. The implication was that if one country gave up the production of something for which another country would have a comparative advantage, the capital invested in that production would flow to another area of the economy, with the result that employment was more or less maintained. Despite some attempts to assure us that this is still the case today, the evidence overwhelmingly shows the opposite: free trade is accompanied by massive exports of jobs to areas in the world where the costs associated with the production of goods and services can be minimized and externalized to the greatest extent possible. It has completely transformed the meaning of competition. It now produces "cancer cities" and the like, with little or no pushback from cultural values. This trend is now rapidly expanding to include all kinds of services, including those dealing with medical diagnosis, telephone and internet support, information technology services, warranty service and other technical support, architectural design, engineering, management, and accounting. When a government announces new jobs created in our economy, these are overwhelmingly part-time "McJobs" with ever fewer benefits and rarely a pension. As a result, free trade is actually a race to the bottom. First, for a growing portion of the population, the only available participation in the economy is through underemployment or unemployment. Second, free trade has made it much easier for corporations to avoid paying their fair share of the costs of running a society and maintaining its ecosystems on which we all depend. Third, it steadily increases income inequality, which is completely undermining people's sense of belonging to a civil society. Fourth, the steadily growing poverty, particularly among senior citizens, may in the near future undo many of the gains we have made in the last two hundred years. Fifth, by allowing the World Trade Organization to rule over free trade, we have given up much of our political sovereignty to an unaccountable body, for little in return. What is democratically controlled is increasingly minimal and excludes almost everything important for our lives. This has meant the weakening of a variety of standards and regulations that have protected working conditions, employment, health, social security, secure retirement, environmental quality, and much more. Free trade has been turned into forced trade for citizens and communities, who have lost all control over it. In comparison, the free trade that occurred a hundred years ago was mutually advantageous and could thus be freely entered into. Today, anyone who carefully studies the trends that have resulted from our free trade deals is not likely to give consent. Free trade has made the world safe for an ever-expanding technique-based connectedness but increasingly unsafe for people, communities, and ecosystems. We will return to this subject in a later chapter.

In sum, if the roles of the economy, markets, and free trade agreements have undergone substantial structural changes with far-reaching consequences for human life, society, and the biosphere, talking about them in more or less the same way as in the past amounts to a level of overgeneralization that has surely caused us to loosen our intellectual grip on what is happening to human life in the world. Our current ineffectiveness in dealing with a growing range of serious problems may well testify to this.

It is not difficult to add many more concepts to this list. For example, from a cultural perspective, a mass society has little in common with a traditional society.[21] This is equally the case for education, law, politics, morality, religion, and the arts. Table 1.1 provides an overview of some

Table 1.1. Technique and Culture

Dimensions of mediation	Civilization based on culture	Civilization based on technique
Science	Embedded in experience and culture, local, culture-specific	Separated from experience and culture, non-local, universal
Technology		
– knowledge	Embedded in experience and culture, local	Separated from experience and culture, universal
– organization	Non-systematic	A system
– regulation	By means of cultural values	By means of non-cultural performance ratios
Economy	Embedded in and inseparable from a way of life and regulated by its culture	Initially a separate and distinct economy ruled over by the universal institution of the Market and based on the economic approach to life involving the maximization of utility or profit. The remainder of society ruled over by the cultural approach. Later, the economic approach mutates into the technical approach and rules over all of society.
Social structure	Tradition-based societies	Mass societies with weak cultures supplemented by integration propaganda
Political structure	A small and distinct political sphere because most of human life is self-regulating by means of a culture	Everything is political. The state is involved in almost every area of human life.
Legal institutions	Laws are a strong expression of, and an extension of, the metaconscious values of a community. Common law tradition.	Laws have an organizational character reflecting the technical orientation of society
Morality	Group morality anchored in the cultural unity of a society	Individualistic, statistical morality; what most people do is normal, and what is normal is normative.
Dominant personality type	Tradition-directed	Other-directed
Religious institutions	Traditional religions	Secular political religions
Art and literature	Symbolizing the cultural unity and changes within it; what is the most profound in a civilization	The arts depicting non-sense, non-cultural
Relations with the biosphere	Generally sustainable	Unsustainable
Overall characteristics of a way of life	Culture-based connectedness dominates technology-based connectedness	Technique-based connectedness dominates, permeates, and envelops culture-based connectedness, universal

of the changes. It points to the far-reaching consequences for human life, society, and the biosphere of our growing reliance on discipline-based science and technique at the expense of culture.[22]

In terms of assessing how well we understand our lives and the world, the above discussion points to two different findings. First, in the double referencing system where the meaning of words is closely related to the differentiation of our daily-life experiences, thus functioning as a meta-language, symbolization appears to have no difficulties striking a good balance between overgeneralization and undergeneralization (as we saw in the case of tables and games). In terms of what really matters for our lives, superficial differences are not mistaken for fundamental ones, and vice versa. Second, in a mass society where many experiences of the world are mediated by the mass media and social media, the meaning of words is linked to our meta-languages in the most tentative way, with the result that their "meaning" derives from their being differentiated from all other words in our daily lives, but especially those used on the media.[23] As a result, their meaning is greatly affected by their use as "tools" as opposed to bearers of meaning. Images dominate the words that accompany them, and their use is carefully shaped by the many public relations, human relations, advertising, and other psychological techniques to create the desired effects.[24] Manipulating language by a variety of these kinds of techniques has led to integration propaganda, as a complement to a highly desymbolized culture.[25] In these cases, it would appear that symbolizing this technically mediated language causes a substantial loss in how well we understand our lives and our world. Fundamental differences with far-reaching consequences for human life, society, and the biosphere are overlooked. This is clearly illustrated by our uses of words such as the economy, the market, free trade, mass society, and the law.

What can we learn from experts who are interviewed by the media regarding a particular issue? When this happens, all experts face a difficult dilemma. The issues about which we are being interviewed rarely correspond to the domains of the disciplines in which we have gained a level of expertise. If we decide nevertheless to use the technical vocabulary of these domains, we are consciously or unconsciously transferring this issue out of the symbolic universe of our culture into the domains of our disciplines. As a result, much of the rich context of the issue will be lost. On the other hand, if we rely on our experience outside of these domains, we speak our minds much like others who have thought about the issue, except that they may not have the benefit of the authority that comes with a certain expertise, even when it cannot be applied directly without a great deal of intellectual impoverishment. We are now penetrating to the very heart of the dilemma of our civilization. Our growing reliance on discipline-based science and technique at the expense of culture has shifted our knowing and doing related to human life and the world to a knowing and doing related to the domains of disciplines without any ability to reintegrate all this into a comprehensive understanding available through symbolization. The implications, however, are not abstract in the least.

The above situation is not limited to specialists being interviewed by the media. We face the same dilemma when we are engaged in our scientific or technical work. Our discipline-based knowing and doing cannot deal with issues characterized by many different categories of phenomena intermingling and evolving in relation to each other. When this evolving of human life, society, or the biosphere presents us with an issue, we cannot grasp it. We can only deal with those situations in which one category of phenomena dominates all the others, thereby approximating what happens to what occurs in the domains of disciplines. As noted, the latter kind occur mostly outside of life, in the "world" of scientific laboratories or technical entities, processes, and systems. Even though this is obvious, we have handed our ways of life over to absolute knowing and doing at the expense of their relative counterparts based on symbolization, experience, and culture.

There are several practical and far-reaching consequences of all of this. First, everything we touch is dealt with as non-life. The fact that what is happening outside the domains of our disciplines is more complex may well be externalized in our discipline-based knowing and doing, but this does not mean that it goes away. Whatever is externalized will be affected by relations and networks of relations being first intellectually distorted and then being distorted by the application of the disciplines. As a consequence, the relationships or networks of relationships will be strained and even ruptured. Each advance is accompanied by a whole range of negative consequences, which are inevitable due to the way we have organized our knowing and doing by means of disciplines. It is not the result of incompetence, greed, or any such negative personal attributes, but the consequence of how our knowing and doing are structured. For example, if we organize a number of electronic components by wiring them together in one way, they may collectively behave as an amplifier. If we wire them in another way, they may collectively behave as an oscillator. It is the structure of the circuit that determines what happens, whether we like it or not. Why is this so difficult to understand for an economy, markets, free trade, society, and more? When we regard our industrial, economic, trade, social, educational, or environmental policies, we can immediately predict what they will achieve and all the problems they will create, provided that we have the kind of understanding of the situation derived from the equivalent of symbolizing it. Fossil-fuel-based energy policy has a major impact on employment, investment, the economy, resource depletion, the biosphere, trade, north-south relations, the global distribution of production, security, and peace.[26] For over a hundred years, we have had energy policies that have essentially ignored all these major relationships, with the result that we now have systems that provide us with the energy we need but at the same time cause many preventable problems. The same is true for our economic policies, which affect almost everything in our lives and beyond. Nevertheless, we entrust these to economists who are suspended in a triple abstraction, just like all other specialists.

Once we create a variety of problems, our discipline-based approaches to knowing and doing are incapable of going back to their roots in order to greatly reduce or prevent them in the future. The reason is again derived from the structure and organization of our discipline-based knowing and doing.[27] The domains involved in the creation of a problem are different from those that must be applied to deal with it, but no institution comes anywhere close to dealing with this structural dilemma. Consequently, we are stuck with end-of-pipe approaches: the structure first creates a problem in one area and then must address it in another in an after-the-fact manner. As a result, our ways of life are layered with endless technical compensations that never get to the root of any problem and thus must transfer it from one domain to another.[28] Only by means of something analogous to negative feedback can any issue be definitively dealt with, by preventing it in the future.[29]

Our civilization has created a technical order with absolute knowing and absolutely efficient doing. It feeds on human lives, societies, and the biosphere by incorporating some connections of everything being related to everything else and externalizing many others, first in knowing and then in doing. Almost all of the significant problems we face today are thus related. They are the collective symptoms of a structure and organization of our ways of life that treat all life as non-life, and in the process do tremendous harm. It is impossible to think that we can deal with widespread depression in one way, unemployment and underemployment in another way, unsustainable health and social costs in still another way, global warming and other environmental issues in yet another way, and so on. First, in this one-at-a-time approach to the issues we face, we make them amenable to discipline-based approaches, which necessarily will solve any problem by creating several others. Second, by not understanding how all these issues are the symptoms of a common underlying problem, we cannot get at their roots and are thus guaranteed to fail. We have created the perfect structure and organization of human knowing and doing, only to threaten the destruction of all life and the planet. We are out of touch with what is really happening with our lives and the world, and anxious because deep down (metaconsciously) we know something is going fundamentally wrong, but we have no idea what it is.[30] If we examine almost any issue that the so-called industrially advanced societies have faced during the last hundred years, in many cases we would likely come to the conclusion that this description fits all too well. No matter how dazzling the increases in the performance of everything around us may be, most if not all is gained at the expense of everything living.

If we wish to strengthen our intellectual grip on our lives and the world, we will have to learn to connect the mesmerizing successes of increasing the performance of everything touched by technique to the numbing failure of effectively dealing with everything it externalizes. Our inability to do so is deeply rooted in the division of human life and society that accompanied the process of industrialization. The technical division of labour made the lives of industrial workers an externality. The growing market economy further reinforced this by compelling them to behave like *homo economicus*. When knowing and doing separated themselves from experience and culture, these developments spread to the highest socio-economic strata of society: one part of the lives of specialists was lived in the domains of their disciplines and the remainder in the increasingly desymbolized symbolic universes of their cultures. The only way all this could be reintegrated was through secular myths. Since these were hidden in our metaconscious, and since according to each and every ideology myths were a thing of the past, keeping an intellectual grip on all of this became impossible. As noted, a technological approach developed in industry, an economic approach in the economy, and the cultural approach came under increasing pressure in the remainder of society. The "arts" (music, painting, sculpture, literature, and theatre) were moved out of the mainstream of society to become the icing on the social cake, as it were.[31] In the evening, people could attend a play or visit an exhibition, and they might be deeply moved by the message of what was happening to human life and the world. They might even shed a tear, but then they had to get back to "real" life. What the "arts" symbolized became largely disconnected from the mainstream of life, and the emergence of the word "culture" in the English language portrayed this very well.[32] There were significant differences in how the cultures of the industrializing societies responded to all this, but ultimately, this division of human life and society had to be faced in one way or another.

Humanity had always maintained its intellectual grip on human life and the world by symbolizing all experiences in a manner that was individually unique and at the same time typical of the culture of people's communities. Skilfully coping with others and the world was based on metaconscious knowledge built up in the organizations of people's brain-minds as a consequence of neural plasticity. How fundamental this neural plasticity is for a symbolic species was discovered at a time when cultures were being desymbolized under the pressure of technique.[33] The need for a concept of culture followed closely on the heels of the need for the creation of phrases such as "appropriate technology" and "sustainable development," when these could no longer be taken for granted because they were being displaced and undermined by the emergence of technique. No such concept of culture existed as yet in the nineteenth century, and embryonic forms of it gradually emerged in the twentieth century as a consequence of desymbolization. Nevertheless, the vast implications of neural plasticity are far from being worked out in the relevant disciplines. Without this neural plasticity, the metaconscious knowledge required for skilfully coping with our lives and the world could not evolve. At the same time, this development had been entirely obscured by the notion that the brain was much like any other organ: it was first formed in the womb by progressive cell differentiation, after which it matured when children became adults. When metaconscious knowledge became severely weakened by desymbolization, our difficulties of coping with our lives and the world were multiplied, and are most likely at the root of phenomena such as our current epidemic of anxiety and depression. Similarly, the weakened cultural support for contemporary ways of life is in a large measure offset by technical integration propaganda: the bath of images we receive from the media, which collectively portray the ways of life we must live in almost every detail.

From the above perspective, adopting the word *culture* with the meaning outlined in this and previous works[34] represents an attempt at reintegrating the division of human life and society that began in the nineteenth century. When, in the twentieth century, the technical approach began to displace the cultural approach throughout human life and society, a new word was needed to designate what technique will never be able to accomplish, and which babies and children continue to rely on as they grow up as members of a symbolic species despite high levels of desymbolization. The issue will likely dominate the twenty-first century: technique desymbolizes human cultures, but it can ultimately not survive without them.[35] A biological analogy would be that of a parasite killing off its host. We need to have a word for what we are losing, but this will only happen through attempts to improve our intellectual grip on our lives and the world.

The other word we need to improve our intellectual grip is *technique*. The philosophy of technology during the last half-century has done us an enormous disservice by arguing that either technology or *techne* is sufficient. It is not an accident that the French language has no word equivalent to technology. The closest word, *technologie*, refers to the discourse about and the study of technique, in the same vein as sociology is the study of society. From the beginning, there have been influential voices that argued for the building of a rational society, of which industry was but one component.

Testing the Concept of Technique

The importance of technique can be confirmed by means of the following analysis. First, the previously noted definition requires that it is co-extensive with culture, and thus that it can be applied in any area of human life. That this is the case is obvious from our difficulty of coming up with lists of things that discipline-based science can never know, and what discipline-based technique will never be able to accomplish. Our exclusive reliance on this kind of specialized knowing and doing says it all. Jacques Ellul developed the concept of technique on the basis of his comprehensive analysis of almost every aspect of contemporary ways of life, finding that almost everywhere technique had displaced culture. These and related studies have been compiled in a bibliography of his complete works.[36] Everywhere he found human life oriented by technique, which limits the context taken into account to what is necessary for achieving absolute efficiency; whereas human life oriented by culture takes as much of the context into account as possible by means of symbolization. Technique and culture thus represent diametrically opposite orientations for human life, and today very little can be understood without this tension.

A second step in the evaluation of the applicability of technique to human life is related to what individual techniques share and in what ways they differ. If what they have in common is of fundamental importance to human life, society, and the biosphere, and if what divides them is not important, technique is essential for improving our intellectual grip on our lives and the world. In this respect, what they share goes to the very heart of our being a symbolic species. Individual techniques separate knowing and doing from experience and culture by transferring whatever they examine into the domains of various disciplines. There it is studied for the purpose of improving it. This is accomplished within a triple abstraction that leaves no option but to equate "better" with absolute efficiency. Models are made, the parameters of each model are varied, and the resulting forms are correlated with performance. The "best"-performing model demonstrates the possibility of technical improvement, which is then imitated in the world by reorganizing whatever was studied through this technical approach in order to replicate the absolute efficiency of the "best"-performing model. As noted, this is the diametrical opposite of how humanity has always proceeded through symbolization and culture.

Moreover, this technical approach separates the knowers, who carry out the study and determine the absolute efficiency to be obtained, from the doers, who execute this prescription, and from the supervisors as well as the managers, who mediate between knowers and doers. As a result, a human being can participate only as a knower, a doer, or an intermediary, but never as an entire person. In addition, both the internal integrality and external compatibility are degraded because the models incorporate only what is relevant to the attainment of absolute efficiency. In sum, the consequences of technique for human life, society, and the world could not be more fundamental in scope and depth.[37]

Finally, the concept of technique must be tested against the theoretical competition. As noted, during the decades following the Second World War, many new and important phenomena arose in industrializing societies. Many of these phenomena were so important that they were taken to be the key to understanding what was happening to human life and the world because they affected so many aspects of individual and collective human life. Some observers continued to hold on to industry or capitalism as the key phenomenon, although they acknowledged that these also had undergone substantial transformations. The theory of technique has three decisive advantages over all of these. A first advantage is that all the other theories were centred on phenomena that derived directly or indirectly from the transition from a primary reliance on symbolization and culture to a primary reliance on discipline-based approaches and only a secondary reliance on symbolization and culture. For example, in the former case, ways of life depended essentially on metaconscious knowledge, while in the latter case, the technical approach generates and manipulates enormous quantities of information, to the point that a bottleneck began to occur (which has been eliminated by the computer and information technology). Another advantage of the theory of technique is thus its ability to explain the relationships between all these new and important phenomena, which none of the other theories has been able to accomplish. After all, why did all these important new phenomena spring up at more or less the same time? This should have alerted observers that they might possibly be symptoms of a common deeper underlying development.

A final advantage of the theory of technique is its clearly stated goal: to examine what happened to human freedom and alienation at this juncture in the human journey. None of the other theories acknowledges a clear and explicit goal by which the unintelligible complexity of human life and the world can be reduced to an intelligible one. The other theories appear to be the theoretical articulation of the convictions of their authors, but when these are neither acknowledged nor critically examined, there is a considerable risk of developing ideologies instead of theories. What did these theories leave out of the unintelligible complexity of human life and the world to end up with the images they did? What values were being used? It is only in this way that it is possible to decide which developments ought to take a central place in a theory, which ones should be given a secondary place, and which ones can simply be ignored. If we are not critically aware of which values we are following in the development of our theories, it is highly likely that we will uncritically allow ourselves to be guided by our secular myths. Here the warnings of Georges Devereux[38] should be taken seriously. Anything that does not conform to our secular myths will threaten our existence, with the result that there will be powerful tendencies to unconsciously reinterpret what we observe in order to reduce our anxiety.

Having previously tested the concepts of technique in relation to that of culture,[39] I must emphasize one detail. There is no such thing as an absolute objectivity in science.[40] Every theory requires values of one kind or another for its development. When we are faced with the unintelligible complexity of our lives in this universe, it is impossible to make any theoretical model of it without the guidance of values. They are necessary to discern between what is important, peripheral, or irrelevant in order to make the required decisions to include or exclude details from the theory. Without these simplifications, theories would be as complex as the reality they represent, which would get us no further ahead than our daily-life experiences. Even the model disciplines – physics, chemistry, and biochemistry – require these kinds of values in order to abstract their domains from human life and the world and commit to a discipline-based approach. In turn, these values imply others, as will become evident later.

As noted, the discipline-based approach to the study of human life and the world is unscientific when the situations being studied are characterized by multiple categories of phenomena making non-negligible contributions. A society is not a mechanism made up of parts that correspond to our disciplines. These disciplines must rely heavily upon our new secular myths to hide their limits. Elsewhere,[41] I have begun to explore the role our secular myths appear to be playing in some of our leading disciplines.

The claim that any theory is dependent on values is likely going to meet with a great deal of opposition. I am not merely thinking of the Chicago school of economics, which has so effectively provided the ideology for people to serve the economy. The lack of any effective opposition shows clearly how both the political centre and the left are firmly in the grip of the cult of the fact. Furthermore, the mass media so effectively play out our secular myths that the situation is not likely to change in the near future. If any serious attempt were made to interpret what is happening in terms of what it means for our lives and those of our children, people would be out in the streets. Given the discipline-based approaches to understanding our economy, it is highly unlikely that we are going to come up with theories that take responsibility for the economy. Instead, we have the cult of economic facts. I would have thought that this debate had been decisively settled during the debates between Karl Marx and the physiocrats of his day.[42] He pointed out that the regularities they observed in the economy were not laws in the sense of the law of gravity. The economy was a human creation for which we must take responsibility. To submit to economic statistics and trends in the name of being "realistic" is to serve the economy as opposed to the economy serving us. Economics as a discipline must be subjected to the equivalent of symbolization: that is, exploring each element in terms of the meaning and value it has for human life, society, and the biosphere.

It is the conservative politicians who have been the most able in using the cult of the fact in formulating economic policy, but the remainder of the political spectrum is not far behind. The cult of economic facts leaves us with no economic policy other than economic growth at almost any cost: economics has no limits, more is therefore better, and growth can thus overcome everything. The wealth it creates can then be distributed according to our political values. It amounts to an end-of-pipe politics, which renders us impotent in the face of our deep structural economic crisis. All politicians are more or less singing from the same song book.

Another kind of attack may come from the left. First and foremost we should be concerned with the victims of technique and our present economic system: the poor, the dispossessed, the vulnerable, the unemployed and underemployed, the South, and so on. The fact that everyone in our civilization is alienated by technique is not very appealing to the left. We do not want to worry about the rich and the powerful, but only about their victims. While this is understandable, it nevertheless blocks the road to reaching the root of the problem. We need to understand how the system we have created operates, and how on the deepest levels it affects the rich and the poor in much the same way. It is true that the rich and the powerful have many more means to compensate for their being alienated by the system, but their humanity remains alienated nevertheless. Simon and Garfunkel expressed it well in their song about Richard Cory.

A true mutation of the system based on the equivalent of symbolizing it will be beneficial to all. There is such a thing as a common good, even though this term no longer appears to cross the lips of any politician. With the desymbolization of our cultures, we appear to have given up on the possibility of a "common sense" to achieve the "common good" based on shared meanings and values derived from symbolizing everything.

Humanity has declared our contemporary ways of life to be unsustainable by the biosphere. We have also declared slavery an unacceptable form of human life. It would appear, therefore, that we ought to continue the Western tradition of placing human freedom at the centre of the values involved in making sense of our lives and the world.

The theory of technique based on the value of freedom points the way forward. By revealing the greatest threats to human life resulting from alienation and reification, it points our attention to where we ought to engage the struggle to make human life more liveable and the future more sustainable. The core value that guided the development of the theory of technique is thus consistent with what most of us believe about slavery and alienation. The Christian right, which has been so influential in North American politics, ought to be more mindful of the Jewish roots of Christianity. Freedom is the prerequisite to all relationships of love, but this freedom is always compromised by the service of false gods. The Jews, therefore, had to live as secular a life as possible. Since Christianity is grafted onto Judaism, its practitioners need to understand that Christianity is an anti-morality and an anti-religion because its God desires relationships of love, to which we cannot be committed unless we are free.[43]

Our social and historical situation appears to be characterized by a tension between technique and culture. The former is impossible without the latter, but the latter can exist without the former. If our secular myths could be removed, a dialectical tension could be established between technique and culture, where each one prevents us from going beyond the limits of the other. The accomplishments of our civilization are all related to technique, and our failures to the limits of technique as well as to its desymbolizing influence on all cultures, thus preventing them from doing what they did so well in the past. Until the nineteenth century, cultures generally ensured that economies were integral and functional in the service of society (the exception being market-driven commerce), that technologies were mostly appropriate, and that ways of life could be sustained by the biosphere. It must not be supposed that this implies the restoration of traditional cultures. Even if this were possible, the results would be highly unsatisfactory, since these cultures have always alienated their members through an enslavement to their gods and their myths. This alienation also resulted in substantial distortions of how everything was related to everything else, and how everything evolved in relation to everything else. Rather, it is a matter of creating a civilization in which technique is evolved on human terms as opposed to its own absolute terms; this would require the conscious equivalent of symbolizing technique in terms of its meaning and value for human life, society, and the biosphere. In the meantime, the vocabularies of most people, which furnish them with the linguistic models for making sense of their lives and the world, appear to be entirely inadequate; and science appears to be impotent at improving the situation.

The Cult of the Fact

The next stage of our inquiry is an examination of what science itself implies about human life and society. By exclusively focusing on what can be replicated, science reveals how human life and the world are determined by all kinds of "mechanisms" as depicted in the domains of each and every discipline. In the case of human life, if we extrapolated and interpolated all these determinisms, we would arrive at the conclusion that human life is alienated. In contrast, when we critically transcend the findings of the disciplines by means of an intellectual equivalent to symbolization that I have referred to as intellectual map making, it becomes evident that such a view is as true as it is false. What science is unable to show is that, despite the many determinisms revealed one domain at a time, human life and the biosphere are not mechanistic. In the interplay of all these determinisms, there remains a fundamental indeterminism that is obscured by examining human life and the world by means of discipline-based approaches. A fundamental indeterminism would be revealed if we had a "science of the sciences" capable of generating an overall scientific theory of human life in the world.

There is no question that science reveals the depth of our alienation through the many determinisms discovered by the many disciplines. The occasional debate over nature or nurture demonstrates this very well as proponents of both sides gather together the many indentified determinisms in order to argue their case. What this debate really demonstrates is that the determinisms of the "body" (physical, chemical, electrical, and biological ones as well as their collective development over time) and the sociocultural determinisms of the mind and society (psychological, psychosocial, social, economic, political, legal, moral, religious, and aesthetic determinisms as well as their collective influence on the brain-mind over time) add up to an enormous diversity of determinisms playing themselves out in individual and collective human life. Included in all of this are the interplays between the two kinds of determinisms: from the molecular level, genes and DNA, to the symbolic processes of the brain-mind, human experience and the living of lives – all these determinisms are interacting in very complex patterns, with the result that, once again, this "everything being related to everything else" and "evolving in relation to everything else" is not accessible to discipline-based approaches to knowing and doing.

Scientific medicine is a perfect illustration. It succeeds brilliantly when, in an acute crisis, the focus of attention is on certain categories of phenomena to the exclusion of all others because it is a question of life or death. This approach works less well when the illness is less acute. A much larger diversity of categories of phenomena needs to be considered if the treatment is not to be undermined by a great many side effects resulting from connections having been externalized in a scientific study to determine the course of treatment and the development of appropriate drugs. When it comes to an ongoing nurturing of health, scientific medicine is virtually impotent because of its discipline-based organization. In contrast, medicine based on symbolization and a medical tradition was sometimes very effective – provided that the "everything evolving in relation to everything else" was not regularly disturbed by major developments as is the case today, which is why some of these practices are no longer successful. Owing to the training of modern doctors, many of them are completely sceptical of the problems of their patients which defy discipline-based approaches. It is only because of the persistence of a few of these patients that some issues have gained grudging acceptance among medical practitioners. When it comes to "sociocultural" diseases associated with our contemporary ways of life, we are barely scratching the surface. How and what we eat, how we work, how we relax and entertain ourselves, how we sleep, where we live, our educational background – all this and more has a profound impact on our health, as social epidemiology has shown. Nevertheless, when we go to our doctors, it is frequently the symptoms that are treated, and not the root problems. In sum, modern medicine is a microcosm of technique: it succeeds brilliantly in situations approaching non-life, and does much less well in relation to life.

Our alienation and ultimate indeterminism can also be approached by means of intellectual map making. In contemporary mass societies with highly desymbolized cultures, human life is alienated by integration propaganda, resulting in public opinion and statistical morality. We have opinions about everything, yet when we critically examine them it is obvious that we have not arrived at them in the usual ways: experience, discussions with others, informed reading, thought, and reflection. They appear to be induced by the mass media and the social media.[44] As noted, a statistical morality results when we begin to use the organizations of our brain-minds as a kind of radar to scan what everyone is doing in order to belong and "go with the flow." As a result, what everyone does becomes normal, and what is normal becomes normative. When our children want to know why they are forbidden to do what everyone else in their class is doing, invoking traditional values hardly convinces them. Nevertheless, in contemporary cultures, what everyone does may be normal, but that does not make it ethical and just. In the same vein, our political opinions are largely induced by political advertising, public relations, and other techniques of manipulation. We are deeply determined by the secular religious myths of our cultures, even when these are highly desymbolized. Our mass societies possess us through our opinions, morality, political choices, and secular religious myths. Similarly, what we like or dislike in terms of the arts, music, or literature is deeply determined by the social and mass media. Technique now imposes a whole new range of psychological, social, economic, political, moral, religious, and aesthetic determinations to complement those exercised by our highly desymbolized cultures. What really matters for human life is their overall effect as an entirely new sociocultural force that pushes human history in a certain direction.

Let us take the example of consumption. As noted, technique has become a system in which every area of specialization functions as a transmitter and receiver of information regarding technical advances. When an advance is made in a particular area, it is communicated into the system by a variety of means. Among the areas that receive this information, there are usually a few that can utilize it to make further advances that are also communicated into the system, and so on. Technique has thus become a global system connected by flows of information regarding all manner of technical advances. Technical development now has two components. The first is the conventional one composed of responses to human needs and desires. The second component results from the permutations and combinations of countless advances within the system triggered by information flows. The latter component became so important that technological forecasting had to be abandoned a few decades after the Second World War because no leading specialist could oversee the entire system and predict how it would affect that specialty. As a result, it cannot be argued that whatever is technically produced is a simple response to human needs and desires. Generally speaking, many (if not most) consumer goods produced during the last half-century or more did not respond to any needs or wants. For example, people did not have a need for cell phones or the internet because their ways of life were arranged so that they could easily get along without them. I am certainly not arguing that these technical inventions are without significant conveniences. They appear to have advantages at first. However, the "technique changing people" factor fundamentally changes our ways of life, to the point that it becomes increasingly difficult to get along without these techniques. It is now possible for our children to patiently smile at us who once lived in the "Dark Ages," when life must have been very difficult without these gadgets. They cannot envisage that social patterns of interaction were once such that there was no need for these gadgets.

When we look at the phenomenon of consumption as a whole, and no longer as one gadget and development at a time, a disturbing interpretation imposes itself. I recognize that, in many books dealing with the relationship between technology and society, the subject is immediately divided into chapters dealing with biotechnology, nanotechnology, information technology, social media, medical technology, factories of the future, and so on. Methodologically, this is the equivalent of saying that the impact of water on paper is that of hydrogen and oxygen. We ignore how everything is related to everything else. I am not implying that the study of these individual technologies is irrelevant. On the contrary; but we must not confuse these different levels of analysis, as most of these treatises do. Examining the influence technique has on human life one technique at a time avoids the complex positive and negative synergistic effects, especially of small and supposedly negligible ones. When these occur over and over again, they become amplified and often constitute the most significant influences. From this perspective, it may be argued that the phenomenon of consumption is a major determination of individual and collective human life. We can almost say that, in a mass society, we consume what the system can produce; we make money to participate in that consumption; we are what we have; without work we are nothing and become nothing because, unless we consume, we do not exist; it is by means of consumption that we ritually participate in the ultimate goodness of technique and live out our secular myths. This consumption is indissociably linked to working in the image of the traditional or information machine, which makes no sense, yet this is what we are all required to do. We are confronted with organizations, systems, and institutions built up for absolute efficiency and not for making sense. Technique thus superimposes the reification of our lives onto our alienation – from walking into a store that has been carefully arranged to maximize the chance that we walk out with more than we had intended to purchase, to the selection of a casket in a funeral home arranged to have a similar result. We are constantly reified in everything we do, since we are always on the receiving end of one technique or another. We are no longer the masters of our lives, no matter how many choices we may appear to have, since these choices themselves are greatly influenced by techniques of all kinds. To defend ourselves by arguing that things have always been this way is a smokescreen. We are responsible for the present and not for the past; it is time we stop escaping into arguments that this or that is a little better or a little worse than it was fifty years ago. Our inner being is deeply penetrated by techniques of all kinds. Whether this happened in traditional cultures in other ways but to the same extent is an interesting question, but irrelevant to our current responsibility for our own creations.

As noted, we live in a technical order of non-sense. It has made us lose trust in our institutions and in our ways of life, which increasingly are abstract and without sense, being under the constant reorganization of techniques of all kinds. Within this order, we experience our alienation in the following ways.[45] Many moments of our lives are characterized by a loss of control in our ability to make sense of them and therefore in our ability to effectively respond to them. The order of non-sense undermines our motivation because we no longer understand why things are the way they are. It makes it next to impossible to live a meaningful life. This loss of control and lack of sense stand in the way of our ability to adapt and evolve our lives in a manner that is meaningful and purposeful. The situation is further aggravated by the overload of disconnected information we receive from the mass media and social media. It has a paralysing effect on our ability to cope. All this adds up to a strong sense of meaninglessness, hopelessness, and abandonment.[46] Many people are unable to cope without the compensations of alcohol, drugs, violence, or antidepressants.

Once again, I wish to emphasize that alienation and reification are moral concepts because they are based on a conviction that human beings are more than cogs in a mechanism and that an enslavement to any system is an unacceptable form of human life. To varying degrees, we expect meaning, direction, purpose, and motivation. The theory of technique is therefore not scientific, like any other theory of this kind, because it ultimately depends on values, by which some features of human life are put centre stage and others on the perimeter, and still others are completely ignored.

The example of consumption is indicative of the fact that human life in contemporary mass societies relies extensively on an order created by technique, thus relegating the highly desymbolized cultural order to our individual lives, where we can claim to be spiritual. Despite all the determinisms examined by science and integrated through intellectual map making, the concepts of technique and culture provide us with the strongest evidence that all these determinisms do not make human lives and the biosphere into mechanisms. There is an ultimate indeterminism in anything living. If this were not the case, I would not have had to write this book, because our civilization would have been unable to construct the most efficient way ever devised by humanity to undermine and eventually destroy the evolving tissue of relationships that constitutes all life and the biosphere.

To develop and illustrate this claim, take human work as a microcosm of technique. Its portrayal by Charlie Chaplin in his film *Modern Times* wonderfully illustrates the mechanistic aspects of contemporary ways of life. Lewis Mumford took this a step further by arguing that our civilization resembles a megamachine.[47] If these portrayals were accurate, the history of human work for the last two hundred years would be incomprehensible. Technique integrates our hand or our brain, but what has not been integrated remains nevertheless connected, thereby becoming a significant source of disorder. Employers have responded to these disorders by means of one end-of-pipe approach after another: job rotation, job enlargement, ergonomics, human factors, industrial relations, group dynamics, and quality of working life approaches. None of these has ever solved the problem of technically divided physical or intellectual work. The only exceptions were socio-technical approaches that went back to the root of the problem.[48] We cannot exclusively focus on the mechanistic side of our civilization and ignore what has been externalized in order to achieve absolute efficiency. The two must be considered together as stemming from the same technical approach, which makes the hand or the brain contribute to this efficiency and externalizes everything else to become a source of disorder. If ever we required evidence of how everything is related and evolves in relation to everything else, in all life in general and in human life in particular, technique provides us with this evidence in abundance. Otherwise, the successes of technique would not be undermined by its failures. Technique has helped us understand, on the one hand, the extent to which we have embraced non-life, and on the other, the limits to that non-life. It is what technique in general and individual techniques in particular have shown us for well over half a century. If people had been nothing but mechanisms, technique could have made them cogs of other mechanistic systems without creating the disorder that now reigns everywhere. The technical division of labour marked the beginning of dealing with life as non-life, the economy as little more than mechanisms of supply and demand, and society as a cumbersome appendage in need of rationalization.

It should be remembered that throughout its development, technique has had a close affinity with warfare. Discipline-based approaches, ideally suited to the development of technology, industry, and the economy, are also well suited to warfare. There are no longer any cultural constraints on this activity, making it purely a question of efficiency. In the design of weapons, battlefield strategies, supply lines, and everything else related to war, the only categories of phenomena that matter are the ones that will determine who lives and who dies. War is the summit of absolute efficiency. Our weapons have now overshot their mark because their all-out use would be so efficient that there would be neither winners nor losers. The First World War was supposed to be the first great technological "war to end all wars" because of its unprecedented horrors. Shortly after, Hitler's army became the summit of absolute efficiency. The Second World War was a laboratory for the proliferation of all kinds of techniques. Possibly the most notable was that of operations research, which has now spread to many disciplines, to the point that many professionals act as if the simulations on their screens represent reality itself. It is leading to the most unbelievable attitudes in many professional faculties, including my own.

If until now there has always been an ultimate indeterminism in human life and the biosphere, does this mean that our alienation and reification will never result in our becoming mechanisms? The conclusion I reached in a previous study is that technique represents a war on everything living, but especially on humanity as a symbolic species.[49] Technique ultimately depends on human cultures in order that each next generation become cultural beings before they can become *homo informaticus*; hence, if technique destroys our being a symbolic species, it will likely disappear with us.

This is where the cult of the fact comes in. In painting such a stark picture, I am appealing to those who live in hope. Hope is the will to live because we have an open future: we could destroy ourselves and the planet, but we can also transcend all the evidence to this effect by symbolizing it as a warning and striving for a liveable and sustainable future. If enough people interpret our so-called reality in this way, they will help destroy the cult of the fact and the reality created by our scientists and our economists. This priesthood of the cult of the fact must be overcome in hope, against all evidence to the contrary. It reminds me of the descriptions of the people who survived concentration camps and nuclear bomb attacks.[50] Even when all the evidence suggested that there was no way out, those who gathered up enough hope that things did not have to be this way and that another future was possible had a much better chance of making it through these "valleys of death."

I recognize that the easiest way out is to simply dismiss my descriptions as overly pessimistic, one-sided, too left-wing or right-wing, or whatever else we can easily dismiss. Devereux[51] has shown that, on a smaller scale, these reactions are all too typical in the social sciences. The most common way out is to escape into one of our many contemporary philosophies that take human freedom as a given by ignoring our social and historical condition. The widespread influence of these philosophies (including post-modernism) is almost certainly a compensation for what science reveals about how deeply we are determined. It is probably an unconscious reaction and an escape, as is characteristic of all of human history. We give ourselves over to all manner of ideologies to make life liveable and bearable. We then surrender to our contemporary cult of the fact and the secular myths that underlie it. We do not even have to make a choice because this is always the fallback position.

If we decide to challenge the cult of the fact, the way forward is clear. We must confront the limits of discipline-based science by recognizing that it cannot go beyond what we know and live to what remains unknowable and unliveable because of our myths. It does not possess absolute knowledge in the forms of facts and theories. We must create a distance between ourselves and science and symbolize it in terms of the meaning and value it has for human life, society, and the biosphere. In this way, a new sphere of possibilities will open up and a small measure of freedom with respect to what alienates us becomes possible. We will then discover that science is very good for knowing certain things, marginally relevant for knowing other things, and completely irrelevant for still others. It will restore science to its true status: a unique and highly powerful human creation which nevertheless has limitations. We can thus begin to live with our new secular gods in the hope of a more secular future. Unfortunately, those who ought to understand the significance of all this are too busy going to heaven while condemning others to hell.

Making Sense of Science

Modern science is unthinkable without technique. Both approach human life and the world in terms of domains in which these are represented one category of phenomena at a time. Technique is entirely dependent on the scientific approach to knowing, while science is dependent on technique to construct the increasingly complex infrastructure required to simulate its domains. As a result, science basks in the glory of technique, and technique in the glory of science. Each one depends on the secular myths of the other to help cover over its limits. Conversely, each one could potentially help to reveal the limits of the other.

From a cultural perspective, science must be interpreted as a unique human activity with its social structures and institutions, like any other. Doing so began around the middle of the twentieth century. Science had defined reality in terms of what could be divided, separated, represented by closed definitions, cast in mathematical form, and have a simple complexity built up one element at a time. Its methods and approaches could directly reach this reality and grasp its details as facts. The specially educated observers were interchangeable because nothing of their lives and cultures would have any effect on their work. Consequently, the philosophy of science was supposed to be entirely adequate for its understanding since it had been completely abstracted from society and history. As such, science had become the victim of its own success. It had blinded itself to the social and historical determinations from which no human activity can escape. When Thomas Kuhn began to open science up to sociological and historical investigations, this philosophical view of science as a kind of mechanism that would lead humanity to reality and its truth had to be abandoned.[52]

In the first edition of his book *The Structure of Scientific Revolutions*, Kuhn essentially focused on the historical development of scientific disciplines such as physics, chemistry, and astronomy. In the postscript of the second edition, he acknowledged that a great deal of light can be shed on science by means of a sociological analysis, beginning by identifying and examining the members of invisible colleges who collaborate to advance the frontier of a particular specialty in order to sociologically investigate what they have in common. Invisible colleges are networks of specialists usually located in different institutions, but whose collaboration advances a part of the frontier. These invisible colleges are to science what technostructures are to technique. In his postscript, Kuhn elaborated his understanding of science from this sociological perspective.

Kuhn's original proposal was that what the members of a discipline or specialty had in common was a paradigm. This concept was extensively criticized for having been used in many different contexts with different meanings. In the postscript, Kuhn replaced the term paradigm with that of a disciplinary matrix having the following four primary elements: symbolic generalizations (the term symbolic is, I think, inappropriate), commitments and beliefs, values, and exemplars. A complementary analysis would examine the process of "socialization" into the "culture" of a scientific discipline. Had Kuhn undertaken such an analysis, the concept of a paradigm might have turned out to be more defensible than a disciplinary matrix. This kind of analysis could begin with examining what happens in high school science classes and then track the development of some students through several degrees in their becoming the practitioners of a discipline. If Kuhn had examined the process of socialization into a discipline or specialty, he would have been able to advance much further in his understanding of what is unique to scientific activities, and which features they share with other human activities.

Let us briefly examine what happens in the science classes of our high schools from a cultural perspective. When students enter these classes, the organizations of their brain-minds have already considerably converged with those of the other members of their culture. Even so, their physics classes do not begin with the extensive metaconscious knowledge of physics that they have acquired in all the daily-life activities involving physical phenomena: crawling, walking, running, cycling, climbing trees, playing ball games, jumping, eating with utensils, and much more. Doing so would have amounted to a kind of Aristotelian approach, which had been abandoned following Newton. Instead, students are introduced to the domain of physics. The first constituents of this domain include disembodied forces, point masses, infinite frictionless planes, and the movement of these masses across these planes at a uniform speed, or under acceleration or deceleration as a result of a force being applied. These earliest constituents and events of these domains are connected by Newton's laws of motion. Next, several problems are demonstrated on the blackboard, following which the students are encouraged to solve others at home. Going from problem to problem is an essential part of learning about this domain of physics. The problems already solved become the exemplars for solving subsequent ones.

If a problem can be conceptualized as being different from all previous ones but somewhat similar to one of them, it can be solved by using this problem as an exemplar.

As a result, interpreting what is happening in a physics class in this way is somewhat comparable to culture acquisition. Learning to distinguish dogs from cats was learning something about dogs relative to cats and about cats relative to dogs. No absolute knowledge was involved. This development has been explained by the processes of differentiation and integration. Just because students have walked into a high school physics class does not mean that the organizations of their brain-minds begin to function differently now that they are doing science. The "experiences" of the different situations within the domains depicted in various problems can initially be expected to be differentiated and integrated exactly like daily-life experiences.

The above developments are inevitably accompanied by the teacher's explanations of what is happening in the physics class. The students are learning science and thus the scientific method. They must adopt an objective approach to studying the world, as illustrated by the fact that when they solve a problem they will either get the right answer or the wrong one. All this and more will make some sense to the students because everything is so different from their daily-life experiences. They will have a sense that this new "world" (really a domain) is very different from their daily-life world: that the entities and events in this "world" are so disembodied as to correspond to nothing in their daily-life experiences. This new "world" is nice and tidy because everything can be defined on its own terms and connected to everything else through mathematical equations; no diverse interpretations are possible because the students either understand it and get the right answer or they do not. Moreover, they are not entirely unfamiliar with this kind of "world." I have suggested that the characteristics of the "world" we apprehend by seeing are different from those of the symbolic universe of our culture. Furthermore, in contemporary mass societies, the image dominates the word, which corresponds to a significant desymbolization of this symbolic universe. Some students may intuit this quickly, while others may experience considerable difficulty making sense of these new "experiences." In any case, the explanations of the teacher will confirm that something very different is going on, but what these differences are the student will discover through the processes of differentiating and integrating the "experiences" of doing science.

The fundamental question is: What will the organizations of the brain-minds of students "do" with these "experiences"? I believe that the acquisition of language may provide us with a clue. The onset of the development of language occurs once the usual experiences of a baby's life in the world become differentiated from the experiences that have a language gestalt in the foreground. A toddler's language experiences are different from his or her other experiences in several ways. The first difference relates to the foreground or the focus of attention. A language gestalt belongs to the symbolic universe a toddler learns to enter by means of language, while the foregrounds of all the other experiences have at minimum a visual dimension. At least in Western cultures, this is probably the dominant dimension of experience. This difference is closely related to a second one, which involves the relationship between the foreground and the background. A language gestalt as a foreground has a direct relationship to the background only if what is being said (and for toddlers this is usually the case) is directly related to the situation at hand. In contrast, when a toddler is told a story, there is no relationship between the foreground (occupied by the language gestalts of the story) and the background (constituted by the setting in which the story is told). Eventually, a third difference will emerge in terms of the kind of metaconscious knowledge built up from the child's usual experiences and that built up from the experiences with a language gestalt as foreground.

Using the previously developed model of culture,[53] it would be expected that within the organizations of the brain-minds of toddlers, the onset of language corresponds to the development of a cluster of differentiated language experiences as being essentially similar to all the others except for the language component. With the entry into a symbolic universe by means of language, this cluster expands rapidly, and likely spills over into another part of the brain-mind when its radical difference becomes clear. As noted previously, this marks the development of a double referencing system. Initially, the meaning of a vocal sign is entirely determined by the other experiences functioning as a metalanguage. With the onset of language, these words and the phrases into which they are embedded become directly differentiated from each other, thus transforming them from signs into symbolic concepts. This development is confirmed by the existence of two distinct forms of memory, referred to as the episodic memory and the semantic memory. A person can lose one without losing the other. Furthermore, I have also suggested that without this double referencing system, the turning of words and phrases into tools by the techniques of advertising, public relations, human relations, group dynamics, and others would be impossible.

Learning science essentially appears to repeat the development of language, with one important difference. Again, the "experiences" of learning physics are different from all the other experiences of teenagers, in three ways. First, the foreground is exclusively related to the domain of physics while the background is that of the setting in which physics is learned. Second, the domain of a discipline has completely different characteristics from that of the symbolic universe of the students. As noted, a domain is constituted by what can be divided, separated, defined on its own terms, represented in mathematical terms, and made to participate in a simple complexity that can be built up one element at a time. In contrast, the symbolic universe of a culture is a dialectically enfolded whole with all the opposite characteristics. Hence, in the "experiences" of learning physics, there is a complete discontinuity between foreground and background. This discontinuity does not occur in the case of language experiences because the visual dimension of experience is symbolized as being integral to the symbolic universe of a culture, even though this has been much weakened in contemporary mass societies. This discontinuity between foreground and background goes to the heart of what I have referred to earlier as the separation of knowing and doing from experience and culture. The third difference between the "experiences" of learning physics and all the other experiences of the students is that the metaconscious knowledge built up with the former will also be separated from experience and culture, while in the latter this is not the case. It will take some time before the processes of differentiation and integration begin to metaconsciously detect these differences, but there is considerable evidence to suggest that it does eventually occur. The result will be an entirely new cluster of "experiences" corresponding to the domains of different disciplines, which will eventually break up into several clusters each corresponding to a domain. There is a great deal of evidence to suggest this, but I will limit myself to the following discussion, since I have reviewed this evidence in detail in a previous study.[54]

A distinction between intuitive physics and school physics has become an accepted concept. The former is essentially based on the metaconscious knowledge of physical phenomena in daily life, and the latter is based on the metaconscious knowledge of the domain of physics learned in physics classes and from textbooks. In the former case, physical phenomena are always intermingled with other categories of phenomena according to the daily-life activities in which they occur, while in the latter case all other categories of phenomena are excluded. It is for this reason that, when a high school teacher tells the students that a science studies the world, they are expressing the usual ideology of science (which has little bearing on what really happens). It is true that the domain of physics gradually approaches the world as students learn to take into account phenomena such as friction, inertia, air resistance, and the weight of strings. However, the domain of physics will never resemble the world in the sense that the former is exclusively populated by physical phenomena. Consequently, the applicability of physics to a great many daily-life situations is exceedingly limited.

What this brief discussion illustrates is that the interpretation of the learning of a scientific discipline from a cultural perspective allows us to metaconsciously differentiate the associated activities from daily-life activities. There can be no question of a discontinuity in the functioning of the organizations of the brain-minds of students just because they enter a science class. What is unique about the activities of a science class becomes evident in their differences from daily-life activities: the former relate exclusively to domains by one category of phenomena, while the latter are incomprehensible without the mingling of many categories of phenomena making non-trivial contributions, with the result that these cateagories of phenomena cannot be neglected.

The separation of knowing and doing from experience and culture as it occurs in discipline-based approaches thus likely corresponds to the development in the organization of our brain-minds of a "paradigm" or a "substructure," which I have compared to an insert into a road map or a bifocal lens for observing the world.[55] Just like the organizations of our brain-minds, the paradigm or substructure manifests itself in many different ways (including eye etiquette, conversation distance, an arrangement of life in space and time, values implicit in our behaviour, mental "images" of the social selves of others and ourselves, our way of life, and our culture). So also a paradigm can manifest itself in the many ways that Kuhn described in his original edition, but which he reduced to four elements of a disciplinary matrix in his second edition. The insurmountable problem with the concept of a disciplinary matrix is that it avoids spelling out the relationships between these elements and how these operate and evolve in different situations. In any case, neither concept is required other than as a reference to a unique development in the organizations of the brain-minds of people who have undergone considerable scientific education. It is likely to be a relatively distinct part of our semantic memory, or possibly a development at arm's length from both the semantic and episodic memories because it is built up from "experiences" separated from all the other experiences, including the culture built up with them. Because of this it is impossible to speak of scientific disciplines as having a universal subculture. It is this development in the organizations of the brain-minds of people having taken advanced degrees in a particular discipline, or who are self-taught from reading the textbooks and journal articles of that discipline, that the members of invisible colleges have in common. Being separated from experience and culture, the paradigm (or disciplinary matrix) acquires a set of characteristics that correspond to what a discipline can achieve as well as its limited applicability to human life and our world. It is in this manner that we can develop a psychology, sociology, cultural anthropology, and history of scientific disciplines, beginning with a detailed examination of how students are socialized into a discipline. This study must consider both the formal and the hidden curriculum in the context of the modern university and its place in contemporary mass society.

Doing so inevitably raises the question of the secular myths associated with the discipline-based approaches to knowing and doing, and with the remainder of contemporary ways of life. Secular myths are necessary to reintegrate our lives and our world in terms of what has been separated from experience and culture, and what remains embedded in it. Elsewhere,[56] I have shown that these secular myths reintegrate the two by the creation of a hierarchy that removes all limits from the former, with the result that the latter are to be excluded from our ways of life, to be delegated to our personal and private lives as vestiges of a religious and cultural past.

These secular myths are built up from and cover over all the things that are never explicitly dealt with, beginning in high school science classes and well beyond. The intellectual journey to the domain of a discipline does not include, let alone formally treat, how that domain is abstracted from human life and the world. What has to be overlooked, intellectually eliminated, and distorted in order to arrive at the intellectual promised land? Why is this land so very different from the symbolic universe of a culture and its representation of what is being studied? By concentrating on what can be replicated, are we affirming that there are nothing but mechanisms or that the non-mechanistic elements are not worth bothering with? Does all of this imply that what really matters is the non-life in everything? Can there ever be a scientific understanding of life and how it is different from non-life? What such questions imply is that the process of socialization into a scientific discipline involves beliefs and values that are so profound and far-reaching that they manifest the secular myths of the symbolic universes in which the future practitioners of science live. The absence of a clear awareness of the limits to discipline-based approaches to knowing is reinforced by this process of socialization. It is in this manner that the domain of a discipline is integrated into the symbolic universe of a culture, on the terms of the former as having no limits in contrast to the latter. Our learning to live within these new secular myths thus involves two parallel developments in young people: their entry into the symbolic universe of their culture, and their taking science classes in high school.

By intellectually forcing everything into what can be replicated and thus into the form of rules, algorithms, mechanisms, systems, and theories that accompany non-life, we leave little or no room for creativity, imagination, or genuine adaptation, which always introduces something new, in design, art, craft, and the like. I am not suggesting that in science these elements of practice have been eliminated. On the contrary, scientific practice is enormously constrained by its dependence on prior accomplishments. What I am referring to are the accusations against Kuhn and others of introducing into our understanding of science elements of irrationality, subjectivity, belief, and myth. However, scientists as human beings are alive, and any description of what they do cannot be entirely captured in scientific terms. Perhaps it would be more accurate to say that what is most essential to science and what makes it intellectually fascinating and compelling is precisely what the discipline-based approach to knowing will never understand. As a result, transforming high school science by making students more critically aware of how they are coming from their symbolic universes into the domains of physics, chemistry, biology, and applied mathematics is the key to reintegrating the scientific creativity and imagination essential for a mutation of science, thus allowing it to enter into a dialectical tension with the cultural approach in which each approach exposes and helps to go beyond the limits of the other. Subsequent chapters will shed further light on this subject.

This description of science exposes a second value to which it is committed. Given what we know, and considering the tremendous difficulties our civilization faces, it is not responsible to assume that there is an intellectual bridge from non-life to life. The reality defined by discipline-based science cannot be assumed to be heading for a qualitative change once the simple complexity of this reality crosses a particular threshold related to its scope. There are situations where quantitative changes lead to qualitative ones, but so far there is no evidence to suggest that an abundance of non-life will mutate into life.

The above description of discipline-based approaches to knowing exposes yet another limit to contemporary science. No scientific discipline can take into account the reciprocal relationships its practitioners have with their surroundings. This eliminates all possibilities of scientists being detached observers free from all cultural influences. These cannot be avoided, and yet they cannot be incorporated into any domain because they involve very different categories of phenomena. Scientific activities change the domain of a discipline, these changes affect the world, which in turn changes the practitioners through its vast influence on human life. In order to be genuinely scientific, scientists must practise a relative, and not an absolute, objectivity. Doing so requires a critical awareness of the influences of their discipline and of the world beyond on their scientific activities. All this lies beyond the domain and the corresponding category of activities of their disciplines. Especially when a scientific discipline studies human beings and other living entities, the reciprocity of the relationship thus established implies that we ought to avoid doing research *on* them and instead we ought to do research *with* them. It is the essential difference between looking at other people and staring at them. In the former case, we acknowledge them as human subjects and expect reciprocity. In the latter case, we deny their status as human beings and treat them instead as objects, which is why we become so uncomfortable when people stare at us.

The reciprocity the practitioners of science have with what they study extends to their laboratories. As they design and work in these laboratories, they affect their surroundings, which at the same time affect them. A scientific experiment ought to be regarded as a complex whole comprising the observers, what these observers bring to the experiment professionally and culturally, the experimental apparatus, what this apparatus seeks to study, the results obtained, and the formal and informal interpretation of these results. In contrast, if something is studied directly in the world, the influence the practitioners of a discipline have on that world derives from treating it as their domain, and the ability to do this has changed them during their education and will continue to change them as practitioners. Denying all these different kinds of reciprocity has put science in a position akin to conducting intellectual warfare against human life and the world by treating everything as a prisoner of the scientific method. Elsewhere,[57] I have begun to make some suggestions of how a variety of disciplines could be transformed to reduce the above kinds of limitations.

The implicit value in contemporary discipline-based science that ultimately all life can be understood in terms of non-life is thus closely related to a discipline-based approach being incapable of dealing with reciprocal interactions. It is another aspect of not respecting how everything is related to everything else, and evolving in relation to everything else, by intellectually ordering this interdependence one category at a time. It amounts to our civilization having exchanged a spiritual journey with the gods for a secular spiritual journey, that of negating everything living in the quest of gaining absolute knowledge – which, in a secular sense, is god-like.

2 The Cult of Efficiency: What Technical Means Cannot Accomplish

The Technical Approach

Everything in our lives and our world is constantly being improved, and yet our experiences of these "improvements" tell a different story. The gap corresponds to the way we improve everything on the basis of the technical approach and the way we used to do so by way of the cultural approach. Prior to the emergence of technique, everything was adapted and improved on the basis of symbolization, experience, and culture. As a result, every improvement was evaluated in terms of its contribution to the whole of everything being related to everything else and evolving in relation to everything else. Improving something on its own terms in isolation from everything else was unthinkable, and a concept of efficiency was absent in a number of Western languages. The evidence is abundant. Archaeology could not do its work if the constant improvements to artefacts were not culturally appropriate, that is, did not bear the stamp of a specific time, place, and culture. Traditional technologies undergoing constant improvements remained extraordinarily appropriate in most cases. The steadily evolving traditional ways of life generally maintained a good fit between themselves and local ecosystems, and the latter generally had no difficulty in sustaining them. The notable exceptions correspond to situations where a society's knowledge base embedded in experience and culture was unable to get an intellectual grip on what led up to a level of unsustainability, or where the symptoms were simply ignored. For example, some irrigation-based civilizations gradually destroyed their soils and turned them into deserts, and Easter Islanders allowed their religious zeal to get out of control and eventually cause the destruction of their environment. These are the kinds of exceptions that prove the rule that making improvements on the basis of the cultural approach amounts to bettering the contribution to human life, society, and the biosphere.

Improving things on the basis of the technical approach is essentially the diametrical opposite of the cultural approach. It begins by studying something for the purpose of making it "better." As much as possible, the findings of the study are cast in a mathematical form as some kind of model because it facilitates the next step: finding the particular form of the model that produces the desired result. Finally, what has been learned is applied to reorganizing what was originally studied in order to improve it. In a civilization that has entrusted all knowing and doing to discipline-based approaches, this technical approach can do nothing else but improve things on their own terms in isolation from everything else. In other words, an area of human life, society, or the biosphere cannot be improved as such. It must be divided and separated into a number of aspects commensurate with the domains of different disciplines. Rarely are all possible disciplines involved in the process, even when it is carried out by a team of specialists. In any case, there is no scientific method for integrating their findings. In order to fully appreciate how different the technical approach is from the cultural approach, we will examine each of the former's four steps in terms of the constraints imposed on it by discipline-based knowing and doing.

The first step in the technical approach begins by studying something for the purpose of making it "better." What is involved will be illustrated by the following three examples. The improvement of a building can be parcelled out to specialists in foundations, structures, building skins, heating and cooling systems, wiring, plumbing, elevators, and any other special equipment it may require. Each of these elements can be represented by one or several domains. The primary function of the foundations is to spread the weight of the building over a large enough area so that it can be supported by the soil. As a result, these foundations can be represented by a continuum with the same geometry, which is a mathematical model of a real material with its properties uniformly distributed throughout space. The footings can then be analysed and optimized by means of the discipline of stress analysis. The primary function of the structure is to carry the weight of the building and whatever forces may be imposed on it. It too can be represented by a continuum of the same geometry and can be analysed and optimized by means of stress analysis. The primary function of the building skin is to control exchanges of light, energy, air, and water. It also can be represented in a similar way to open it up for analysis and optimization by such disciplines as heat transfer, fluid mechanics, and materials science. A similar argument can be made for all the other systems because, in each case, one category of phenomena dominates all the others. In this way, the study of the building is opened up to discipline-based approaches. The improvements are thus limited to its components and their corresponding domains.

When all knowing is surrendered to discipline-based approaches, there is no possibility of examining the building on the level of experience in order to evaluate its liveability. Nor is there any discipline-based approach for evaluating how well it fits into and contributes to its physical and natural surroundings. What has been lost is what could be achieved by means of the cultural approach. It would shift the focus away from the "performance" of the building to its contribution to everything else. In traditional cities built up with the cultural approach, buildings tended to complement and enhance each other to create beautiful, harmonious, and liveable neighbourhoods. These cities were also generally sustainable by local ecosystems. This example is a microcosm of the successes and failures of our civilization. It is also typical of any technical artefact, process, or system built up of elements that can be represented by means of the domains of various disciplines.

A second example deals with situations that cannot be represented in terms of what we have referred to as reality because of their dependence on human beings, groups, and organizations. Consider the situation of an engineer who has been charged with resolving the quality control problem of one of her employer's manufacturing plants. This assignment has nothing in common with anything she encountered during her undergraduate engineering education. There is no problem statement with all the relevant information – not one item too few or too many. I failed to get this message across to my students when I assigned them an advanced fluid mechanics problem and deliberately added one piece of information that was not required to obtain the "right" answer. Many of them became rather upset with me, and they were not very reassured when I told them that in their professional lives it would be up to them to decide which of the myriad of available details were relevant and which could be neglected. In the case of our example, where is the root of the problem? – in the technology of the plant; the interfaces between this technology and the human operators; the shop floor organization; the supervision and management of this organization; the back-and-forth flows of information between the plant and the remainder of the corporation; adjustments made as a result of customer complaints and feedback; maintenance; management-labour relations; modifications made in response to community complaints and feedback; responses to regulatory agencies and their inspections; advice from insurance companies or lawyers; problems encountered with supply chains; difficulties encountered in distribution, including packaging, handling, and shipping; or temperature and humidity control in the plant? The dialectically enfolded "world" of this plant does not respect the boundaries between these categories, making it highly unlikely that the problem can be isolated in this manner. There is a high probability that the problem has multiple roots, and positive or negative synergistic effects between them cannot be ruled out. Since no manufacturing facility is ever perfect, it is likely that the engineer's attention will be drawn to those aspects that are commensurate with the disciplines she is familiar with. There are bound to be things that could benefit from improvements, but it will be next to impossible to answer the question as to whether such improvements are the key to solving the quality control problem.

Consequently, the first task the engineer is faced with (and for which she has received little or no preparation during her education) is to translate an area to be improved in terms of its quality control aspects into the equivalent of the kinds of problem statements encountered during her undergraduate days. This translation of an area of human life and the world into the domains of the disciplines of a technical specialty while ensuring that nothing essential is lost in the process is challenging. Doing so successfully would require what our secular myths cover over: a collaboration between the technical approach and the cultural approach in which each one goes beyond the limits of the other. It presents similar difficulties to the teaching of design in engineering. Once again, this represents a microcosm of the spectacular successes and equally vast shortcomings of our civilization.

The structure of our undergraduate engineering curriculum implies that the answers to most if not all of our professional questions lie in the application of the disciplines of the appropriate specialties. Doing so is erroneously called design and analysis. Nevertheless, these two components in the curriculum hardly prepare future engineers for the kinds of tasks they will face. Nor is the solution to be found in what in some countries is referred to as complementary studies, composed of social science and humanities courses. The technical core, design, and complementary studies components operate in mutually exclusive "worlds."

An extensive quantitative investigation of North American undergraduate engineering education revealed that the "world" of technical disciplines and design was full of technology and little else, while the "world" of the social sciences was full of everything else but little technology.[1] The study also demonstrated the complete lack of a synergistic relationship between the three components. The investigation showed that future engineers learn almost nothing about how technology influences human life, society, and the biosphere, and even less about how to use this understanding in a negative feedback mode to adjust engineering design and decision making in order to achieve the desired results but at the same time prevent or significantly minimize harmful effects.[2] These scores are more than a little embarrassing for a profession that claims that its primary duty is to protect the public interest, unless this is understood in minimal terms, such as making sure that bridges do not collapse and machinery is relatively safe. When the professional accreditation boards in Canada and the United States, the deans of most of the North American engineering schools, and my faculty were informed of the results, there was a deafening silence. In my own faculty, the chair of one of its largest and most influential departments had its undergraduate curriculum committee draft a so-called review of a first-year compulsory course in preventive engineering that had introduced the above type of missing negative feedback. The review was extremely negative. He readily admitted that the review was based on no evidence of any kind, but that the department had to deal with a situation in which the English literacy of many of its students was so poor that it dragged down their grades and put scholarships at risk. When threatened with being brought before the ethics committee of the profession, the chair of the department backed off, only to try again later. Eventually, the only course dealing with the above problem was removed as a compulsory course, and the program in preventive engineering was shut down when the founding director retired, even though qualified successors were available. Colleagues attempting to implement equivalent ventures at other schools frequently received similar treatment. It would be a mistake to attribute these kinds of incidents to the personalities involved. It is a widespread structural problem that the profession is reluctant to address because of its enslavement to the secular myths of our societies and civilization.[3] The cultural approach has been so thoroughly eliminated from the curriculum that the common sense it is able to contribute has all but vanished. For example, engineers ought to know better when they talk about zero emission vehicles, hydrogen economies, life cycle analysis, industrial ecology, and other such ideas as if they were perfectly feasible. Two weeks of a thermodynamics course show that the first example is impossible, and that the second example would vastly increase our primary fuel requirements unless hydrogen was made from fossil fuels, in which case it would only accelerate fossil fuel depletion.[4] Life cycle analysis has never been able to settle the simplest question, such as: Are paper, styrofoam, or washable ceramic cups the environmentally best choice for a cafeteria, or are cloth diapers environmentally better than disposable ones? When these life cycle analyses are done by means of input-output analyses, the numbers generated become completely meaningless because of our globally integrated economy.[5] Finally, industrial ecology retains very little of its original preventive orientation.[6] Yet we continue to waste public money on tenure track positions that make no sense. Again, it is a microcosm of the dilemma of our civilization. Beyond the domains of the disciplines of their specialties, technical experts of any kind have to rely on their highly desymbolized cultures, which causes them to participate in our "war against ourselves," as I have examined elsewhere.[7]

The above problem is even more extreme when external consultants are called in, whether they are consulting engineers or MBAs, and this frequently has the same mediocre results. They have no experience of the plant and thus even less "common sense" than the employees. Whether they like it or not, their suitability for the task at hand is limited to the extent that the issues to be grappled with can be captured in terms of discipline-based approaches. These specialists walk in, apply their disciplines, walk out with the cheque, do not have to live with the consequences, and hence rarely learn from their mistakes. Reading the Semco story may be sobering in this regard.[8]

A third example deals with a situation in which technology plays only a minimal role. It is based on a case study reported by Donald Schön.[9] An interdisciplinary team of experts is recommending how to resolve a food shortage in a valley in Colombia. It further confirms the above-mentioned tendency for technical experts to place whatever corresponds to the disciplines of their specialties in the foreground and everything else in the background. The situation is comparable to a group of artists sitting side by side painting the same landscape, but placing different things in the foreground according to their interests and focus of attention. From these paintings we would be unable to discover what the original landscape was like. We would have to go and experience it for ourselves. In the same vein, dealing with the food shortage meant different things for each member of the interdisciplinary team.

As might be expected, the nutritionist made the lack of an adequate diet the central problem. The specialist in public health pointed out that this was a peripheral issue because many people suffered from diarrhea owing to contaminated drinking water and lack of sanitation, with the result that they could not absorb a nutritious diet even if it were available. The economist smiled politely as he pointed out that both these issues were peripheral. If the people's standard of living were raised by economic development, they would be able to afford a nutritious diet and adequate water and sanitary systems. According to the agronomist, a more direct approach to the shortage of food would involve modernizing the agricultural practices of the community. After all, the people lived from agriculture, and increasing the amount of food a family could grow would create a surplus that could be sold for cash and would thus directly improve the standard of living and everything else that comes with it. At this point, the political scientist on the team protested by pointing out that none of this came close to the root of the problem because the reason why the farmers in the valley were so poor was that they had no political voice. The landowners and their politicians controlled everything, mostly to their own advantage. If the farmers had an effective political voice, things would surely change for the better, and such changes would be deeper and more lasting since they would affect the distribution of power and influence in the valley. The sociologist insisted that this would not solve the problem created by the custom of dividing a family's land when the oldest son married. In the course of generations, this had had the result of reducing each family's amount of farmland to a size too small to make an adequate living. The only real solution to this issue would be to organize cooperatives through which costly farm implements necessary for more modern agricultural practices could be purchased and shared. "Nonsense," blurted the demographer of the group. None of this would ever solve the problem. Non-governmental agencies through immunization and improved health care had succeeded in reducing infant mortality and slightly increasing life expectancy. As a result, the population had grown to a point that it could no longer be sustained by the ecosystem of the valley. The systems expert on the team walked away in frustration, muttering that the situation was hopelessly deadlocked until a systems model was built. However, doing so would inevitably fail: a systems model translates a dialectically enfolded complexity into a reality based on the principles of divisibility, separability, closed definitions, applied logic, and a mechanistic complexity. In a civilization dominated by technique, the members of the team acted in the only way possible. Their diagnoses and solutions all corresponded to the domains of the disciplines of their specialties. They all had the answer but no one knew what the question was because that would involve a great deal of experience of living in the valley in combination with applying discipline-based approaches. Once again, the technical approach must be complemented by the cultural approach so that the limits of each can be transcended by the other. Moreover, the people of the valley needed to be involved in the decisions that greatly affected their lives. There is an obvious conflict between democracy (what little is left of it) and absolute knowledge separated from experience and culture and thus from people's lives.

These three examples also illustrate why so many of us have a sense that in our society nothing appears to work any more. From the assessment notices we receive after filing our tax returns, to making inquiries about errors on a bill, making sense of undecipherable cell phone bills, having repairs made on our homes which may solve one problem yet create two others, acquiring technical objects that perform efficiently but do not fit into our lives, being unable to repair a growing number of items, building prisons when crime is receding, declaring war on drugs without bothering to understand what drives people to take them, building weapons systems that can no longer defend anyone but only destroy everyone – a great deal of what is being done no longer appears to fit the situation. This also reflects a civilization in which technique has so desymbolized our cultures that very little makes sense any more because our personal lives face the mirror image problems of our professional lives.

The second step in the technical approach to making something in human life or the world "better" takes the findings of the study conducted in the first step in order to represent them in a mathematical form wherever possible. If a formal mathematical model cannot be made, a set of correlations between the important variables constitutes the next best representation. In a non-trivial number of cases, doing so is not possible, and the "model" may be largely metaconscious, having been spurred by intuitions. These may be of a kind where, whenever certain difficulties are encountered, or when something appears to work unusually well, there seems to be a coincidence with other things, and these observations can be followed up by further explorations. Wherever discipline-based approaches are relied on, such metaconscious "models," intuitions, and further explorations are separated from experience and culture.

When a technical approach is applied to human life, groups, or organizations, its second step presents insurmountable difficulties because this involves the translation of something living (which must constantly adapt and evolve to everything changing around it) into an element of "reality" (which can be represented by rules, algorithms, statistical correlations, and theories, and thus excludes any creative and imaginative adaptation). Doing so is possible only in the case of routine behaviour, which is the closest we can come to mechanical repetition, but as a consequence of necessity or external force. The most common source of this in our civilization is rooted in the technical division of physical and intellectual labour. The greater the necessity or external force, the more a person or group is compelled to act as a "mechanism" having as much life as possible squeezed out of it. In the social sciences, the observation of any regularities that can be expressed in terms of statistical correlations or theories represents a lack of creative and imaginative adaptation and evolution, thus indicating the presence of alienated or reified human behaviour. As noted, it is the reason why artificial intelligence has been unable to capture daily-life knowledge and skills because we live the dialectical tension between this alienation and freedom. It is for these reasons that the cultural orders of traditional societies were able to support alienated individual and collective human life and why the technical order can support only alienated and reified individual and collective human life. We shall return to this issue throughout this work.

The preference for representing the findings of a study of human life or the world in mathematical form wherever possible is closely related to the third step of the technical approach, which uses the results to achieve improvements. What has been divided, separated, quantified, and mathematically represented to the greatest extent possible is now going to be improved. The kinds of improvements that can be achieved by the practitioners of the disciplines involved are highly restricted, however. The knowledge being applied is separated from experience and culture and suspended in a triple abstraction. Intellectually, it is impossible to connect the domain of any discipline to human life, society, and the biosphere. Consequently, when the practitioners of the relevant disciplines prepare several possible courses of action for improving what now represents a portion of reality (as opposed to a part of human life or the world), the determination of which one is actually better for human life, society, or the biosphere is out of the question. It would require a scientific way of integrating the domains of all disciplines to create a discipline-based equivalent to what was formerly accomplished by means of symbolization, experience, and culture. In the same vein, all the technical disciplines would have to be integrated to achieve an overall improvement from the efforts undertaken one domain at a time. In their absence, the only measure for evaluating alternative courses of action compatible with the discipline-based organization of knowing and doing is what is accessible to the practitioners involved: the inputs into a domain, the transformation of these inputs into a desired output within the domain, and the transfer of this desired output to another domain. The only possible improvements are of the kind that can be measured by means of efficiency and similar input-output ratios, but certainly not by means of cultural values.

Since there is no possibility of improving anything on the level of human life, society, or the biosphere, improvements that make sense and which matter to us cannot be made. If a formal model of the conversion of the inputs into the desired output is possible, the parameters of the model are varied and the associated forms are correlated with performance. This procedure converts making an area of human life or the world "better" into a problem of optimization. If a set of correlations between the most important variables had been established, an optimum arrangement could be determined for obtaining the desired improvement. If no formal model or set of correlations between the main variables could be established, experimentation guided by intuition based on metaconscious knowledge separated from experience and culture may still get good results. It is astonishing that discipline-based approaches have led their practitioners to believe that ultimately everything can be improved by means of optimization. The notion that this is clearly impossible where an area of human life or the world is in need of a better fit with everything else appears to be almost unthinkable.

The fourth and final step of the technical approach for improving human life or the world is to apply the best performing model to whatever was originally studied. The area under consideration should be reorganized to imitate the model in order to achieve the absolute efficiency that the model has demonstrated to be achievable. It is by means of this technical approach that contemporary societies make use of discipline-based approaches to "improve" every aspect of contemporary ways of life. The result is a technical order erected on the disorder of whatever in human life or the world has been externalized in the process but has not just gone away. Absolute efficiency is the only possible goal commensurate with discipline-based approaches to knowing and doing, and anything that cannot be intellectually grasped by them is externalized, with the result that human life, society, and the biosphere are being strained to capacity.

The use of the technical approach has four significant consequences. First, improvements are not made within and in the context of our lives and the world. Whatever is to be improved is first translated into what we have referred to as reality, constructed on the basis of five principles that are diametrically opposite to the dialectically enfolded order of human life in the world. It is a question of improving, not the cultural order of our lives, but the technical order built by discipline-based approaches.

Second, the technical approach divides us into knowers, doers, and supervisors or managers (those mediating between knowing separated from experience and culture, and doing embedded in experience and culture). The introduction of the technical division of labour, the assembly line, and scientific management illustrates the separation of brain from hand and a new technical mediation between the two. Symbolization, experience, and culture are delegated to a secondary role.

Third, the integrality of what is improved by means of the technical approach is first misrepresented by being cast in terms of reality; then distorted by the separation of what is relevant for the improvement of performance from what is not; and, at the end of the process, it is replaced with the best performing model. These distortions are then translated into human life and the world by reorganizing them in the image of this best performing model.

Finally, the external compatibility between what is being improved and everything else will almost certainly be degraded. Since nowhere in the process does this receive any critical attention, there is a possibility of (accidentally) doing no harm, but the likelihood of this is extremely small. It is the third and fourth consequences of the technical approach that lie at the root of our human, societal, and environmental crises.

The technical approach directly contributes to the previously explained characteristics of technique uncovered by an exhaustive analysis by Jacques Ellul.[10] The technical approach contributes to the rationality of technique by imposing the goal of absolute efficiency. It also contributes to the necessary linking together of individual techniques within a technical order because each and every domain receives its inputs from another, transforms them into a desired output with absolute efficiency, and transfers this desired output to another domain. It eliminates any need for human values by creating an automatism of technical choices because the greatest possible efficiency is always best.

The technical approach also contributes to the self-augmentation of technique in two principal ways. By translating everything to be improved from the cultural order into the technical order, it takes on the characteristics of reality built up by means of the previously noted five characteristics. As a result, in principle all specialists speak the same technical "language." Flows of technical information can thus be transmitted and received by every area of specialization, creating the global system of technique. Each area of specialization advances in response to the technical information received, and only indirectly responds to human needs and desires. Based on this internal mechanism, technique is self-augmenting, functioning at arm's length from human life and society. The other source of this self-augmentation of technique is the lack of negative feedback from within it. The technical approach for improving anything is the equivalent of driving our cars by concentrating on their performance as indicated by the gauges on the dashboard and only glancing out of the windows when a major disturbance in this performance takes place. This was illustrated for the case of engineering education, but it equally applies to many other professions.[11] The practitioners of any technical specialty do not know the consequences of their decisions and actions because these fall mostly outside of the domains of their disciplines, where they cannot intellectually "see" them. Hence, the equivalent of steering and pedal corrections cannot be made, with the result that "collisions" can be dealt with only in an end-of-pipe fashion. As a result, technique augments itself by feeding on the very problems it creates, resulting in layer upon layer of compensating techniques for dealing with the problems created by earlier technical improvements. Because of its reliance on discipline-based approaches, it is impossible to get to the root of any problem, thus making the system more and more top-heavy and uneconomic, as we shall see later.

Finally, the technical approach contributes to the universality of technique by removing everything to be improved from human life in the world that has been built up with symbolization, experience, and culture, in order to integrate it into a universal reality. No reference is made to a particular time, place, or culture, thus making everything inadapted to the local context. A universal technology has replaced the traditional locally appropriate ones. It has thereby made them unsustainable by the biosphere.

The technical approach is thus the entry into the technical order in which human beings participate almost exclusively on the terms of technique, and only to a secondary extent on symbolization, experience, and culture. As a result, technique is no longer guided by human values, or, to put it another way, the effect people have on technique is overshadowed by the effect technique has on people through its vast influence on human life and the world. It creates a condition of alienation and reification. Generally speaking, we have ended up serving the system we created in the expectation that it would serve us.

The Cult of Efficiency

The application of the technical approach to improving every area of contemporary ways of life constitutes a cult of efficiency. Accordingly, absolute efficiency is good in itself, endows everything with unlimited technical perfectibility, and, through it, economic growth. Our business reports inform us whether the markets were smiling on us that day, or whether they frowned at our behaviour. Much of this reminds us of traditional religious cults, except that the consequences for human life, society, and the biosphere are now incomparably greater.

Regardless of whether the markets are smiling on us or not, technique has fundamentally altered the nature of competition. It has essentially eliminated everything that was traditionally referred to as design. Prior to the emergence of technique, design was carried out on the level of experience and culture. Its aim was to end up with something that would enhance human life in some fashion. It was on the level of experience that people judged a design to be good, mediocre, or poor.

The technical approach can satisfy the aims of design only in the most trivial manner. As noted, when it is applied to the design of a building, the technical approach is capable of ensuring that the foundations sufficiently spread out the weight of the building, including any forces exerted on it, over the soil to prevent the building from settling, which could cause structural damage. The structure can be "designed" to carry the weight of the building and any loads imposed on it to avoid a partial or complete collapse. The building skin can be "designed" to control the exchanges of light, energy, air, and water with the surroundings to reduce heating and cooling requirements, prevent mildew and mould, avoid leaks, admit an acceptable amount of daylight, and so on. The point is that this kind of "design" only ensures what is absolutely necessary without doing anything more. In order to achieve a well-designed building, real design has to leave the level of the domains of the relevant disciplines in order to move to the level of experience. What impression does the building make on those who approach it? Does it enhance its surroundings? Can it be entered easily by everyone? Is it simple to find your way or orient yourself within the building? Does the building appear to be a sustaining environment for all the activities carried out within it? The bottom line is: would we really like to live, work, shop, or relax in this building? These questions are decided on the level of symbolization, experience, and culture.

It is important to remember that when we live in an urban habitat, we develop a great deal of metaconscious knowledge about it. We have no difficulty deciding that we would not want to live on a certain street, but that we would be very happy in another neighbourhood. Similarly, when renting or purchasing an apartment or a house, we have little difficulty deciding whether the unit suits us or not. No disciplines are required. Living in an urban habitat results in a great many experiences that are differentiated from each other, permitting us to make meaningful distinctions. These experiences are also integrated and yield a great deal of metaconscious knowledge. The tragedy is that this metaconscious knowledge is not critically evaluated, used, or drawn on in any way in architecture, civil engineering, planning, landscape architecture, and so on because these are all built up from a variety of disciplines. Moreover, disciplines such as urban sociology, urban geography, and urban ecology are not built up to complement what we already know metaconsciously. When it comes to choosing a place to live, what we have learned from various disciplines may be helpful, but it is entirely insufficient to achieve the kind of good design that many people admire because it fits their lifestyle and life in general.

Another way of approaching what constitutes good design is the recognition that it is not only the specific features in themselves that we appreciate in an apartment, house, street, or neighbourhood, but how these features interact together with the human activities they sustain. Ensuring this is entirely beyond the abilities of the technical approach. Why is this not recognized in schools of architecture, engineering, urban planning, and landscape architecture? It would appear obvious, because the clients of their graduates are going to make their own decisions. To put it more strongly: if architects had to live or work in the buildings they design, if engineers had to work on the assembly lines they analyse and optimize in relation to the product to be assembled, if doctors had to be regular patients in the hospitals they help run, and if landscape architects had to walk only in the spaces they design, it can reasonably be expected that things would significantly improve. They would gain enough experience to recognize what is so frequently lacking in their "designs," which are often a reflection of the poor education they have received.

It may be argued that the "common sense" necessary for evaluating a design is too subjective for "real" professionals. If by this term we mean practitioners who work exclusively with what we have referred to as reality, which is associated with the building of a technical order, the objection is valid. If, on the contrary, the aim of design is to make a genuine contribution to the enrichment of human life, society, and the biosphere, then professionals will not make much of a contribution, since they cannot operate on the level of experience and culture.

Consider some of the most promising developments that have taken place in architecture and urban planning during the last fifty years, such as urban villages, pedestrian pockets, transit-oriented design, and the new urbanism. These improvements are unthinkable without a synergy between knowing and doing embedded in experience and culture and knowing and doing separated from experience and culture. Before such synergy can be created, it is essential to systematically and comprehensively build up our experiences using real streets and neighbourhoods as our laboratories. Doing so should begin as an integral part of professional education and should include preparing students for these observations, supervising how they make them, synthesizing their findings as "laboratory" reports, and organizing group discussions to make them aware that a high level of consensus can be achieved. As in the learning of a culture, the development of a great deal of metaconscious knowledge will begin to imply metaconscious values as to which streets appear to be "good" for all kinds of people and which appear to be "bad" in sustaining street activities. In "good" streets, people are at ease, they stroll and talk to strangers, children are allowed to play on the street, and a non-trivial number of people appear willing to take some responsibility for this "good" order by getting involved in situations that could threaten it. In "bad" streets, there are usually extended periods of time when there are hardly any people on the street, and when people do encounter strangers, they tend to avoid them. No children play in these streets, and most people avoid walking there at night, taking a taxi to their door if necessary. During this stage of the students' development, analysis should be avoided as much as possible. Students should simply share their impressions to complement their observations. These will be no different from the observations made by any inhabitants of a neighbourhood, which allow them to make decisions about renting or buying, and on which kind of street they would like to live.

Without any analysis, the development of the experience of streets and neighbourhoods permits the differentiation of "good" streets from "bad" streets and, eventually, much in between. This development of metaconscious knowledge goes hand in hand with the emergence of metaconscious values. These make an overall assessment of the liveability of streets and neighbourhoods. It is important to recall our earlier discussion of tables, games, snow conditions, and illnesses. Every street and neighbourhood has countless details open to analysis. The metaconscious processes of differentiation and integration permit us to learn to make distinctions that are meaningful and valuable for our lives. Initially, these distinctions may be highly individual and thus subjective, but through the above kinds of processes, a great deal of convergence will occur (as we noted with children acquiring the culture of their community). As a result, the development of metaconscious values regarding the liveability (or lack thereof) of streets and neighbourhoods will be widely shared.

At this point, the students have achieved something that cannot be accomplished through discipline-based approaches: they are able to make overall assessments of the liveability of streets and neighbourhoods. They are now ready to complement this ability with discipline-based approaches to knowing and doing. Doing so can begin with asking the question as to what very liveable streets have in common, what less liveable streets share, and so on, followed by a comparison of the characteristics of different degrees of liveability. The results ought to converge to some extent with the findings of Jane Jacobs,[12] which took the form of her well-known design principles for liveability: a mixture of functions to bring people into the streets, shops, restaurants, offices, services, living accommodations, etc.; short city blocks to bring as much of this diversity within easy reach; a mix of older and newer buildings on each block to provide a variety of niches for these functions; and a significant density of dwelling units to ensure an adequate number of people on the streets. These design principles are expressed in the previously mentioned designs for new urban forms.

The process can then be repeated for older and newer suburbs to cover the entire range of density of dwellings per net acre. Significant variations will occur that correlate with the number of strangers one meets on the street and how many familiar faces there may be among them.[13]

The above procedure may well be similar to the one Jacobs followed when she began to reflect on her extensive experience of streets and neighbourhoods in preparing her landmark work, *The Death and Life of American Cities*.[14] Although the book appears to identify with systems thinking, I believe that it implies a synergy between knowing and doing embedded in experience and culture and knowing and doing separated from experience and culture.[15] However, the process must be repeated over and over again in our professional schools and universities. There is no guarantee whatsoever that the findings today will not be somewhat different because of the explosion of box stores, which draw business out of the downtowns with disastrous effects (social and environmental); the proliferation of the social media, which greatly reduce the need for face-to-face contacts, with far-reaching negative consequences; the growing gridlock on our streets, since we cannot find a feasible way of shifting the balance between private cars and public transportation; the out-of-control proliferation of "throwaway" buildings, especially condominium towers, and so on.

By going into some detail, I hope that I have made the lack of design on the level of experience and culture more evident. Elsewhere,[16] I have made the same point for the design of the Toronto streetcars, kitchen refrigerators, and production facilities, and it is possible to extend this approach to design on the level of experience and culture to everything around us.

In sum, technique has essentially eliminated design on the level where it counts: the way everything fits into and ought to enhance our lives, our habitat, and our biosphere support. Because contemporary ways of life are in the grip of technique, competition is now almost exclusively based on performance achieved by gaining absolute efficiency in each and every domain of the disciplines relevant to the "design" of everything. Our box stores are full of items claiming enhanced performance, and our lives are full of things which neither fit into, nor effectively sustain, the quality of our lives, communities, and habitats. This new kind of competition does not serve our purpose, but simply enslaves us to technique.

This new kind of competition on the level of discipline-based approaches contributes to our race to the bottom in another important way. It is no longer a question of competition among products, services, and artefacts in terms of their meaning and value for human life and society. Instead, the competition is based on absolute efficiency, which is indissociably linked to externalizing as many costs as possible. That this is the case follows directly from the above description of the technical approach. Improving anything on the basis of discipline-based approaches to knowing and doing means trading performance for an increased disorder in human life, society, and the biosphere. Competition now has no other reference than the building of the technical order by undermining everything else. When we speak of increasing competition through free trade, we are primarily referring to the ability of corporations to move to places where human life, society, and the biosphere are the least protected so that costs can be externalized to the greatest extent possible. Any protective standards have now been turned into trade barriers, and development has often been turned into technical and economic growth for its own sake with little regard for human needs, wants, or desires.

It should now be obvious that what is referred to as the design for X (DFX), where X can be any aspect such as manufacturability, reliability, quality, and serviceability, remains entirely on the level of discipline-based approaches to knowing and doing. In any case, the list can be extended, and still we would not be anywhere near what genuine design on the level of experience and culture can accomplish. The only exception is design for environment, which is a genuinely preventive approach.[17]

There is one final aspect to present-day competition that needs to be mentioned. As noted, Adam Smith already warned that the technical division of labour would make people as stupid as they could possibly become.[18] The information revolution has extended the technical division of labour to include a great deal of intellectual and professional work. The so-called re-engineering of the corporation involved the redesign of its organization in the image of the computer.[19] As much as possible, the intellectual work required to sustain our institutions is divided and distributed over a number of domains by what is called enterprise integration. These domains are subdivided into other domains corresponding to business processes, and these are further subdivided into smaller domains until eventually everything is expressed in terms of rules, algorithms, subroutines, and programs. From the perspective of experience and culture, this means that the use of computers and information technology has pushed all human skills back to the equivalent of the third stage of the model of human skill acquisition, which is based on problem solving.[20] All metaconscious knowledge on the level of experience and culture becomes externalized along with the ability to reach the levels of proficiency and expertise.[21] Again, competition based on technique "dumbs everything down," as we all know when we have the feeling that the "system" doesn't work any more. As a result of the influence of technique, some politicians claim that one of our main problems is government waste and inefficiency, thus opening government up further to technique. Essentially, what they propose amounts to bringing in more technique: doing more of the kinds of things that produced the problem in the first place.

The situation that has existed for a very long time on assembly lines is thus spreading to all intellectual "assembly lines" based on enterprise integration. From our discussion of the failure of artificial intelligence, it follows that there will be an initial gap between the flows of information as they occur within an institution and their representation in the enterprise integration software. The gap will continue to widen, since this program cannot evolve along with the society in which an institution operates. A growing number of situations will no longer fit the categories of the software, and the people working with it will become increasingly frustrated because they are unable to do a good job. Many of them will have no understanding of the consequences of arbitrarily choosing one category or another because neither fits the situation. This is equally true for people interacting with the institution from the outside, who may also have a sense of frustration as nothing appears to work very well any more. The situation is exactly the one faced by a conscientious assembly-line worker. Confronted by a deviation, the worker will have no idea of the significance for subsequent production steps and for the product as a whole. Is this a deviation that really matters, or is it one that is not going to have any significant effects? Being unable to make sense of anything, one can best protect one's humanity by daydreaming and longing for the weekend.

As a result, the cult of efficiency is enforced by means of an entirely new kind of competition. Unlike competition in the past, it is no longer based on what is significant for human life and society. It is a competition based on performance to grow and strengthen the technical order, and thus a competition to disorder human life, society, and the biosphere. This is the future prosperity made possible by computer and information technology. It is what I have called "our war on ourselves."[22]

Why Nothing Works Any More

When I use the terms design and the art of good decision making, I refer to all approaches seeking to transcend the limitations of discipline-based approaches. Can this still be accomplished, given our nearly complete dependence on discipline-based approaches to knowing and doing, which most of us are expected to practise in one way or another, at the very least through our work? From the outset, we must be as frank as possible about what we are attempting to face. From a cultural perspective, our task is nothing less than breaking the cult of the fact and the cult of absolute efficiency. Unless we critically understand this, our good intentions, especially towards environmental responsibility, will soon be thwarted by our secular myths, and we will be reassimilated by technique. Much of the environmental movement today has already fallen into this trap. Environmentalists do not understand the crisis as being inseparable from our ways of life, which have been structured by discipline-based approaches. As a result, it is possible to pick a favourite issue such as alternative green energy production, the protection of wildlife and its habitats, responsible shopping, and above all, education. In the case of the latter, most efforts amount to little more than sprinkling the words "sustainable development" throughout the curriculum as a kind of secular magic, or adding end-of-pipe environmental options to the curriculum (as my own and many other university faculties have done). Our secular myths ensure that our good intentions remain enslaved to technique through discipline-based approaches. These initiatives are not without value, but they will not even come close to getting us to first base. As noted, the dilemma we face is the result of the shift we have made from a primary reliance on cultural approaches to a primary reliance on the technical approach, with only a secondary reliance on cultural approaches, essentially limited to people's personal and spiritual lives.

This shift from culture to technique is accompanied by the disappearance of what in daily life we refer to as common sense. Attempting to revive this concept as designating something we have lost is a risky venture. The term has been much abused, and in philosophy the claim has been made that common sense does not exist. Lacking the social and historical context, the conclusion of philosophy is correct, but that does not mean that common sense did not exist prior to the emergence of technique. By common sense I mean the body of meanings and values of a culture by means of which a community lives and finds its way in an ultimately unknowable universe. It must be remembered that, prior to the emergence of technique, the working culture of a community represented the most objective forms of knowing and doing available. Because no situation in anyone's life ever repeats itself in quite the same way, everyone is busy adapting the culture of their community. These adaptations are individual and subjective. Some of them attract people's attention as being particularly effective, and they may adopt them for themselves. If this process continues, what began as an individual adaptation becomes an objective element of the working culture of a community. It is this objectivized culture that is transmitted from generation to generation as the common sense by which people live.

As contemporary cultures became highly desymbolized under the pressures of technique,[23] this kind of common sense all but disappeared. Because everything cultural is now regarded as the vestige of a religious and mythical past, any claim of common sense in the face of discipline-based approaches is easily dismissed. Nevertheless, if meaningful and purposeful design and decision making are to have a future, and if the race to the bottom is to be reversed, the re-establishment of a body of common sense will be a necessary condition.

The difficulty in re-establishing a place for common sense ultimately lies in the fact that we have confused what I have referred to as reality (constructed by discipline-based approaches to knowing and doing) with what is left of our symbolic universe and cultural order. The latter are as close as we can still get to the common sense we share, even though it is highly desymbolized. Our discussions of the way we have learned to differentiate and integrate our experiences of tables and games, how other cultures have learned to do this for snow and ice conditions, and how medical students learn to diagnose diseases illustrate that common sense continues to play an unacknowledged role in our lives. It is important to make a distinction between what is *real* and what is *true*.[24] We have already shown that what is real is constructed by dividing and separating everything, excluding anything contradictory, defining things on their own, measuring and mathematically representing the results, and integrating all of this in a complexity built up one element at a time. We have also seen that the way babies and children learn to make sense of others and the world is based on the diametrically opposite orientation. Everything is differentiated from everything else and integrated into a life, and via that life, into the working culture of a community. The result is a dialectically enfolded complexity which is indivisible, inseparable, contradictory, definable only in open-ended terms, and not open to measurement and quantification, and which grows through progressive differentiation and integration. This complexity represents what is true for our lives by differentiating and integrating everything in terms of its meaning and value for our lives.

Unfortunately, as technique has taken hold of our lives, most practitioners of discipline-based approaches no longer feel a need for the common sense that ought to guide their knowing and doing. Without this kind of guidance, these approaches take on an absolute and unlimited character because their users can no longer discern when and where to use them and when and where to put them down when their limits have been reached. I will give some simple examples, but in order to protect the identity of some of my former students, I will limit the details.

It is no exaggeration to claim that the students who specialized in nanotechnology and biotechnology in my faculty had for a time a great zeal for saving the planet by helping to bring to an end the environmental crisis. When they were taken on a "reality check" (literally checking the meaning and value of these disciplines, including the limits of what they could do for human life and our planet), they had to face the question as to why these new and exciting disciplines would accomplish something different from all their predecessors. Is it not more likely to expect that their positive and negative effects will follow the same pattern as all others? They will draw matter and biomolecular life into reality, make them perform better, and thus disorder their context. In sum, based on a great deal of prior experience, it may be argued that the likely long-term effects of nanotechnology and biotechnology will be to add two new frontiers of pollution: the pollution of matter and the pollution of the DNA pool of all life on the planet. The reactions of many of the students was deeply disturbing because they were unwilling to even consider the possibility that these two disciplines, to which they were committing themselves, could not possibly deliver what we all desire. The very possibility was dismissed as technology-bashing. This is a perfect illustration of the kinds of issues that, as Georges Devereux has shown, exist in the social sciences.[25]

Many students taking human factors engineering found themselves in much the same situation. They had correctly learned to recognize that a great many of the problems we face today stem from a poor fit between technology and society. To address this situation, they were learning to focus on the interface between technology and human life in order to improve it. They had become convinced that the kind of future humanity will likely have depends fundamentally on the success or failure of this endeavour. Human factors engineering was the new and exciting discipline charged with this task, and it would accomplish it by drawing on psychology. Once again, a reality check was undertaken by drawing on the experience of human history. Why in the social sciences is psychology not the only discipline we need? Are all the issues faced by earlier societies and civilizations reducible to issues to be dealt with at the interface between people? Can we reduce economics, sociology, political science, law, morality, religion, and much else to psychology? Should the limitations of human factors engineering not be evident, and should the students therefore not learn the limits, to avoid using their intellectual hammers to treat everything like nails? Most of the students who had an unwavering belief in human factors responded by dropping the course that questioned it. It was simply too threatening to their careers. Again, the work of Devereux helps us understand the situation.[26]

The same thing was true for the students taking the end-of-pipe environmental option offered by my faculty. End-of-pipe approaches are absolutely necessary because their preventively oriented counterparts are not practised on a wide scale, and it is much better for all of us to have end-of-pipe approaches than to do nothing. Nevertheless, these approaches do not go to the roots of the problem, with the result that they trade certain problems for perhaps less serious ones.[27] Again, this is useful, but it is not economically viable in the long run.

The use of computer simulations also represents a shocking loss of common sense. The reality constructed by these simulations is taken for what is really happening. I experienced this first-hand when, as editor-in-chief of an international journal, I was approached by one of my editorial board members and asked if I would consider publishing some articles about the health effects of wind farms. At first, I did not understand why the question was being asked. After all, the topic was clearly within the scope of the journal, and if articles were submitted for publication, they would be refereed like all the others. Why was this a special case? I quickly learned something new. I live in a province where the health effects of wind farms have become a significant issue, one that is even blamed for a Liberal government losing its majority in the legislature. There was a great deal of controversy and enormous vested interests. I attempted to do my due diligence as an editor, spending a great deal of time exploring the issue. I encountered a team of eminent professionals who had been drawn into the controversy because they were themselves affected, or knew people whose lives had been deeply affected by these health problems. In their own free time they had assembled what evidence they could by initiating pilot socio-epidemiological studies.

I also learned that intellectual games were being played by many different vested interests. Even Greenpeace was on record as stating, in a program on the CBC (Canadian Broadcasting Corporation), that there was no scientific evidence of any health effects from wind farms. The reasons for this were never elaborated on. Was there no scientific evidence because the required socio-epidemiological studies had been done and no evidence was found? Was there no scientific evidence because these studies had not yet been undertaken, or because they had been blocked by lobbying efforts? I became more and more upset at what I was discovering. The provincial government apparently had even used the Ontario Provincial Police to intimidate people engaging in peaceful and legal protests. I myself was served with a letter of demand (one step short of a libel suit) in my own backyard by a (trespassing) lawyer representing the wind farm industry. This was clearly designed to intimidate me and prevent me from publishing further articles on this issue. My publisher's lawyers had little difficulty defending what we had published, but it did create a very stressful time for me because neither the university nor the faculty association would commit itself to aid in my defence. I later published an article reporting acoustical measurements that had been taken by some technicians, one of whom actually developed the typical symptoms some people experience when living near wind farms.[28] I suppose the above type of behaviour can be expected when powerful vested interests are challenged, but I was not prepared for the confidence specialists placed in their own acoustical simulations. How is it possible to simulate something that we clearly do not understand well enough? The occurrence of the typical symptoms was undeniable, but the pathways by which they were caused were under intense debate. In the meantime, some of the engineers and their employers acted as if what was not represented on the computer screens by their simulations simply did not exist. It is a deep structural problem rooted in our inability to identify and locate the limits of our discipline-based approaches to knowing and doing.

There are many other situations of this kind. A public transportation company under financial pressure engaged consultants to eliminate waste and cut costs. The usual approaches led to the usual results: greater performance for the company and greater disorder for its drivers and passengers. The company did not realize that their long-term viability was being negatively affected. Since then, the quality of driver training has plummeted, the quality of the people applying for that job has significantly gone down, and the passengers noticing these problems are beginning to contemplate alternative transportation wherever possible, with the result that in the long term this vital public transportation has been put into a downward spiral and anyone who has any alternative simply does not consider using it.

The same approach is being taken to reduce costs and waste and to improve the performance of our hospitals. From time to time, there have been voices in the literature claiming that approaches based on operations research have significant limitations.[29] One of my colleagues has quit the field altogether.[30] Nevertheless, the consultants arrive, they use their discipline-based approaches to make changes, they leave with the cheque, and everyone else in our hospitals is frustrated by all the problems these reorganizations create.

I could continue with more examples and easily fill the remainder of this book with them. The situation is now completely out of control. There no longer is any common sense that allows us to understand the limits of our discipline-based "tools." Collectively we are like a contractor who has not yet gained enough experience to understand that not every problem can be solved in the same way. It amounts to a deep structural problem that does not depend on the validity, or lack thereof, of what is presented in this work. Even if my diagnosis is completely faulty, there is no way of getting around the issue that our discipline-based approaches need to be guided by something broader, unless we can accept that, unlike all other human creations, these approaches are absolute and have no limits. In that case, we are simply completely alienated by our secular myths, and nothing can be said that will make any difference. However, if we are willing to admit that these approaches, being human creations, must have limits of some kind, then what can we use to keep their applications within these limits, and what shall we do for situations that fall outside these limits? If it is not common sense derived from a cultural approach, what else shall we use?

If contemporary cultures were not so highly desymbolized by technique, and if so many people had not been deprived of their common sense by desymbolization, our experiences that things do not make sense any more and that the system is "broken" would have caused us to question where we are headed and to fight back a long time ago. We admire other societies when they do so because they have simply had enough, but we do not appear to recognize that we are conducting a war against ourselves as a consequence of technique.[31] All the hype regarding the role the social media are playing in other parts of the world to aid people in standing up for a better future is a wonderful distraction from what these technologies are utterly failing to do at home.

We have built universities almost exclusively on disciplines. Doing so makes sense if these disciplines are absolute and without limits. However, if we are unwilling to make these disciplines into our secular gods of knowing and doing, we need to organize a compulsory course that every entering student must take to do a check on the reality we are constructing by means of these disciplines. In the meantime, the public university no longer serves the public interest, and we should have no illusions about this.[32]

Design and Decision Making That Make Sense

Our civilization has come to a fork in the road. We can decide to serve technique at the expense of our own interests and those of our children as well as of all life on earth. Alternatively, we can attempt to make decisions that make sense for human life, society, and the biosphere. Making sense of things includes our employers as well as the economies of our societies (which now act as anti-economies by extracting more wealth than they create). It is not a choice against anyone, but a choice against the secular myths that hold us captive as combatants in our war against ourselves.[33] What it requires is to do a check on the reality we have constructed by symbolizing it in terms of the meaning and value it has for our lives and the world.

Reintroducing symbolization, experience, and culture into technique begins by organizing our professional activities in the same way as we drive our cars. Keeping in mind where we are going, we look out the windows to see how getting to our destination is placing us in relation to all the other occupants of the road. As a result of what we observe, we make steering and speed corrections to get to our destination safely by preventing collisions with others. All this includes an occasional glance at our dashboard to check up on the performance of the vehicle. In other words, the performance of the vehicle is not an end in itself, and not keeping it in check may prevent us from reaching our destination.

As noted, a detailed study of undergraduate education quantitatively showed that future practitioners are not taught to look out of the "windows" of their disciplines to experience how technology is influencing human life, society, and the biosphere, and they learn even less about how to use this to make corrections to achieve the desired result but at the same time prevent or greatly minimize harmful effects. It would not be difficult to change this situation, provided that we are willing to recognize our secular idolatry of technique as absolute knowing and doing in order to experience what it is truly doing to human life and the world. Doing so can be undertaken by design and decision making that make sense.

How can design make sense? It begins by recognizing that what ultimately matters is to design something that enhances its contribution to the evolution of everything being related to everything else. This is not as abstract as it may sound. It begins by taking what are regarded as state-of-the-art designs and benchmark accomplishments in a field. These accomplishments act as design exemplars. For example, lean production is the leading exemplar for a manufacturing system. Agribusiness is the design exemplar that is replacing the family farm. The low-floor streetcar common in Europe is a vastly superior design exemplar to the high-floor streetcar of Toronto. The common household refrigerator is a design exemplar well adapted to the needs of manufacturers, but a poor design exemplar for our homes.

Any attempt to make sense of these design exemplars can begin by systematically examining all the relationships established by the people using them as well as the relationships of reciprocal interdependence with their surroundings. By way of an example, let us attempt to make sense of the high-floor streetcar in Toronto. How do these streetcars make sense to their drivers, passengers, people on the sidewalk, people living along streetcar routes, other users of the road, the supervisors and managers of streetcar lines, the people who service and repair them, and so on? Among these categories, the people who travel by streetcar, those who drive them, and the other users of the road form a good starting point for the analysis. In the case of passengers, their experiences of streetcars involve boarding, arranging payment, finding a seat or place to stand, keeping track of where they are in order to not miss their destination, and finally leaving the streetcar. These experiences can be assessed for different kinds of passengers, including mothers with strollers or toddlers, the elderly, people with motor or sensory limitations, people with luggage, bicycles, or dogs, visitors unfamiliar with the system, people escorting small groups of children, and people engaging in anti-social behaviour. In each case the experiences can be rated on a scale of highly user-friendly to virtual exclusion from use. Repeating the process for other categories of people will build an inventory of experiences rated on how much sense the design exemplar makes for them. Based on this inventory of experience, the meaning and value of the features of the streetcar become apparent.

Next, this assessment can form the point of departure for the modification of a design exemplar. It is the art of modifying a design in such a way as to eliminate most or all of the negative experiences and enhance as many positive ones as possible. In addition, a comparative approach can usually enrich the analysis by repeating the process for other design exemplars and correlating different features with the relevant experiences. Even the leading design exemplars can frequently be improved on the basis of this approach. In many instances, students studying them quickly conclude that these types of exercises have not been undertaken as they discover how easy it is to prevent negative experiences by making modifications to the design exemplars. In the case of the current Toronto streetcars, students soon came to the conclusion that one of the most important and most unacceptable features of this design exemplar is its high floor (apparently originally intended for the typical European train station platform). The number of steps not only makes boarding and exiting needlessly slow but also impedes widespread use by all people. In addition, the weight of these cars (especially the articulated ones) makes no sense whatsoever. The weight of a train car is not such a critical factor because it tends to move at a uniform speed as much as possible, but for a streetcar, which is constantly accelerating and decelerating, weight is an important design parameter. A heavy weight makes these streetcars much noisier than they need to be, increases wear and tear on the tracks, transmits excessive vibration to the surroundings, and requires more energy than necessary even though some of it is regenerated as the vehicle brakes. From a safety perspective, these streetcars take too long to stop, which is significant in heavy traffic. The students concluded that these streetcars should never have seen the light of day, and this opinion is reinforced by people who have visited European cities.

Once the students are convinced that they have arrived at the best possible design exemplar, discipline-based approaches can be applied by dividing the design exemplar into its primary constituent design exemplars, including the frame, the "skin" with its windows and doors, the interior design, the driving controls and instrumentation, the wheel sets with motors and brakes, the heating and air conditioning systems, the electrical systems, the communication and security equipment, and the payment devices. These can be further subdivided until they reach a level where one function dominates all others, with the result that they can be treated as a domain. The relevant disciplines for analysis and optimization can be applied, and the results on this level must then be synthesized on the next level, back down the "tree" to the root design exemplar. As a result, discipline-based approaches to knowing and doing are guided by our experiences of the streetcar, in contrast to the usual approach in which the use of discipline-based approaches determines our experience. The results that students can achieve in a few tutorials are encouraging, and a sign that a great deal of improvement is possible.

A similar experience-based approach to the design of low-floor buses capable of accommodating wheelchairs leads to somewhat similar results. It is not difficult to improve the experiences people have of these buses, whose accessibility for wheelchairs is achieved by avoidable inconveniences to many other people. There are far too few seats, especially in an aging society, and the expanse of standing room is poorly served by supports designed for passengers to hang on to while increasingly poorly trained drivers make it difficult for many people to hold on and remain safe. This kind of design exercise does more than simply introduce experience back into design in order to control and guide discipline-based approaches. It also reintroduces negative feedback to support minor or major design changes guided by experience. Design features can be correlated to the ratings of the experiences of everyone involved. I cannot sufficiently emphasize how quickly students learn to make significant improvements.

The next stage in this design exercise is to have the students recognize that streetcars are merely one component of what ought to be a seamlessly integrated public transportation system. Additional factors need to be considered. Should streetcars or buses be used on a particular route? How much weight should be given to studies that suggest that the more positive experiences on streetcars tend to build ridership to an extent that buses do not? In addition, how flexible is the overall design to accommodate significant changes in ridership? Is a new design exemplar for a streetcar compatible with the existing track system? If not, should the design exemplar be changed or the tracks upgraded? What kind of upgrades should be made so as to increase the likelihood that future designs can be accommodated? What can we learn from innovative experiences such as those of Curitiba, a city that could not afford the high costs of subways and light rail transport, thus necessitating the invention of interesting alternatives? Again, comparative approaches can be most helpful here. I have spent many hours in my office listening to former students tell me how dysfunctional many public transportation systems and private bus companies are. It is more than a little depressing, given that the only solution to the gridlock in many metropolitan areas lies in convenient and rapid public transportation.

The design of homes and buildings can be guided by our experiences of them in much the same way. We need to design them from the "outside in" by means of experience as well as from the "inside out" on the basis of discipline-based domains. As noted for streets and neighbourhoods, living in an urban habitat leads to a great deal of experience of homes and buildings and thus to metaconscious knowledge that allows us to effortlessly distinguish between those homes we would like to rent or buy and those we would not. Although this may appear to be little more than a subjective assessment of their liveability, these kinds of evaluations are likely to be widely shared among certain groups of people, much as the experiences of our culture have a dominant objective component despite desymbolization. Today, this body of experience is not well used by those who design our homes and buildings. As a result, it usually does not take more than a few days of living in a home to discover that even a few small changes here and there could have made a world of difference in its liveability.

Again, students can be taught to have human experience guide the design of homes and buildings. For example, at a first tutorial, students can be handed a floor plan of a family home in order to assess its liveability. They are instructed to begin by determining the kinds of families who are likely to rent or buy it. They can then begin to imagine living in this home as a member of one of these families. The assessment of its liveability can best be started by adding the footprints of furniture to the floor plan. The students can then imagine carrying out a variety of daily-life activities and begin to superimpose the trajectories people would follow as they carry them out. Different copies of the floor plan can show groups of these trajectories that are likely to be carried out at more or less the same time, so that possible interactions can be imagined. For example, one copy may show how the home is used in the morning when everyone gets up, gets ready, eats breakfast, and goes to school or work. Another copy may show its use when people come home in the late afternoon or evening. The students will have to experiment with how to best group these activities into meaningful sets, to help them imagine the ways in which the layout of the home sustains, interferes with, or obstructs a particular set. Special scenarios may also be imagined. For example, one copy of the floor plan may show what happens at a birthday party with an influx of friends and family. How well will the entrance hall, kitchen, dining room, living room, and washroom facilities sustain the celebration? As people are seated in the living room, how easy is it for them to move in and out with snacks and drinks, leave to make a phone call, answer the front door, visit the powder room, and so on? Difficulties with the layout may be identified, such as a room frequently serving as a corridor to get from one place to another. Based on the imagined experiences of these kinds of scenarios, students can learn a great deal about the liveability of the home with different types of furniture arrangements and a variety of families with different lifestyles.

The evaluation of the home should next turn to how well it is integrated into its surroundings. Do the placements of windows and doors take full advantage of all the good views and provide outdoor access to them? What is the level of privacy from people passing on the street, neighbours sitting on their back patios, and people looking out of nearby windows? Are the locations of doors and windows such that advantage can be taken of prevailing wind patterns to create cross-ventilation when the weather is warm? What exposure does the house have to sunlight, and what are the energy implications for summer and winter? Which outside areas are likely to be sunny and shady, and can various activities be accommodated during various times of the day in hot weather? Is it possible to plant trees to shade the house during the summer? Although all this initially appears as a daunting task to the students, a little practice soon allows them to focus on the most important questions. In this way, they quickly learn to make improvements to furniture arrangements as well as the floor plan, to enhance the liveability of the home.

The next stage in this exercise is to assess the suitability of the floor plan for the intended building site. If they can visit the site, they will be able to stake out the house's footprint and mark the rooms to be located on the first story. Windows and doors can be identified by additional stakes connected by tape to mark their tops and bottoms. Small platforms can be built to accommodate folding chairs, and these can be placed in the various prospective rooms to evaluate the views offered by the windows.

The students are now ready to apply what they have learned regarding the liveability of homes by designing a new house to be built on a particular site. If an actual site cannot be used, a detailed map showing how a hypothetical building lot is situated in relation to others in the street may be provided along with some pictures of the major views. Students can begin to imagine the appropriate building forms that would fit the neighbourhood, while at the same time considering possible floor plans appropriate for the most likely inhabitants. The most promising designs can then be drawn on different copies of the site plan for an evaluation of the above kinds of details. Of course, visits to an actual building site would greatly enhance this exercise because the students could then walk around and imagine different kinds of building forms, narrow down the possibilities, and stake out the most promising ones as described. Views from windows can be carefully checked and further adjustments made. It should be noted that during an actual construction, this procedure can be repeated for the second story once the first has been framed and the flooring for the second story laid down. The location of the interior walls can be marked and the location of windows simulated in order to "experience" the second story. If necessary, minor adjustments can be made. In most jurisdictions, building permits may not have this kind of flexibility, and it may delay the framing of the second story by a few hours. Generally speaking, the results are likely to be worth it. As a university student, I followed the above procedure when designing a home for my parents, only to discover later that others had developed similar experience-based approaches.[34]

Once a design exemplar has been finalized, the analysis and optimization of the building by means of discipline-based approaches can readily be carried out as suggested earlier. The design exemplar established by experience-based approaches again guides the discipline-based approaches used for analysis and optimization of the details. In this way, the usual relationship between culture and technique is reversed by the former guiding the latter. It may be objected that these kinds of exercises are appropriate for schools of architecture but not for engineering. These kinds of objections will ensure that technique continues to dominate culture. Instead, these exercises help students to establish a new type of relationship between experience and what they learn in their disciplines. This helps them to understand how the latter can serve the former to create what makes sense to us, instead of pursuing absolute efficiency. It creates an alternate interpretation of what engineering is all about, and this is essential in the education of future engineers who can help us create a more liveable and sustainable future.

It is possible to extend the above approach for assessing the liveability of houses to larger buildings in order to arrive at design exemplars that make sense for people. Software programs that can assist students in making three-dimensional representations of a building that allow them to "experience" it may be of some limited value. Their usefulness would be greatly enhanced if arrangements of the appropriate furnishings could be added.

In conjunction with what students have learned regarding streets and neighbourhoods, the above kinds of exercises can lead to the development of a great deal of metaconscious knowledge that would be invaluable to anyone involved in the design, evolution, and redevelopment of our urban habitat. It may even be possible to teach some of this in a more elementary form in high schools as a way of making students more literate regarding the importance of our urban habitat and the way that it sustains or undermines our activities within it. The challenge is enormous, as the equivalent of mass production is putting its stamp on the urban habitat. The repetition of story after story in tall buildings may be absolutely efficient, but it is not very liveable. The repetition of minor variations on the same design exemplar over large subdivisions may save some money now, but will inflict incalculable harm as a monotonous and uninteresting habitat. Surrendering our urban habitat to the cult of absolute efficiency will surely become recognized as a failure to design an urban habitat that makes sense to people.

Taking human experience into account can also make a fundamental difference in the design, organization, and management of work and workplaces. Since our work occupies a significant part of our lives, we all acquire a great deal of metaconscious knowledge about it, including the metaconscious values of our society regarding work and workplaces. We have no difficulty deciding what defines a good job or a bad one. If our students have been so lucky as to have had some summer work, they will quickly build this type of metaconscious knowledge. Unfortunately, when students study engineering, business administration, management science, or industrial accounting, little or no use is made of this experience. Because their fields of study are built up with discipline-based approaches, they quickly learn to discount anything they have learned from their own work experience. The path of industrialization would almost certainly have taken a different course if the people involved had been able to make work-related choices based on their experience. However, owing to skewed power relations, this was rarely the case. During the first phase of industrialization, the pool of unemployed workers was large, with the result that, for most of these people, any work was better than nothing. Those who did have work in the factories experienced enormous difficulties adapting to the order imposed by the technical division of labour. Draconian measures were required to impose and enforce this order.[35]

When these factory workers began to defend themselves by organizing unions, wages and benefits began to rise along with standards of living. A portion of this may be interpreted as an end-of-pipe compensation for what many experienced as meaningless and dehumanizing work. A great deal of anecdotal evidence confirmed by research suggests that many people recognized the enormous negative influence their work had on their lives and persons and how this in turn affected their families and communities.[36] By this time, they no longer had any choice in the kind of work they did: almost without exception, their work was a part of a technically divided hand or brain. In the case of the former, they merely carried out one of the production steps in the horizontal division of labour; in the case of the latter, they could participate in a level of control, ranging from shop floor supervision to senior management. Eventually, both management and unions took this situation for granted, with the result that there was no longer any discussion regarding the kinds of workplaces that should be designed and built. Everything had become a struggle over who had what power in these workplaces, and who would get what slice of the benefits. During times of near full employment, following the Second World War, unions were able to drive up wages and benefits as compensation for the mostly unfulfilling work their members had to do. Because economies were strong, their employers could afford it; for decades, the opportunity was lost to act on what we knew about work from experience.

The situation began to change when economies turned into anti-economies, which Keynesian policies were not designed to prevent (more about this in chapter 3). Decades of economic growth and near full employment had been accompanied by the raising of labour, health, social, and environmental standards, which appeared to be affordable at the time. Since these standards were mostly met by means of end-of-pipe approaches, they quickly imposed huge costs with modest benefits. It put enormous pressures on corporations. Many of them resolved these by moving manufacturing operations to countries where they could externalize costs to the greatest extent possible because wages were extremely low, benefits were negligibly small, and labour and environmental standards were almost non-existent. To make this work, free trade agreements became necessary, which further reinforced what became one of the most significant exports for many countries: good jobs that enabled people to raise families.

Despite the previously noted early warning of Adam Smith, the enormous negative role the technical division of labour played in industrialization did not receive much attention. Nevertheless, many major and highly successful corporations quietly experimented with ways of reorganizing work to prevent or greatly minimize the harmful effects of the technical division of labour.[37] Some of these experiments created genuine choices for people.

The Scandinavian countries learned a great deal from the mistakes made by those who industrialized before them. Moreover, living in a cold climate had given their cultures a highly cooperative orientation, and these and other factors led to some of the most cooperative relations between government, industry, and labour. A situation developed in which work experience was able to play a significant role, thereby beneficially affecting the distribution of power. Some time after the Second World War, under conditions of near full employment and high labour standards (which included the principle of equal pay for equal work), Swedish citizens working on automotive assembly lines voted with their feet, gradually refusing such work. Since financial or other end-of-pipe compensations could not legally be offered as incentives to accept assembly-line work, the situation became so serious that Volvo decided to begin assembling vehicles without an assembly line in its newly built Uddevalla plant.[38] In this plant, there were two kinds of workshops in which work teams assembled cars without repeating any operation, for eighty minutes in the first type of workshop and for a hundred minutes in the other. As the organization perfected the system, assembly costs steadily dropped, retooling costs for model changes were dramatically reduced, and vehicles could be built in response to customer orders. Self-managing work teams enjoyed high levels of method and timing control, which, according to the demand-control model based on a great deal of socio-epidemiological evidence, translates into much healthier work and fewer negative spillover effects into people's lives.[39]

The Uddevalla plant eventually became the victim of rising unemployment levels, growing competition with other European car manufacturers (who could afford to ignore the experiences of guest workers from the Middle East and North Africa since they were on temporary work permits), and growing free trade (which made it more difficult to protect high labour standards). Nevertheless, this attempt showed that, had the experiences of workers been taken seriously from the beginning, alternative manufacturing systems could have been designed. These facilities would have transferred far fewer costs to society, thereby reducing tax burdens. Even if these kinds of production systems had not been able to reach the same level of labour productivity as the lean production systems (and I believe this result would be highly unlikely, as we will soon see), there would almost certainly have been a net benefit to corporations and societies. Hence, to dismiss Volvo's initiative as uneconomic or an experiment in socialism represents an uncritical submission to the cult of the fact and the cult of absolute efficiency. It should be remembered that the lean production system took a very long time to perfect and is an integral part of our contemporary anti-economies, which extract more wealth from societies than they produce.[40]

In order to critically assess the Uddevalla project, we should recognize that it divided the Volvo organization against itself. In the Uddevalla plant, an organizational culture was being created in which the hand and the brain were reintegrated to the greatest extent possible: where everyone was as self-managing as possible while participating in work teams that had a great deal of autonomy, and where negative feedback loops were shortened accordingly. Workers learned to assemble vehicles without sacrificing their own lives. In the remainder of the Volvo organization, the hand remained fully separated from the brain, according to the principles of scientific management and bureaucratic organization. In order to reunify the corporation, Volvo would have had to undertake the kind of complete top-to-bottom reorganization that will be described next.

In Brazil, Semco reintegrated the hand with the brain throughout its organization.[41] All dead-end jobs were eliminated by reassigning the "undesirable" tasks to the people who needed to have them done, thereby making everyone as self-reliant and self-managing as possible. Negative feedback loops were shortened everywhere in the organization. People could take the experiences of their work into account in planning and organizing it. Everyone was taught to read and interpret the company's balance sheet, and financial information was openly discussed, with the result that everyone understood how they affected and were affected by the company's bottom line. The owner was adamant that the benefits of capitalism should be for everyone. Hence, all employees could vote on major business decisions that would affect the organization and themselves. There was also a profit-sharing scheme, which employees helped to manage. People at the centre of this organization were so highly self-managing that they even determined their own salaries, made decisions about when they needed additional training or a mini-sabbatical to rethink their work, and arranged all their own travel. The values of freedom, individualism, and competition created an organizational culture which, by the standards of most managers and CEOs, would be unthinkable. Many of these people paid Semco a visit to see for themselves.[42] This reorganization of the company made it five times more profitable and increased its sales sixfold, in spite of operating in an economy going through some very hard times.[43]

It should be noted that what gave the lean production system its competitive advantage over the Fordist-Taylorist system was a partial reintegration of the hand and brain and a substantial shortening of some negative feedback loops.[44] During the decades when lean production was being perfected, Japanese engineers recognized that the extreme technical division of labour already used by Ford in the United States, before the Second World War, would not work in Japan because its market for vehicles was too small and segmented. To everyone's surprise, they discovered something that to the engineers of those days was completely counter-intuitive, namely, that making small batches of parts just in time for their use on the assembly line created a substantial advantage. This was because the losses in economies of scale were more than offset by the advantages derived from shorter negative feedback loops. These made it possible to catch errors earlier, quickly eliminate them at their source, and then use the same negative feedback loop to continuously improve quality.[45] Even though people on the lean production assembly lines were supposed to use their brains as well as their hands, the lines almost never stopped once the system had reached a certain level of development. As a result, the lean production system does not even come close to resolving the problem of technically divided work rooted in the hand-brain separation. The country that invented lean production is now also the one that recognizes death by overwork as a disease.

If Volvo had reunited the company by reintegrating the hand and the brain to the greatest extent possible throughout the organization, much more could have been achieved. It could have combined the many advantages of the Uddevalla assembly system, the organizational advantages of Semco, and the shorter negative feedback loops in supply chains by lean production, as well as Toyota's design for ease of assembly. Volvo's system might have become unbeatable, even at a time when the race to the bottom was well under way.

The creation of these types of organizations is just the beginning of what is possible. In addition to becoming a much better fit with human life and society, these organizations could also have learned to improve their fit with the biosphere because of their short negative feedback loops. To do this, car manufacturers could turn themselves into transportation service providers, which represents a further development of the leasing process by making it comprehensive. Under such an arrangement, a customer could purchase a transportation service for a certain number of kilometres with the desired fuel economy, safety, comfort, and luxury according to the model selected. The transportation company would manage the vehicles from beginning to end by engineering a system that has many advantages and would realign the interests of these companies with those of their clients because profits would now depend on providing these services with the lowest possible throughput of matter and energy. Companies would begin by manufacturing high-quality vehicles designed to be refurbished at regular intervals (with the use of remanufactured parts as much as possible), and at the end of the vehicle's useful life, a substantial portion of the value it represents could be recovered.[46] Elsewhere, I have developed a set of design principles based on a great deal of experience described in the literature.[47] These kinds of reorganizations would pave the way to building a service economy based on consumers buying the services rather than the products that provide them.[48] Because the company's and the consumers' interests would now be realigned, a measure of trust could be reintroduced to make these arrangements work more cooperatively. It has been estimated that, by carefully exploiting a hierarchy of preventive approaches, corporations could significantly reduce the burdens they impose on the biosphere with little or no sacrifice to their bottom lines.[49]

When governments become aware of these possibilities, an entirely different type of economic strategy becomes feasible. It would undertake to rebalance the productivity of matter and energy with the productivity of labour, which became increasingly distorted as industrialization advanced. At the beginning of the Industrial Revolution, economists correctly argued that the scale of the human economy was small compared to that of the biosphere, with the result that the former would not deplete the "natural" capital of the latter, and therefore natural resources did not have to be priced. Consequently, their costs were essentially those of extraction and refinement, which made these resources inexpensive compared to labour. Entrepreneurs seeking to keep their production costs as far below market prices as possible therefore focused on economizing labour, with the result that labour productivity grew enormously. Today, this imbalance has reached a point where we overuse nature and underuse human beings, which makes no sense in a world facing an environmental crisis and growing underemployment and unemployment.[50] Governments have aggravated these problems by taxing what we have too little of (good jobs) rather than what we use too much of (materials and energy). By gradually reinventing the tax system, governments will be able to promote a shift to a service economy, and new kinds of organizations are likely to benefit from this. It will not be easy, but by creating organizations that can learn and respond to new situations by the shortest possible negative feedback loops, careful experimentation is likely to be highly successful. There is enormous scope for preventive approaches.[51] However, because of their professional education, engineers, managers, economists, and policy makers simply do not understand how preventive approaches hold the key to creating a more liveable and sustainable future.[52]

The new organizations could realize another substantial advantage by ensuring that knowing and doing embedded in experience and culture will work hand in hand with knowing and doing separated from experience and culture. It involves challenging our secular myths, but if we persist, each approach to knowing and doing can complement the others by dealing with those situations that fall beyond its limits. As we have seen, what makes for good design and decision making is that experience is used in directing discipline-based approaches, and this relationship should be extended to all sectors of these new organizations. As energy costs are likely to continue to grow because the costs of extraction tend to increase with the level of depletion, the concentration of global production is likely to become uneconomic.[53] When that happens, it will be possible to repatriate a great deal of manufacturing activity by maintaining and improving social, labour, health, and environmental standards through preventive approaches. Because we have thus far done this in an end-of-pipe manner, many observers erroneously assume that it cannot be done. In fact, preventive approaches are usually highly cost-effective because a little investment in extra design time can substantially increase the ratio of desired to undesired effects resulting from our activities.[54] The new kinds of organizations described above will be able to take advantage of these preventive approaches because of their greater dependence on negative feedback loops.

The above kinds of developments can benefit still further from the possibility they offer to substantially increase human skill levels. Owing to the current exclusive reliance on knowing and doing separated from experience and culture, the levels of skill at which people participate in our current organizations often do not exceed the third stage of the five possible stages to expertise level, according to the human skill acquisition model described previously.[55] Moving from the third to the fifth stage requires the growth of metaconscious knowledge built up from experience. As a result, creating a more synergistic relation between the cultural approach and the technical approach can substantially increase the level of expertise of the members of these new organizations. It will also enhance their ability to learn, which is essential for the difficult transition we need to make.

In conclusion, good design and decision making, as well as the organizations able to sustain them, appear not to be limited by anything other than the cults of the fact and absolute efficiency and their exclusion of symbolization, experience, and culture. Science and technique need to be exclusively used within their limits. As all of human history shows, human freedom demands a greater distance from the primary life-milieu, which in our case means symbolizing science and technique according to their meaning and value for human life and society. Our secular gods must be reduced to means serving human ends. It is a question of creating a civilization that includes science and technique but which is not enslaved to it. Our participation in such a civilization will be a great deal more satisfying – and a great deal less destructive – as we begin to bring under control the disordering effects of science and technique.

One final warning: we will not succeed if we politicize these kinds of considerations. We are dealing with a technical order based on the pursuit of absolute efficiency. It does not depend on any human values, and thus does not respond to our politicians. The only policies that will make a real difference are those that will restructure the technical order itself.

It is next to impossible for us to accept that the problems of our workplaces and economies are not subject to political control. The reason is that we regard everything as political, that is, we ultimately believe that the state is omnipotent (another of our secular myths). There are no correlations between the policies of the right, middle, or left and the values by which people work and live, although elections do periodically eliminate governments that are corrupt or that identify too closely with the "system" regarded by most of us as problematic for one reason or another. Perhaps the time has come for everyone across the entire political spectrum to recognize that the introduction of the technical division of labour and its perfection by current computer and information technology has been an enormous mistake in terms of human freedom and democracy. If we rebalance the productivity of matter and energy with that of labour, create a service economy, and modify the tax structure, the possibilities for a different kind of economy may open up. What stands in the way are our secular myths, which cover over our enslavement to science as the god of all human knowing, technique as the god of all human doing, and the state as the god that organizes everything.

The following anecdotal evidence suggests how difficult the struggle against our secular myths may be. I began my career in a Department of Industrial Engineering which, prior to the re-engineering of the corporations and the institutions, was the locus of almost unlimited devotion to the technical division of labour and what we have referred to as reality. From time to time, colleagues would seek my advice about how to incorporate social factors into their mathematical models. The way they posed the issues and the kinds of solutions they expected me to provide made me realize that, by and large, these colleagues had no idea that there was more to human life and the world than what we have referred to as reality. Unfortunately, I was still too young, and my ideas not developed well enough, to help them understand that operations research cannot work when people are involved.

Before I left the department, I received an astonishing paper in one of my graduate courses from a visiting Chinese student who had been a factory manager in his home country. He had always been told, and this was confirmed by his experience, that the peasants who worked in his factory were lazy and irresponsible, and there was nothing to be done except to maintain a harsh discipline. When he was exposed to the extensive literature on healthy work, including the demand-control model,[56] the reward-imbalance model,[57] and the preventive approaches for the design of more effective work organizations based on a variety of case studies,[58] he recognized the possibility of an entirely different interpretation of his experience. For his final paper he was required to develop a strategy for diffusing preventive approaches in an area of his choosing. I will never forget his paper. He developed an economic strategy for China based on healthy workplaces capable of producing high-quality products, which would also help reduce the vast social and environmental problems his country faced.

Some years before, I had been asked to join an advisory board of the new Hong Kong University for Science and Technology, which the British were developing just prior to handing Hong Kong back to China. On my first visit, I inquired whether the school was prepared to learn from the mistakes we had made in the West in order to develop new strategies. To my surprise, there was no interest whatsoever in doing anything of the kind. What mattered was the stimulation of economic growth through the unlimited application of discipline-based approaches. Since I had nothing to contribute to that enterprise, I withdrew after the first meeting.

I am certainly not the first, and I hope I will not be the last, to observe how completely North American undergraduate curricula are out of touch with anything beyond our current generation of Tayloristic interpretations of what is happening to human work. The considerable literature dealing with healthy work and the very successful experiments conducted by many major corporations (always in a piecemeal and hence contradictory manner to everything else in the organization) are not even brought to the attention of the students. Because human beings do not conform to what we have referred to as reality, they are the problem; what does conform to reality, such as traditional and information machines, is the answer. We refuse to accept that in a democracy corporations ought to be communities of stakeholders or people who are citizens; that they ought to respect the values of society, including democracy, human freedom, and the dignity of the individual; that the growing gap between the lowest and highest salaries is evidence of a slavish devotion to technique at the top, which makes no sense, and does not reflect the values of most other employees; and that it is unacceptable to treat human beings essentially as resources. The real challenge to human freedom and democracy is coming from our workplaces. To varying degrees, in our factories and offices, we are treated as items on a budget line who cannot be trusted and thus must be given as little responsibility as possible, only as hands or brains. Our jobs are restricted as much as possible to discharging one step in the horizontal or vertical division of labour. From time to time, we go through soul-searching exercises of inquiring why so many people do not bother to vote, as if it is not obvious that the values of democracy and the benefits of capitalism are increasingly restricted to a small portion of society.

The Technical Approach as Mediation

How do good design and decision making transform the technical approach? Or should we ask: will the technical approach transform design and decision making? I have suggested that, first of all, a design exemplar has to be created or a decision taken in terms of its meaning and value for human life and society, following which it can be analysed and optimized by discipline-based approaches. This latter step cannot be undertaken on the level of the design exemplar or the decision. Each one must be broken down into ever-smaller constituents until these begin to resemble domains because one category of phenomena dominates all the others. The results of this analysis and optimization must be synthesized on the next higher level, and this process must be continued until it returns to the design exemplar or decision. John Kenneth Galbraith came to the same conclusion when he examined the constraints imposed on the design of modern products using advanced highly specialized knowledge.[59] Will this analysis, optimization, and synthesis undermine the symbolization that preceded it? In order to answer this question, it is necessary that we examine the technical approach as mediating the relationship between its user and what it is being applied to. We are in the habit of thinking about technique exclusively in terms of means, but it also constitutes a non-neutral intermediary which acts as a kind of filter that passes some aspects, transforms others, and blocks out still others.[60] Can this reduce the design exemplar or decision to its "reality" and thus to technique?

The first step of the technical approach is now applied to a design exemplar or a decision. Its study by means of discipline-based approaches will abstract it from a highly desymbolized symbolic universe of a culture and reduce it to its reality. Doing so translates the design exemplar from what is true (the world of sense) into what is real (based on the principles of divisibility, separability, non-contradiction, definability on its own terms, mathematical representation, and symbol complexity).[61] As a result, the analysis and optimization of a design exemplar or decision by means of the technical approach begin by filtering out everything that does not belong to what is real.

The second step of the technical approach uses the findings of the first step to create some kind of model which is now the reflection of the design exemplar or decision in terms of what is real. The design exemplar or decision is thus modelled on its own terms as its reality and not in terms of its human significance.

The third step of the technical approach takes this further. It is the reality of the design exemplar or decision that is being optimized. Will this not turn everything into an element of technique when the results are used to reorganize the object of the technical approach? It would appear that the optimized design exemplar or decision will have to be resymbolized to ensure that its meaning and value for human life have not been altered. If a design exemplar or decision was strictly related to a technical artefact, process, or system, no difficulties will arise because discipline-based approaches are well suited to their reality. What happens, however, if a design exemplar or decision involves human beings, as is the case for human work in a factory or office? The technical approach can deal with these kinds of situations only in terms of non-life as their reality. Here we have once again encountered the limits of the technical approach. It must not be applied to any entities that involve technique as well as human beings, communities, and ecosystems. What happens if these have been subjected to the technical approach in their evolution for some time? Can a design exemplar or a good decision be arrived at under these conditions?

Let us return to the design exemplar for a manufacturing facility. Symbolizing the current design exemplars for these facilities makes it evident that they do not serve human life, but that instead the people involved serve them. It is thus necessary to create design exemplars that reverse this relationship. By appealing to a considerable literature on past accomplishments in this area, including the previously mentioned Uddevalla plant and the Semco organization, doing so appears to be possible, but the window of opportunity is rapidly closing.[62] There are fewer and fewer of these kinds of successful alternatives as the technical order aided by free trade disorders everything else. In another decade or two, the above proposal for alternative manufacturing systems may well be greeted as something so far in the past that it cannot be taken seriously. In this case, democracy will be reduced to an even greater shadow of its former self, and unemployment (including underemployment) will continue to rise.

Good design and decision making thus face a double limitation. First of all, the internalization of elements of human life filtered out by their representation in terms of reality will become more and more difficult as technique continues to disorder what lies beyond reality. Second, the desymbolization of experience and culture under the pressure of technique will make it more and more difficult to imagine good design exemplars and decisions. It is another consequence of our war on ourselves.[63] It is entirely possible that in a few more decades, technique will have triumphed over the kinds of efforts described in this work.

Technique as Life-Milieu and System

Until recently, humanity lived either in the life-milieu of nature (during the period erroneously referred to as prehistory) or in the life-milieu of society (during history). In both situations, these life-milieus had such a profound influence on human consciousness and cultures as to create a situation of alienation. The enslavement to these life-milieus did not soften until they were symbolized. During the last century, technique gradually became our primary life-milieu via which we related to others, our communities, and the biosphere. As a result, this technical mediation filtered out almost everything that could not be expressed in terms of reality. As noted, this has now developed to a point where a great many academics have difficulty understanding that something essential about all life is being filtered out in our experiences and thus barred from cultures. Their desymbolization threatens everything that has made us human until now. As a result, the window for good design and decision making is rapidly closing. The day will come soon when wake-up calls of this kind will be received by polite smiles because they will be regarded as well intentioned but "unrealistic." The specialists who build the technical order will be unable to connect what they are doing to the disorder being created, while the people on the receiving end of this disordering will have no idea of its source. In this way, the life-milieu of technique will alienate humanity in a manner from which symbolization can no longer offer a way out. It is a wake-up call Jacques Ellul issued more than fifty years ago.[64] There is still time to profoundly change the situation through major structural reforms to our educational institutions and workplaces in order to transform the system of technique.[65] The critical issue of this century will undoubtedly be how reversible the extensive desymbolization of human cultures will turn out to be if we attempt to reduce our servitude of technique. Our present zeal for economic growth and free trade makes this highly unlikely, but there is nothing less predictable than human history.

3 The Cult of Growth: The Anti-Economy

Economics as Our Secular Theology

For the discipline of economics to have scientific validity, one or both of the following conditions must be met. The influence of economic phenomena on human lives, society, and the biosphere must overshadow the influences of all other kinds of phenomena, to the point that these can be neglected. Alternatively, the influences of all non-economic kinds of phenomena on human lives, society, and the biosphere must change so little that they are essentially static, with the result that the forces of change are economic in character. In other words, either technological, social, political, legal, moral, religious, and cultural developments must have trivial influences on human lives, on society, and on its relations with the biosphere in comparison to those resulting from economic changes, or their combined influence must be stable. If either or both of these conditions were met, what in artificial intelligence has become known as the "frame problem" would disappear, because human lives, society, and the biosphere would no longer be the inseparable context in relation to which economic phenomena take on their meanings and values. Under such conditions, economic phenomena could be understood on their own terms, and economic "facts" would be determined by an economic rather than a political-economic theory. On the other hand, if these conditions are absent, economics cannot take on the form of a discipline without losing its scientific validity. It will therefore not be possible to create a domain and "populate" it exclusively with economic phenomena and processes, as if these were inert to all other influences. Only the gods of the past had these kinds of qualities; hence economics would turn into a kind of secular theology.

From the above perspective, it is difficult to identify a historical period which was more turbulent and produced more massive and lasting transformations of every aspect of human life, societies, and their relations with the biosphere than the last two hundred years, which brought industrialization, urbanization, secularization, and extensive desymbolization. Vast economic changes were indissociably linked to equally vast technological, social, political, legal, moral, religious, and cultural transformations. To be sure, there have been historical periods during which economic phenomena dominated all the other kinds of phenomena in terms of their influence on individual and collective human life, especially during times of crisis of one kind or another. For example, during severe famines people are willing to pay almost anything and with almost anything to procure food in order to survive. There have also been historical periods during which economic phenomena essentially behaved as only one dimension of human lives and societies. Consequently, any genuinely scientific economic theory capable of helping us understand economic phenomena and guide policy making would have to be able to explain under which historical circumstances economic phenomena can become dominant, and when they must be interpreted as integrally contributing to the historical evolution of human lives, societies, and their relations with the biosphere. Such economic theories cannot be constructed without reference to all non-economic categories of phenomena.

Since the process of industrialization and its consequences have been examined in four previous volumes, and since a brief overview of these analyses was provided in the Introduction, I will limit myself to highlighting a few details to show that during this latest historical period there was very little possibility of economics constituting itself in the form of a discipline, and yet it did exactly that. To begin with, the conditions in Western Europe just prior to the beginning of industrialization were historically unique and constituted what may regarded as the "resources" necessary for sustaining the process of industrialization once it had begun. These resources were a unique mix of technological, economic, social, and demographic conditions strengthened by a technical orientation in the cultures of what were about to become industrializing societies.[1] From the outset, the influences of non-economic phenomena could therefore not be neglected.

Next, the introduction of the technical division of labour, followed by mechanization during the first phase of industrialization, completely upset the dynamic equilibrium of the technology-based connectedness of the ways of life of the societies involved. It quickly became evident that this situation could be managed only by economic rather than cultural approaches, thus causing the economy to become the locus of the former and the remainder of the society continuing to be the locus of the latter. As a result, distinct economies appeared that had to be reintegrated into their societies. This was done by giving economic phenomena a limitless potential by means of the myths of capital, progress, work, and happiness.[2] Thanks to these myths, economic development could now achieve everything. The omnipotence of economic phenomena overshadowed the importance of all other kinds of phenomena, with the result that economics could take on the form of a discipline and political economy could be abandoned. Nothing as ultimate as all this could be achieved without myths. There was very little opposition to this transformation of economics. Nevertheless, it is impossible to defend it scientifically. Economic phenomena thus became the "base" or "core" of society and *homo economicus* the new secular "soul" of its members. It also became thinkable that the massive technological, social, political, legal, moral, religious, and cultural transformations could be interpreted as secondary phenomena.

In addition, the introduction of the technical division of labour to pave the way for mechanization (and later automation, computerization, and enterprise integration) necessitated the reorganization of human work in the image of the machine, to be regulated by market mechanisms. Economies were built up by dividing human work into distinct and separate production steps that were endlessly repeated and connected only by flows of inputs and outputs. Doing so set this sphere apart from the remainder of society and the world, in which everything evolved in relation to everything else – a world in which it was next to impossible for anything to repeat itself. The closest we can come to repetition in human life is in our routine activities, but this is an entirely different matter.

Since our lives are constantly working in the background, we are able to anticipate almost all situations we encounter in our daily activities. This enables us to delegate to the metaconscious level everything that appears to be routine, thus freeing up our conscious attention for those details that are unusual and require special attention. If a situation turns out to be different, we may be surprised or even startled. We may have to pause for a moment to think it through. All this is very different from machine-based repetition, be it of a classical or information kind.

Human work can become repetitive only through alienation that excludes as much as possible our lives working in the background, thus leaving few possibilities other than escaping the situation by thinking of other things or by daydreaming. In the same vein, routine activities cannot be explained by cybernetics or negative feedback control, since these are organized by means of domains in which "inputs" are transformed into intermediary "outputs" until the desired result is obtained. These kinds of mechanisms are a far cry from our lives working in the background, with their multiplicity of meanings, values, and beliefs that are vastly more complex than a "set point" or a set of "reference values." Our civilization still does not know how to divide a human activity related to work into two components, one that is open to disciplines and thus may be delegated to classical or information machines, and another that must remain under human control.

The partial replacement of the cultural approach with an economic one was the beginning of reorganizing human life and society in the image of the machine (i.e., non-life). It involved a strategy of separating and dividing everything from everything else in a manner that is the diametrical opposite of how human life in the world up to that point had dominated everything by means of symbolization, experience, and culture.

Nevertheless, on the level of daily-life experience, "shadows" of life continue to be experienced. For example, some people are troubled by the ambiguities related to our understanding of what it is to be human in the face of computers and information technologies. Differentiating the associated experiences from all others has proven to be far from simple. Moreover, doing so has greatly changed over time, beginning with personal computers entering our lives and our homes, followed by a growing dependence on the internet. What we have referred to as "technique changing people" began to make its mark in the late 1970s and early 1980s. At this point, technique had sufficiently created the conditions which necessitated the adoption of information technology, and its growing use helped to transform technique into a life-milieu and system into which were embedded other technologies of all kinds. The research of that time is of particular interest because many people could still differentiate their lives before computers from their lives after computers. It did not take very long, however, before life without computers became almost unthinkable and supposedly unliveable. This may well be the reason why this kind of research has become relatively rare.

The research of Sherry Turkle revealed the growing effects of "information technology changing people" over several decades.[3] Joseph Weizenbaum argued that we ought to exercise our responsibilities towards these new technologies by deciding what we should entrust to them and what is best left to human beings.[4] Hubert Dreyfus drew our attention to the impossibility of a *homo informaticus* because all human experience, knowledge, skill, and expertise are embodied, and this implies that we live our lives in the world from the vantage point and in the context of being members of cultural communities.[5] Our lives are constantly working in the background. He also showed that for many decades the successes and failures of artificial intelligence research could be explained in this way.[6] Craig Brod explained how technostress is generated at the interface between life and non-life.[7] As far as human work is concerned, Peter Brödner showed that it could be evolved in two ways: one in the image of the machine (the technocentric alternative) and the other in the image of life (the anthropocentric alternative).[8] Shoshana Zuboff examined the difficulties people experienced when their work was being computerized.[9] Stuart Dreyfus showed that human skills and intelligence cannot be reduced to problem solving that knowledge engineering can then transform into rules and algorithms, thus opening skills and intelligence up to logical and mathematical representations.[10] These and other researchers have attempted to trace what happens when we need to make sense of computers and the internet, which in some ways appear to be alive and in other ways not. Young children especially are faced with the task of making sense of these technologies by differentiating them from everything else in their lives.

This type of research on how "technology changes people" has fallen far short of reaching a comprehensive understanding. There are two important methodological issues which have not been satisfactorily resolved. The influence of individual technologies (in this case mostly computers and the internet) on human lives and societies cannot be understood apart from technique as a whole, any more than the influence water has on paper can be understood in terms of the separate effects of hydrogen and oxygen. Moreover, the influence of an individual technology or of technology as a whole cannot be understood apart from the kind of human life and the kind of society into which it is introduced, just as the consequences of exploding two identical bombs, one in a crowded square and another in an open field, will be entirely different. During the process of industrialization, people changed technology, but technology changed people at the same time, eventually creating different kinds of societies. These are referred to as mass societies, which (as we will see) have little in common with all earlier societies. These two methodological issues were rarely dealt with in a comprehensive manner, thereby leading to premature and erroneous conclusions as to what was happening to human life. Moreover, all discipline-based approaches to knowing and doing involved the separation from experience and culture, and these approaches have little in common with knowing and doing in earlier societies.[11]

In terms of the possibility of economics taking on the form of a discipline without violating scientific principles, two conclusions may be drawn from the above discussions. First, *homo economicus* cannot be the secular soul of the members of contemporary mass societies. If no clear boundary can be drawn between those aspects of human lives that can be modelled in terms of classical or information machines as non-living and whatever escapes this as the living "core," neither can the mechanisms of *homo economicus* be divided from what is governed by symbolization, experience, and culture. Consequently, economic modelling of human behaviour is necessarily limited to situations in which most influences remain relatively stable. Such situations are relatively rare, if they exist at all, given the constant turbulence that is created by "people changing technique" and "technique changing people."

Second, every element of a contemporary economy is constantly modified by technological and technical changes of all kinds that affect the planning, design, innovation, production, marketing, and displacement of each and every product and service.[12] As will soon become evident, it is impossible to think of a modern economy other than as being a subsystem of technique. It is no longer governed by the values and aspirations that societies culturally establish for themselves. Instead, economies are ruled by a striving for efficiency in the service of a cult of growth. Hence, the situation today is very different from what it was during the nineteenth and the first half of the twentieth centuries. This may be confirmed by making a distinction between the internal relationships that occur between the elements of an economy and the external relationships that connect any individual element to its societal context and via it to the biosphere. If the influences of the internal relationships on the elements of an economy are far more decisive that those of the external relationships, an economy takes on the properties of a system with respect to society and the biosphere. If the reverse is the case, the economy is non-systematically organized and its elements may participate either in a culture or in technique.[13] Capitalism was the economic system that occurred historically between the era of traditional preindustrial societies, whose economies were generally governed by their cultures, and the era of contemporary societies, whose cultures have largely been desymbolized by discipline-based approaches to knowing and doing that have also transformed their economies.[14]

The above discussion introduces another reason why economics cannot take on the form of a discipline without losing its scientific status. Any economy is deeply affected by the way human life in the world is conducted. According to the Jewish and Christian traditions which had such a profound influence on the cultures of the industrializing societies in Western Europe, this life was to be conducted by means of a practical wisdom. Karl Marx formulated the secular equivalent, which he called praxis.[15] Its importance can be illustrated by his dispute with the British physiocrats (the forerunners of economists). Like Marx, they were gathering every possible economic fact in order to discover the economic regularities which the physiocrats regarded as natural laws, while Marx argued that these were human creations for which society ought to accept responsibility.[16] Doing so required a praxis, which began by examining the implications of these economic regularities for human lives and society in order to determine whether these were in line with human values and aspirations. An intervention was required, if this were not the case, in order to redirect the economy and bring it in line with such values and aspirations. If the anticipated results of such an intervention conformed to the best possible scientific understanding of the situation, the praxis would be a responsible one. If, on the other hand, the consequences of a praxis were different from those that were anticipated based on a theoretical understanding of the situation, economists would have to re-examine this situation in order to understand the reasons why a gap occurred between what was expected to happen and what actually did happen. The understanding of the situation then had to be modified, as well as the response to it, in order to close this gap. Doing so would improve the intellectual grip economists had on the situation and this would also improve their ability to responsibly deal with it. In other words, a praxis was also a check on people's understanding, since this was the closest economists could get to the work of natural scientists carrying out experiments to validate their theories. It would also ensure that economies would fit into people's lives and societies in accordance with their values and aspirations. However, the possibility of an economic praxis was entirely lost when political economy was transformed into a discipline. This discussion also shows why economics as a discipline cannot create sound economic policies. The equivalent of design exemplars will be required for the development of suitable economic policies.

The possibility of creating an equivalent to practical wisdom has been enormously weakened as a consequence of the desymbolization of experience and culture. Technique limits the context taken into account, reduces the level of participation, minimizes personal commitment, and shifts the balance between freedom and alienation.[17] As a result, the processes of adaptation and evolution essential to life are significantly impaired. This restriction is further intensified by the way our perception of our lives, communities, and the world is being shaped by the mass media and its bath of images portraying the lifestyle we must live and how we must make sense of things in conformity to technique. Much of this makes little sense because technique does not permit any problem to be addressed at its roots by means of self-regulation, thereby resulting in the creation of layers upon layers of compensating activities. In sum, the alienation and reification by technique blocks the self-regulation of all human activities, thus making technique self-reinforcing because of a lack of iconoclastic symbolization designed to create something equivalent to a daily-life practical wisdom.

Desymbolization diminishes our reliance on experience and culture and thus our ability to know and do anything in the context of our lives in the world. Possibly more than ever before, nothing in our lives and the world is in the right place, and every relationship is distorted and truncated (the secular equivalent of what in the Jewish and Christian traditions is referred to as sin). Nevertheless, experts, lobbyists, pressure groups, think tanks, politicians, and governments go about their business as if we can deal with the situation in which we find ourselves one issue or problem at a time. We fail to recognize that all these issues are deeply connected by the way everything is related and evolves in relation to everything else. Worse, this interdependence is now malfunctioning under the pressures of technique. It is therefore impossible to address issues such as global warming, economic crises, free trade, underemployment and unemployment, poverty, depression, and injustice outside of their joint context. From wind farms to tobacco products, the situation is more or less the same: the public denial of any problems, aided by specialists (frequently with vested interests); the lobbying against any adequate studies being undertaken; the intimidation of dissenting experts by such means as SLAPP lawsuits (Strategic Legal Action against Public Participation); the intimidation of whistle-blowers; police pressure against peaceful protests; and so on. It is easy for discipline-based specialists to deny any negative consequences because these almost always fall beyond their domains, and this is compounded by their almost unlimited confidence in their simulations, even though these are rarely adequately tested. Questions whether there are significant negative effects or whether the social epidemiological or other required scientific studies have been undertaken are rarely addressed. The phrase "there exists no scientific evidence ..." is perfectly ambiguous. All this makes the responsible conduct of human life in the world very difficult. In observing how our public universities in general and their departments in particular are preparing the future of our factories, offices, technical systems of all kinds, urban habitat, and much more, we see a picture emerging of discipline-based approaches of knowing and doing being completely out of touch with the conduct of our lives and the kind of future we desire. Nevertheless, alternative preventive approaches[18] could be implemented as an integral part of living more responsibly. In the meantime, the cult of the fact, the cult of absolute efficiency, and the cult of economic growth are blocking our abilities to responsibly live our lives in the world.

Yet another reason why economics cannot take the form of a discipline without giving up its scientific status is the way economic decisions affect the work-life balance. From a historical perspective, it would appear that no civilization has been as devoted to work as our own.[19] This is veiled by our belief that, prior to our civilization, human life was a brutal struggle for survival, and that, thanks to technique, we have achieved a relative state of abundance. Yet the opposite appears to be the case. The place of work in human life, the place of the economy in society, and the place of society in the biosphere were essentially the opposite of what they are today. In the past, people escaped work and consumed less whenever possible in order to relax, play, and enjoy the abundance of nature. There was a very different and delicate balance between human life and the economy based on the greatest economy of effort. A modest standard of living was the means for achieving a certain quality of life. A relative abundance of natural resources was thus associated with the relative scarcity of goods produced by crafts.

It was the West which began to turn this around. There is now a great abundance of industrial goods and a scarcity of many natural goods, including clean air, water, food, game, and fish. The more we produce and consume, the more we deplete what once was a natural abundance. There is no question that a growing population necessitated agriculture, which demanded a great deal of hard work. It is also the case that empires imposed enormous burdens on their people, which necessitated still more hard work.

Our current work-life balance in the industrially advanced nations was gradually developed in the West in the course of some five centuries. Before this time, Western civilization had been primarily influenced by Christianity and its teachings regarding work and life. Beginning in the fifteenth century, all this was reinterpreted into its opposite, giving rise to what Jacques Ellul has referred to as the bourgeois spirit.[20] It regarded work as a virtue instead of a consequence of the break between God and humanity. In other civilizations, work had been associated with servitude, and everywhere work was considered to be far from noble. In the bourgeois spirit, however, work was so essential in life that almost everything else paled in comparison. For example, as long as you worked hard you could be forgiven almost anything, including being ruthless, cheating on your spouse, and exploiting other people and cultures. This interpretation created an ideology serving the interests of entrepreneurially minded circles in society. Christian theologians by and large justified this reinterpretation of work by quoting the well-known text from Genesis in which people are commanded to dominate the creation. As already noted, this was the domination by the word, which ended with the break between God and humanity.[21] Ironically, this glorification of work spread through the many working poor in the nineteenth century, especially those who attempted to make sense of their lives by means of a socialist perspective.

It was Karl Marx who provided the theoretical justification for this glorification of work when he convinced the proletariat that work would eventually liberate them.[22] In this sense, as well as in others, Marx's thought was entirely in the bourgeois spirit of his age.[23] The Nazis inscribed their belief that "work would set you free" above the entrances to concentration camps. Both the communists and the Nazis assured the working people that, despite their poverty and misery, work was their great virtue because it and it alone would transform the world. Hard work would power capitalism. Capitalism's internal contradictions would grow with its development, causing its eventual collapse and the ushering in of the socialist paradise.[24] Marx's thought represents one of the ways in which the great myths of the eighteenth and nineteenth centuries (capital, progress, work, and happiness) were intuited and developed. Not working became a crime that eventually necessitated the re-education camps designed to reorient people in the right direction by means of hard work. Because work was the great virtue, is was to be done with great joy despite the harsh factory discipline.[25] Max Weber examined the role the Protestant ethic played in the rise of capitalism in the West.[26] This ethic was the diametrical opposite of what had been regarded as the Christian message on this subject: work became a necessity following the break between God and humanity, but necessity is the opposite of freedom.[27] The necessity of work is confirmed in both the Jewish and Christian Bibles by statements to the effect that people who refuse to work should not eat. It is better to work than to have to steal what is essential to others, especially in the context of a subsistence economy where everyone has just barely enough to survive. Work is, therefore, neither a virtue nor a pathway to freedom.[28]

Western civilization thus turned Christianity upside down by creating societies dedicated to hard work. As noted, in the course of the eighteenth and nineteenth centuries, newly emerged secular myths implied that humanity had always gone about the creation of a decent life in the wrong way by striving for the good moral and religious culture-based connectedness of human life at the expense of its technology-based connectedness.[29] Nevertheless, the past has not been dominated by a struggle for survival, as contemporary studies show.[30] In addition, it was the very nature of work that was completely transformed, beginning with the technical division of labour, which first gave rise to the traditional assembly line and later to the intellectual assembly line.[31] Work became much harsher because it was robbed of what little sense it had. Moreover, industrialization and technicization created an abundance of industrial goods and a scarcity of almost everything else, social as well as natural, as we will show later. The more we consume, the more we deplete the social and natural abundance. Social abundance refers to the ability of communities to sustain their members, and natural abundance refers to the ability of the biosphere to sustain a rich diversity. Consequently, in the past there was a very different balance between work and life and also between the economy and society.

The above examples were chosen to show that industrialization is essentially a shorthand for one of the greatest historical transformations.[32] As has been pointed out in previous works,[33] our being a symbolic species creates a reciprocal relationship with everything we do and experience. For the first phase of industrialization, this may be summed up as "people changing technology" and the reverse, "technology changing people," or, alternatively, "people changing the economy" and "the economy changing people," where the first term of each pair refers to the way people changed their lives and their world, while the second term refers to the influences these changes had on them. Although most studies completely neglect this reverse interaction, it has transformed almost everything in human lives and societies into its opposite.[34] As the new industrial societies emerged, people's experiences changed dramatically, and since the organizations of their brain-minds were built up with these experiences, substantial legal, moral, religious, and cultural changes accompanied what most studies concentrate on, namely the introduction of the technical division of labour followed by mechanization and industrialization, the liberalization of trade in order to expand markets to gain the full advantage of this technical division of labour, and the building of the new economic system referred to as capitalism. There were a great many other changes, but very few studies put all this together as constituting a radical historical divide.

We need to take into account the experiences of the people who lived during the first phase of industrialization. Every year they were surrounded by more factories full of machines, whose output increased with every innovation. When the organizations of their brain-minds "interpolated" and "extrapolated" these experiences, there was a steady development of metaconscious knowledge. In the course of generations, this implied for people that industrialization meant a constantly growing output of goods and services and that this kind of material progress would soon lead to social progress, because the many problems associated with poverty would be eased. There was nothing on their horizon of experience to suggest that eventually poverty could not be eliminated. This metaconscious knowledge also implied that social progress would inevitably also produce a spiritual progress. As people were saved from a struggle for survival, a great many destructive social relationships would be eliminated, with everything this implied for their lives and communities. In other words, there appeared to be no limits to this progress as it potentially extended to all aspects of human lives and societies. All the average person needed to do to help bring it about was to work hard, and happiness would follow. All this was presided over by capital as entrepreneurs required ever larger investments and as ever more wage earners translated their wages into demands for goods and services. In this way, in the course of several generations, the deepest metaconscious knowledge began to correspond to the myths of capital, progress, work, and happiness.[35]

The suspension of people's lives into a new set of secular myths changed everything. The cultures of all earlier societies and civilizations had created legal, moral, and religious institutions to help people lead good lives according to the metaconscious values that sustained them. The new secular myths of the first generation of industrial societies implied that all that had been a waste of time. Salvation would not come from the gods but from the earth, as it were. All the resources and ingenuity available to people needed to be harnessed to material progress, and the good life would then follow in its wake.[36]

In other words, economists as members of these societies knew metaconsciously that the economic phenomena they studied offered humanity an unlimited potential for progress. Nothing could stand in the way of these phenomena, hence economics could be reconstituted as a discipline thanks to the new secular myths. Like all myths, be they traditional or secular in kind, the "interpolation" and "extrapolation" of the experiences of people's lives by the organizations of their brain-minds were not accompanied by a corresponding formation of metaconscious knowledge related to the reverse interactions, those associated with the effects of people's actions on their lives and communities. These took time to build up and thus lagged in time. It was not until the shocks of the First World War, the Great Depression, and the Second World War that the first generation of secular myths of the first generation of industrial societies was decisively desymbolized. These myths were first weakened and then replaced by a build-up of new metaconscious knowledge associated with all the transformations of knowing and doing separating themselves from experience and culture and being organized by means of disciplines. It gradually took the form of the myths of science, technique, the nation-state, and history.[37]

In sum, the economists, more than any other members of society, used their deepest metaconscious knowledge to intuit that economic phenomena had no limits in terms of what they could accomplish. Instead of recognizing that this meant that all other categories of phenomena would undergo massive transformations that could not be neglected, they created a pseudo-scientific discipline with a domain exclusively filled with economic phenomena. Everything had to be explained in economic terms, and this meant that any major non-economic changes could undermine their theories at any time. We have seen this over and over again. Economic modelling using input-output tables or any other mathematical constructions of this kind is linear in nature and can thus be used to make very short-term extrapolations, but these can hardly be treated as forecasts given their almost complete neglect of all non-economic changes.

The secular myths of the first generation of industrial societies also justified taking a rational approach to the development of economic phenomena because of their unlimited potential. The overriding purpose of human activities thus ought to be the strengthening of these phenomena, to prevent anything from interfering with them. It invited the extension of a rational approach that had taken shape in dealing with the implementation of the technical division of labour and the subsequent mechanization and industrialization. Human work had become divided into a sequence of domains in which a production step takes the output from the previous one and repeatedly and efficiently transforms it into another intermediary output, to be passed on to the next production step until the desired result is reached. There is no meaning or value to be discerned for any of these steps in relation to human life and society. There is only the technical significance of each production step in relation to those that precede it and those that follow it. Jointly they produce a commodity to be priced in the market, but this price is a poor indicator of its meaning and value for human life and society. In all these domains filled with mechanisms of one kind or another, the cultural approach is of little use, since everything must be assessed in relation to the local technology-based connectedness and the economic mechanisms governing it, and not in relation to the culture-based connectedness. People's lives could no longer work in the background, which required the imposition of external goals. The result was a rational approach, which was transformed into a technical one when knowing and doing became organized by means of disciplines.

A rational approach was also required to make everything compatible with the new emerging economic order. The labour force had to be entirely restructured, both in terms of new and old categories of work and in terms of the numbers of people employed in each category. Work was no longer done at home and was separated from the extended family, therefore contributing to its eventual replacement by the nuclear family. The population had to be redistributed and concentrated in the new industrial centres. Governments had to prop up common law systems, which were swamped with new situations that were so fundamentally different that few if any legal precedents could be applied. Governments were also compelled to create a new legal basis for many institutions, including the banks and the corporations with limited liability. The role of the state had to be expanded as the self-organizing character of human life and society based on culture was weakened as a consequence of desymbolization. Political decision making had to shift from the local to the national level, and eventually to the international level as production and trade became increasingly global. As a result of all these kinds of development, what remained of the culture-based connectedness of human life and society became an unstable blend of the little that still made sense and the necessities imposed by the new economic order built up with approaches that were diametrically opposite to those of symbolization, experience, and culture.

Not only did the industrializing societies acquire relatively separate and distinct economies but their social structures, political frameworks, and legal organizations also became more and more "unfolded" from the culture-based connectedness to take on the forms of relatively separate and distinct structures. These gradually appeared to legitimate other social sciences, following the example of economics. Before this development, social sciences all studied dimensions of human life and society whose relative significance could vary greatly from time to time but which nevertheless remained enfolded in the culture-based connectedness of human life and society.

To sum up: reorganizing political economy into a discipline produced all the previously examined changes. Its practitioners would be suspended in a triple abstraction leaving them no decision criteria other than efficiency and growth. The practices of economists would now be characterized by economic techniques that would be fully integrated into the system of technique.[38]

The massive increase in the throughput of matter and energy in the technology-based connectedness of the industrializing societies had a growing influence on the biosphere as well. If this influence had been the result of satisfying everyone's basic needs, it would have levelled off once this had been achieved. All this was implied by the secular myths of the first generation of industrial societies. Since these myths did not take into account "technology changing people," one of the consequences was the gradual turning of people into consumers with unlimited and insatiable needs. Consequently, material progress initially led to a limited social progress in the reduction of poverty but then began to create all kinds of new social issues, and no lasting happiness could be gained from this emerging consumerism. Instead, material consumption became our symbolic participation in the greatest good humanity began to develop following the Second World War: the fruits from a supposed unlimited scientific knowing, technical doing, and economic growing. Relationships with the biosphere were thus transformed through these two generations of secular myths.

The technology-based connectedness made up from overlapping networks of flows of matter and energy and everything built up with them has to borrow all matter and energy from the biosphere, and eventually returns them in altered forms.[39] The biosphere "produces" the inputs of matter into the corresponding network by means of successive transformations that form closed cycles, so that the wastes from one transformation are the resources for another and no pollution occurs. The biosphere produces the energy inputs into the corresponding network from the solar radiation it receives by day, from which different plant formations produce biomass as the food base for life. On the outputs side of the network of flows of matter associated with our way of life, the transformed flows of matter can sometimes be absorbed by these closed cycles. Failing this they are simply stored, while the transformed energy is radiated back out to space as low-temperature heat by night. As a result, the network of flows of matter and energy that make up the technology-based connectedness of a way of life are embedded in corresponding larger networks representing these flows within the biosphere. Consequently, the human economy is a kind of subsystem of the biosphere in terms of its material and energy usage. Initially these subsystems had little effect on the biosphere, but today humanity has agreed that the scale of the global economy can no longer be sustained by the biosphere. In economic terms, "natural capital" is being depleted, and yet it remains unpriced.

The above examples show that human life and society cannot be broken down into categories of phenomena corresponding to the disciplines of our contemporary universities. The way of life of a society is made up from a diversity of human activities that at best may have a dominant aspect that corresponds to one of the disciplines. Furthermore, the process of industrialization involved many different categories of phenomena, of which a significant number made non-trivial contributions. In sum, during the time political economy was being transformed into the discipline of economics, the scientific prerequisites for this change did not exist. Hence, the new discipline of economics did not have a scientific status, and it gradually assumed the role of the secular religious theology of the myths of the democratic industrial world. In the next section, we will confirm these findings by examining how three successive economic theories developed by the new discipline of economics dealt with the non-economic phenomena.

The principal consequence of this transformation of political economy into a discipline was the development of economic techniques solely preoccupied with efficiency and economic growth. Their increasing efficiency and power legitimate these techniques and mask their unscientific status.

The Limitations of Three Successive Economic Theories

Robert Theobald[40] has argued that the following assumptions must be met if neoclassical economics was to have any validity. First, the people at that time had to behave as rational economic actors wooed entirely by economic motivations. Any other human needs, concerns, values, convictions – everything human – had to become an economic externality. The validity of this assumption was beginning to weaken by the end of the nineteenth century and into the twentieth century.

Second, all corporations had to be small relative to the size of the markets they served in order to make it impossible for them to exercise any power over their markets. This assumption contradicts the development of industrial corporations, except in their early days. Already near the end of the first phase of industrialization, technological and economic growth was entering a phase of diminishing returns, with the result that one of the few ways in which corporations could continue to gain competitive advantages was by means of economies of scale, thus necessitating their growth. This limit was removed and even greater growth was necessitated by knowing and doing separating themselves from experience and culture, to make way for discipline-based approaches to knowing and doing highly adapted to evolving the technology-based connectedness of industry and society.[41] In other words, technological and economic growth made this a questionable assumption almost from the very beginning of industrialization. In addition, it is more than a little naive to expect corporations not to exercise their powers wherever this can gain them competitive advantages. Attempts to limit the size of corporations by means such as anti-trust laws were not only difficult to enforce but completely incompatible with the structure and dynamics of technology.

Third, the organization of workers into unions should not occur, in order to prevent workers from having any power over labour markets. Anyone who has read anything about the working conditions during the first phase of industrialization would probably agree that unions were an absolute necessity if some kind of civil society was to survive. The balance of power was already so terribly skewed in favour of corporations that there was little choice. As a rule of thumb, it would appear that the later a country industrialized in Western Europe, the more cooperative and sensible were its management-labour relations, presumably because they had learned from the mistakes of their predecessors. Unfortunately, Canada and the United States inherited the very worst kind from Britain.

Fourth, there must be no government intervention in the economy. In other words, the reorganization of production by means of the technical division of labour and the distribution of the products by markets was automatically in society's best interests. Apparently, human goodness was now built into these mechanisms. This assumption was as socially and historically naive as the previous two.

Fifth, all economic actors had to possess the correct information to permit them to make economic decisions most advantageous to them. This implied that all societies had been wrong in constraining self-interested behaviour by their values. Humanity's experience was turned on its head: forget your neighbour and your community and you will help bring forth a better world. Moreover, this assumption was well on the way towards being invalidated in a process that eventually led to mass societies, which depend on a technical complement to their culture. This consists of a bath of images which portray almost everything equivalent to the roles of tradition and culture: they show how people must live, what they must have and do, and so on. The advertisements embedded in these images, however, have been shown to contain almost no economic information.[42]

Sixth, because of its dependence on economic mechanisms, the neoclassical economic model is essentially static and thus requires an absence of significant scientific, technological, economic, social, political, legal, moral, religious, and environmental change. Undoubtedly, this assumption is the most bizarre of them all, and yet it remains the most enduring. Industrialization plunged human life into enormous turbulence, thus requiring its complete restructuring.[43]

These six assumptions are a testimony to how isolated the domain of the discipline of economics was from those of all the other social sciences. The result was the representation of life by models that were entirely asocial and ahistorical. Moreover, representing an economy in the form of a domain caused it to take on all the characteristics of what we have called reality. Almost everything of human life, society, and the biosphere was turned into an economic externality. The values Western civilization had stood for were brushed aside, first out of economic necessity and then by pseudo-scientific ideologies. From a mathematical perspective, this opened the door to representing economies by elegant and seductive mathematical models. It is tempting to conclude that the assumptions of neoclassical economics have more to do with remaking human life and society in the image of these models than with understanding the actual economies, societies, and ecosystems. Nevertheless, as a result of economic policies and interventions, people changed the economy, but at the same time the economy changed people through its vast influences on human life, society, and the biosphere. Consequently, economic policies based on neoclassical models moved economies towards these models while simultaneously moving human life and society towards alienation and reification, and the biosphere towards commodification.[44] They helped to convince Western civilization that guiding human work and economies by means of measures of efficiency and growth was synonymous with human well-being, viable societies, and sustaining ecosystems. Nothing was further from what was really happening at that time.

It must also be emphasized that economic policies based on neoclassical models did much to undermine Western values. They enormously enhanced the power of corporations and diminished the power of their employees. They also removed the economy from any democratic scrutiny and control, since the pursuit of absolute efficiency and limitless growth were considered to be good in themselves. The greatest threat to democracy did not come from the extreme left or right but from within. What is so terribly shocking about this chapter in our economic history is not the complete lack of scientific validity of the neoclassical models but that anyone took neoclassical economists seriously. Western civilization has paid a very high price for what Herman Daly and John B. Cobb[45] have called disciplinolatry.

The macroeconomic crises of the 1930s helped to discredit neoclassical economics. At this point, John Maynard Keynes, in *The General Theory of Employment, Interest and Money*,[46] showed that there was no automatic balance between production and consumption and that government interventions were required to avoid crises. Governments were to represent the public interest, as opposed to favouring those with economic power. This could have spurred a major effort to ensure that economies serve us as opposed to our serving them. Unfortunately,

Keynes restricted his analysis to situations in which there were no significant changes to the skill and quantity of labour, the quantity and quality of available equipment, and the existing technique.[47] Although this is a partial prerequisite for the application of a discipline-based approach, almost every analysis of the period showed that these represented some of the most dynamic elements in the economies of his day. Moreover, this dynamism was bound up with much deeper changes, some of which we will briefly consider.

Around the turn of the twentieth century, Germany was able to overtake Great Britain as the leading industrial power because it had successfully exploited discipline-based approaches to knowing and doing, especially in the chemical and electrical industries where knowing and doing embedded in experience and culture had encountered extreme limitations.[48] Germany had done the same in many other areas in its way of life, thus giving rise to what Max Weber referred to as the phenomenon of rationality.[49] It demonstrated its industrial power during the First World War, also called the first great technological war. It was able to keep other Western European industrial powers at bay for several years. It later experienced the scourges of industrial civilization, namely, unemployment and inflation, although war reparation payments psychologically and culturally aggravated the situation. Hitler intervened and wrought a widely admired economic miracle by putting the nation back to work and bringing inflation under control in a very short time, but the cost was a second great technological war. In the United States it was President Roosevelt who had the courage to intervene in the economy during the Great Depression. In sum, between the two world wars, industrial societies experienced a great deal of turbulence, which affected their economies. Nevertheless, such effects are difficult to acknowledge by any discipline restricting its attention to one category of phenomena. It is worth speculating that the ongoing focus on economic policies, as opposed to more comprehensive policies, may well have contributed to one of the most deadly periods in human history.

Following the Second World War, what had begun as the phenomenon of rationality evolved into that of technique by displacing the role of symbolization, experience, and culture in almost every area of human life.[50] The shift from a primary reliance on culture to a primary reliance on technique and only a secondary reliance on culture was at the root of many new phenomena that sprang up following the Second World War: the emergence of so-called post-industrial information societies;

the information explosion and the growing computer revolution; the emergence of an entirely new kind of society referred to as a mass society with mass production, mass consumption, and mass media; the tension between the mega-machine-like aspects of human life and society and the disorder of everything externalized by the technical approach; the shift from *homo economicus* to *homo informaticus*; the commoditization and reification of human life; new secular political religions; statistical moralities; democracy ruled by public opinion; and much more.[51] I have shown that all these new phenomena as well as the relationships between them can be explained as the direct consequences of the shift from culture to technique.[52] The role of the state in society also changed dramatically. The displacement of culture by technique robbed societies of their ability to evolve everything in relation to everything else, with the result that a great many activities had to be externally coordinated and regulated. The scope of this development is illustrated by the commonplace: today everything is political.

It is difficult to exaggerate the scope and depth of the shift from culture to technique. As far back as we can go, humanity has always relied on culture. This transition may well turn out to be as significant as the birth of human societies at the beginning of the period we refer to as history. In other words, although this transition was certainly more comprehensive than the one we described for the first phase of industrialization, what industrial societies had to cope with were ultimately thermodynamic constraints, which were more rigid than the ones associated with the shift from culture to technique.

During this time, we might have expected the status of economics as a discipline to come under increasing scrutiny. How was it possible to understand the economy by ignoring all the other categories of phenomena that were jointly transforming everything in human life and society? It is no surprise that Keynesian economics began to fail, given its limiting assumptions, but it is incomprehensible that this opportunity led to monetarism's taking control of the discipline of economics, because it was an intellectual throwback to neoclassical economics.

The rise of the phenomenon of technique transformed all economic phenomena. John Kenneth Galbraith[53] in the West and Radovan Richta[54] in the Communist bloc drew our attention to the transformative role highly specialized scientific and technical knowledge had begun to play in the economy. In order to take advantage of this highly specialized knowledge, corporations had to reorganize, which necessitated the change in their relations with economies and societies. Galbraith showed that these new corporations could no longer rely on markets for many of their direct and indirect "inputs" and "outputs" and that the large corporations jointly had constituted a planning system which was of a technical as opposed to a political nature. Because of the enormous technical constraints imposed on corporations, markets became a kind of appendix to their technical planning, which included extensive consumer testing and the "management" of consumer demand by advertising. Furthermore, the products, processes, and systems built up with technique rapidly gained in absolute efficiency (a shorthand for all output-input measures) at the expense of their compatibility with their social and natural surroundings. For example, during the decades following the Second World War, the contribution technological change made to the burden the US economy imposed on the biosphere was greater than the combined contributions of demographic expansion and the increase in the standard of living.[55] Technology also increasingly collided with human life and society. It is hardly surprising that the costs incurred in the production of wealth began to rise sharply, with the result that net wealth production levelled off or even declined depending on the assumptions made.[56] It gradually turned economies into anti-economies that extract more wealth than they produce.[57]

During these decades, an entirely new kind of economic activity began to grow exponentially: the making of money from money without any intervening economic activity, which Herman Daly argues is a thermodynamic impossibility.[58] A former head of the Belgian Central Bank has calculated that, by the 1990s, only about 3 per cent of the total financial flows around the globe still corresponded to real economic activities, with the rest being speculation.[59] All these developments were superimposed on the usual business cycles and recessions, further aggravated by two significant oil shocks. In sum, the shift from culture to technique required major scientific, technological, economic, social, political, legal, moral, religious, artistic, and environmental adjustments.[60]

The response to the failure of Keynesian economic policies was explained by the monetarists exclusively in terms of the category of economic phenomena. Economic phenomena were implicitly regarded as being inert to all other categories of phenomena, so that the latter could be neglected. I can think of no greater example of pseudo-science than monetarism. It is perhaps instructive to briefly sketch some of the arguments advanced by the monetarists to explain various events. I will give a brief overview of the merit of such arguments, in part based on a non-disciplinary account by Allen Lane.[61]

The above six conditions, which had to met by the first generation of industrial societies in order to validate the neoclassical models of their economies, essentially implied that these economies were equipped with mechanisms capable of guiding them through any turbulence as long as there was no outside interference from corporations, unions, government, or society at large. For example, when a crisis such as the Great Depression arose, economies would simply right themselves. High levels of unemployment meant that the supply of labour exceeded demand, which the labour market could "correct" by reducing wages and creating a new balance that would put most people back to work. For Keynesian economists, such a depression represented a shortfall in demand caused by people being reluctant to spend their money in a time of uncertainty, when they might face layoffs. In order to avoid this triggering a downward spiral, governments had to step in as spenders of last resort by increasing the public sector in the economy to offset the contraction in the private sector. This was done by means of initiatives such as the building of roads, schools, hospitals, or parks. To finance this, governments were to dip into the surpluses created during good economic times. In this way, governments could smooth out economic fluctuations and create more or less steadily growing economies. Such economic strategies had worked very well in the decades following the Second World War. This success came to an end when the shift from culture to technique began to have a massive impact on net wealth creation by vastly expanding the costs incurred in the production of wealth. In addition, this shift imposed many other costly transformations and adjustments. As a result, the creation of surpluses became next to impossible.

Because of their disciplinolatry, the monetarists, led by Milton Friedman, had no choice but to restrict any explanations to economic phenomena. Hence, they argued that the Keynesian fiscal policies were doomed to fail. Money spent by governments had to come from somewhere, and that somewhere was from the private sector, either by taxation or (indirectly) by borrowing and thus deferring the matter to the future. They also claimed that the Keynesians were mistaken regarding their perceived tradeoff between inflation and unemployment. Inflation would not fall as unemployment rose, and vice versa. As a result, there was no point in tolerating modest inflation in return for keeping unemployment in check. As inflation rose, society could expect that wage earners would demand higher wages to compensate for higher prices, and this would cause corporate profits to rise along with their costs. There would be no incentive for employers to hire more workers. The monetarists convinced the politicians that it was impossible to spend your way out of a recession by expanding the public sector and reducing taxes. The overarching priority now became the control of inflation, and if unemployment rose this was the unavoidable price that had to be paid. Hence, fiscal policy had to be replaced by monetary policy because so much depended on the money supply. According to the monetarists, the only solution was to make it easier for employers to hire or lay off workers, which would turn everyone into "just in time" labour. All this was coupled with a relentless push to expand markets through free trade.

In its original conception, trade was free if both parties could gain an advantage, but this required that the capital withdrawn from one area of an economy be reinvested into another area to ensure that no net loss of jobs would result. Such assurances of mutual advantages were made meaningless by the almost unrestricted global flows of capital. The results for all to see have been a massive export of jobs to a few areas of the world where costs can be externalized to a much greater extent, in return for cheap and mostly low-quality consumer goods. It is a poor bargain for countless people who depend on jobs in order to participate in their communities.

Much as a government was supposed to be the spender of last resort, the International Monetary Fund (IMF) created under the Bretton Woods agreement was supposed to act as a lender of last resort on the international stage. There was considerable controversy over how large this fund ought to be because it might lead to a great deal of irresponsible behaviour on the part of banks, in the knowledge that the central bank would stand behind them; failing that, the nation could get a bailout from the International Monetary Fund. It might also discourage nations from exercising their financial responsibilities, and instead to run up large trade deficits in the knowledge that they would be bailed out. Initially, Keynes had supported the creation of a kind of international clearing house for the settlement of trade imbalances similar to the ones in effect with domestic banks in many societies, which, at the end of the day, clear out any imbalances by transferring the necessary funds. Each country would have a limited overdraft option; beyond the limit of this overdraft, its currency would have to be devalued or interest paid to the international clearing house.

In 1971, the Bretton Woods system was abandoned, giving capital markets a great deal of independence. Some argued that this was inevitable because of the policies of that time advocating full employment coupled to fixed currency rates. Currencies were now allowed to float in currency markets, and by 1973 there no longer was any connection between the gold standard and paper money. When a government ran up a large fiscal deficit or a large trade deficit, it would have to borrow money. International investors now acquired the ability to punish it by pushing up interest rates. Because of floating currency rates, governments were no longer required to spend considerable resources to prop up their currencies. According to the monetarists, the markets could be relied on to put everything right.

Instead, in less than three decades, financial markets created the largest speculative bubble ever seen, thus dwarfing real economic activities to about 3 per cent of the daily financial flows around the globe. Banks, hedge funds, and pension funds became some of the main players in this international gambling casino.[62] The scale of this phenomenon could never have occurred if a modest Tobin-like tax had been imposed on this speculation.[63] Within a few decades, these financial transformations helped to turn the United States from the largest creditor to the largest debtor in the world. Trade was no longer the one explanation for currency fluctuations. The crisis of 2008 was a monumental market failure that probably would have brought down the entire global economy, had more people been aware of how large the speculative bubble had become and how little all this finance corresponded to anything genuinely economic. As an aside, the monetarists were remarkably silent at this point. It is tempting to speculate that the massive new economic activity of creating money from money without any real intermediary economic activities had to eventually be matched by quantitative easing, by which governments appeared to complement the creation of money out of nothing by "printing" it.

The fiscal implications of monetarism are clear. First and foremost, inflation had to be controlled, property rights secured, and global markets entrusted to the World Trade Organization. It essentially made the world safe for technique and unsafe for human life, communities, and ecosystems. Capital was now attracted to those areas in the world where almost every possible cost of manufacturing could be externalized because protective standards and regulations were absent or ineffective. As noted, the transnational corporations had constituted a technical planning system which maximized the externalization of these costs, thereby giving a completely new meaning to the concept of competition. These developments greatly accelerated the shift from culture to technique, along with all the devastating adjustments and necessities this imposed.

The monetarists also launched an entirely asocial and ahistorical attack on government. According to them, almost everything government bureaucrats could do the markets could do better. They entirely forgot that even though government bureaucracies are a very poor substitute for the self-regulating capabilities of symbolization, experience, and culture, they became necessary everywhere industrialization began. There was no exception to the growth of the state, even though every political and religious Western tradition (including conservatism, liberalism, socialism, communism, Christianity, and Judaism) had been against large government, for the simple reason that it was undemocratic or oppressive. Either the people decide or government bureaucrats do. Neither the political right nor the left has ever adequately acknowledged this obvious necessity, which came with the shift from culture to technique. With the massive loss of the self-regulating character of culture-based ways of life came a growing role for government as a locus of coordination, thereby making everything political. Nevertheless, without governments' technical coordination, civil societies would have shrivelled even further than they have. Industrialization made the large state a necessity.

There continues to be little awareness that monetarism implicitly advocates an ultimate economic alienation and reification. We are to stay out of our own lives, and to entrust them to infallible and omnipotent market mechanisms. Everything in human life and in the biosphere has to be kept at bay as an economic externality, which essentially closes our eyes to the fact that the kinds of markets served by monetarists with all their hearts and minds have not existed for at least a hundred years. We must also forget that human lives are sustained by communities and that communities are sustained by the biosphere. In sum, we must slavishly follow the dictates of the most extreme form that economics as a discipline has taken on.

When we stand back from all the neoclassical, Keynesian, and monetarist forms of the discipline of economics, one thing stands out. The neoclassical and monetarist versions seem to explain everything in terms of economic phenomena endowed with limitless capacities for self-correction to steer the economic course. This course is good in itself, and thus requires a complete human alienation and reification. The supposed focus is on job creation, growing the economy, and paying down the debt, but since all this depends on free trade, it causes unprecedented job losses, and the strengthening of what have become anti-economies increases the debt. From a social perspective, this economic policy leaves behind billions of people who, for reasons of underemployment and unemployment, can no longer meaningfully participate in the ways of life of their communities. From an environmental perspective, this economic policy will destroy the planet by driving all protective standards to the lowest common denominator.

Unfortunately, Keynesian economics did not escape the limitations of the discipline-based approach to knowing either. As noted, Keynes's analysis was limited to situations in which there were no significant changes to labour, equipment, and technique – contrary to the massive upheavals that had been going on ever since the beginning of industrialization. A discipline-based approach to economics requires entirely unscientific assumptions to justify the neglect of all non-economic phenomena. Since no essential progress had been made in relaxing the fundamental assumptions underlying Keynesian economics, it was vulnerable to any and all non-economic changes. Hence, the failure of Keynesian economic theories was inevitable.

The inescapable conclusion to be drawn from this very brief overview of economic theories is that any discipline-based approaches to economics are bound to fail, since industrialization and globalization go hand in hand in vastly increasing the way everything is related to everything else. As a result, the reform of economic theory will have to become an integral part of reducing our dependence on discipline-based approaches to knowing and doing by reforming the modern university. This will pave the way for a more integrated understanding of our situation, as will be examined later in this chapter. It will then become possible to "design" economic policies operating on the level of experience and culture by addressing the question of how our economies serve us and how we serve our economies, and how to rebalance these two interactions in favour of human life and the planet. Several design exemplars will be presented to which analytical exemplars can be applied for the purpose of analysis and improvement.

We have all read a great deal about the contribution of Christianity towards the growth of capitalism, but what we often forget is that the opposite role should have been expected. In the Jewish Bible, the book of Amos dealt with a time of rapid economic and military expansion, during which the king and the religious establishment were accused of a blatant disregard of the commandment to love your neighbour, by neglecting the poor and the most vulnerable members of their society.[64] Some of the harshest language against political and religious authorities is found in this book. Throughout the Jewish Bible, not protecting widows and orphans (symbolizing the weakest and most vulnerable people) was interpreted as a failure to keep this commandment and thus as a failure to love God, as opposed to being a matter of social justice. The Christian Bible builds on this by saying that the nations will be judged by their conduct towards their weakest and most vulnerable members. The letter of James puts it most succinctly when it states that the only religion acceptable to God is to take care of the widows and orphans.[65] This letter was written when churches struggled with the issues related to wealthy people becoming members; some of the harshest language is used against those who (because of one form of riches or another) exercise power and domination over others.[66] Surely the events in North America would have been very different during the last fifty years had Jews and Christians paid a little more attention to what their Bibles had to say about these issues.

Economics without a Scientific Theory

What are the implications of our lacking a scientific economic theory? First, it affects all economic statistics. The gathering of economic statistics is guided by theoretical frameworks rooted in discipline-based approaches to knowing and doing. For example, the gross domestic product is a measure of gross wealth instead of net wealth, and yet it is used in many calculations as if it were an indication of true wealth. In addition, there are a host of other economic categories whose validity will partly or wholly diminish when they are reinterpreted by a new theoretical framework, which will gradually emerge as we develop an alternative approach. We do not have direct access to the economic dimension of human life. Any so-called fact is relative to a theoretical framework, and the theoretical framework limits and defines what can be related to it and thus constitute a fact. All so-called facts are relative to a framework and can never be absolute. In addition, any theoretical framework must incorporate all non-economic phenomena, whose influence cannot be neglected.

Second, there is no longer a clear demarcation between an economy (where the technology-based connectedness dominates the culture-based connectedness) and society (where the reverse is the case).

Technique-based connectedness has become entirely coextensive with culture-based connectedness. Everything is being re-engineered: matter by nanotechnology, the DNA pool by biotechnology, human knowledge by information technology, the corporation by enterprise integration and the intellectual assembly line, our behaviour and opinions by public relations and communication technologies, our interactions by social media, daily life by medical enhancement technologies, and so on. All this re-engineering by discipline-based approaches involves the taking of elements from human life, society, and the biosphere, reorganizing them to make them internally effective, and integrating them into the technique-based connectedness of a technical order. All these activities simultaneously disorder human lives, communities, societies, and the biosphere. From an economic perspective, the creation of wealth is essentially associated with the building of the technical order, whereas the costs incurred in the creation of this wealth are mostly associated with the creation of chaos in the cultural and natural orders.

If economic theories and policies are supposed to increase net wealth, then virtually every technical intervention must be considered as well as every associated dislocation. The intellectual and practical difficulties arise from the fact that it is impossible to conclude that the economy has now become coextensive with societies and the biosphere. The technique-based connectedness of the technical order has created a new way of relating everything to everything else on the basis of absolute efficiency, without being able to disconnect itself from how everything continues to be (however weakly) related to everything else through the fabric of all life, both cultural and natural.

In other words, the locus of economic phenomena is no longer the economy distinguished from society by the way the technology-based connectedness of human life is related to the culture-based connectedness. Almost every technical innovation for improving the performance of any human activity is harnessed to the making of money, thereby creating an economic dimension essentially coextensive with all of human life and society. A relatively distinct economy was therefore a unique feature of industrializing societies roughly up to the Second World War. This is yet another sign of how our civilization continues to rely on an inadequate stock of concepts for making sense of and living in the world. The dominance of the technique-based connectedness over the culture-based connectedness is taken to be of an economic character and dealt with accordingly. If we did have an adequate stock of concepts, we would be more concerned with the theories of technique and the policies related to it, as opposed to economic theories and policies. The latter now play the role of a secular theology designed to guide secular religious economic practices. For example, the hosts of many radio and television shows will ask business reporters how the markets are doing today much as they might consult a priest about the gods in the past. As noted, these markets correspond to nothing that is economically true, and yet this is one of the most important things most of us wish to know. It simply is not the case that economic phenomena rule over human life, society, and the biosphere because their influence dwarfs those of all other phenomena. It is a question of our civilization conducting human life *as if* this were the case, thus attributing a kind of god-like limitless quality to these economic phenomena.

Treating economic phenomena as if they are omnipotent has long and deep roots in the inadequacies of the above three economic theories. As noted, Adam Smith's description of the economy was fundamentally incomplete. Every market transaction agreed on by two parties would have consequences for human life, society, and the biosphere and thus for everyone else. This is guaranteed by the way everything is related to everything else, and does not disappear just because we act as if everything traded in markets really is fully commoditized and reified, that is, disconnected from the interrelatedness of all life. As a result, when the economies of the industrializing societies had to be organized on the basis of the Market (the organization of the economy as a whole), each and every market transaction generated a diversity of so-called market externalities, which are the unforeseen and usually unintended consequences of any market transaction for human life, society, and the biosphere. All together these little market externalities added up to Market forces, which Adam Smith appears not to have noticed – other than the one referred to as the great invisible hand, to which he did not nearly pay the kind of attention we do today. He also noted that the new economy, based on the technical division of labour and regulated by markets, would make a great many people as stupid as they could possibly become. Here is an embryonic admission of a Market force directed towards human life. We can now add the environmental crisis as the consequence of another Market force directed towards the biosphere. As a result, the force of the great invisible hand of the Market as protected by the World Trade Organization is accompanied by others that appear to mostly undermine the fabric of everything living, as a consequence of the necessity of commoditizing and reifying everything before it can be organized by the Market. In sum, the above three economic theories are fundamentally incomplete with regard to these Market forces acting on human history. In contrast to the Market force of the great invisible hand, all the other Market forces act in the long term. Discounting the future and short-term economic decision making adds to the problem. It is only possible to live this way if economic phenomena are taken to be limitless and all-powerful in relation to all other categories of phenomena, so that the influences of the latter are of little or no concern. We are slowly beginning to appreciate the disaster this represents. Cultures are being pushed back by an economic ordering which necessitates that human life, society, and the biosphere become Market externalities. It slashes through the fabric of relationships of all life.

Following the Second World War, the role of the Market shrank to a shadow of its former self (except for financial markets). John Kenneth Galbraith showed that the transnational corporations could no longer rely on markets for procuring most of the necessary "inputs" or sell their "outputs" when these became the result of technical planning based on the latest discipline-based scientific and technical knowledge.[67] I reached the same conclusions by examining the rise of technique.[68] However, this did not alter the fundamental problem in the least. Instead of market externalities, the technical approach (as examined in the previous chapter) systematically externalizes everything that is not relevant to making everything as efficient as possible. These externalities disorder human lives, societies, and the biosphere.[69] There continues to be no possible compromise between a system building an absolutely efficient order and all life on our planet. For some two centuries, we have put together a way of systematically slicing through the fabric of relationships of all life, with the result that we are now in an all-out race between the problems created by the building of a technical order and the putting into place of compensating techniques to protect that life in an end-of-pipe fashion. Since human life and the biosphere cannot be turned into megamachines, it is a battle that we cannot win. In this respect, discipline-based approaches to knowing and doing have reached their most devastating results through our economic theories and practices. These belong to the category of techniques, to which we continue to ascribe an ultimate importance despite all evidence to the contrary.

Very little remains of the kinds of economies common in the industrializing world during the nineteenth and first half of the twentieth centuries. Few, if any, so-called industrially advanced societies have the three sectors of their economies serve most of their needs. Local agriculture has largely been replaced by technique-based agribusiness.[70] It is concentrated in those areas of the world where turning local ecosystems into mostly monocultures can be done by passing on the externalities to the greatest extent possible. Similarly, manufacturing has flowed to those areas where labour and environmental standards are comparable to those found in Western Europe during the first phase of industrialization. Many of the services forming some of the "inputs" into enterprise integration and the intellectual assembly lines of the transnational corporations have flowed to countries with low protective standards, thus permitting the maximum externalization of labour and environmental costs. How long will China be able to produce consumer goods for most of the world before it is compelled to boost protective standards of all kinds in order to avoid becoming a wasteland? At that point, will Africa become the next locus of this kind of production? Similarly, how long will India be able to provide a variety of information and service inputs into the enterprise integration and intellectual assembly lines of large corporations before it has no choice but to increase its labour and environmental standards? How long will it be before these and other countries will acquire the ability to do what mostly occurs on the technical frontier in California?

If we all have to earn our bread by the sweat of our brows, this export of much of the primary, secondary, and tertiary sectors of our economies cannot continue without creating the levels of unemployment and underemployment that compelled the Germans to pay attention to a few extremists proclaiming the secular gospel of national socialism. It is simply not possible to endlessly marginalize a growing number of people from the ways of life of their societies without creating the kinds of crises that have always led to wars in the past. How do our economic theories and policies take all this into account?

Surely we cannot take seriously that all this is happening because it is what the people want. It is true that the filled parking lots of the McWallDepotCountryMarts of this world are a sign that we are not connecting the dots because of the distortions produced by our secular economic theology, much as the Easter Islanders failed to understand that the religious idolatry necessitating the complete deforestation of their island would collapse their civilization. All this has been explained by Galbraith, and I will simply reinforce it with the theory of technique.[71] We have pointed out that technical advances are rarely responses to human needs and desires, but instead result from the flow of technical information within the system of technique.[72] When the possibility for a new consumer good is opened up at the technical frontier, a need can readily be created if it does not already exist. Production has become an activity that involves adapting human needs and desires to what is technically possible. The evidence is overwhelming. When examining the literature attempting to project how much use past societies were likely to make of new technologies such as telephones or mainframe computers, the estimated potential was always a fraction of the actual eventual use.[73] Mainframe computers are a good example. With the shift from a primary reliance on culture to a primary reliance on technique, the technical approach created vast amounts of technical information, which soon could not be coped with without the use of mainframe computers. This was obvious when the situation was looked at via the lens of the theory of technique but not through the lens of conventional explanations.

The same argument can be made for cell phones. The first users quickly learned that a variety of advantages could be gained from this device, while additional advantages could be gained by slight adjustments to how things were usually done. As others discovered these advantages, they acquired cell phones and added to these kinds of adjustments. It was not very long before life without a cell phone became more difficult as employers and friends now expected rapid responses and as public telephones were gradually being phased out. Our lifestyles have adapted themselves so completely to this device that our children have a great deal of difficulty imagining how life was possible without it. Their lives probably *are* impossible without it because they have been so completely built around the cell phone, which is now used as a mobile computer. Some of us can still remember growing up without cell phones and computers because the lifestyles of that time did not require them. We had slide rules; and when our cars broke down, others would stop to help out because the same thing could happen to them.

The kinds of societies in which we live today are in a large measure the result of our adapting to a variety of technologies and techniques: techniques are changing people. As we will see in the next chapter, the desymbolization of cultures by technique led to the need for a bath of images on the media portraying what we must do and own in order to live the "good life" made possible by technique. Many social functions rendered by cultures in the past have now been taken over by what Jacques Ellul has referred to as integration propaganda, which performs many of the roles previously provided by customs, tradition, morality, and religion.[74] John Kenneth Galbraith's description of the planning system includes its careful examination of human needs and desires as well as consumer trends in order to explore the potential for adapting them to new technical possibilities so as to break the gap between the plan and consumer responses by means of the above bath of images.[75] Moreover, this reversal of economic democracy also suggests that what people really consume is not so much the product itself but a participation in the greatest good we know, which our new secular myth has centred on technique. Our current forms of production cannot be understood apart from mass consumption, mass advertising, and the dependence of mass societies on the media. The dependence of people on integration propaganda in a mass society, and thus their dependence on technical planning, will become more evident in the next chapter.

It is no longer possible to limit contemporary economies to these three primary sectors. Superimposed on them are several others which further complicate the possibility of comprehensive economic theories and policies.

A first additional sector is constituted of entirely new economic activities based on the making of money from money without any intervening traditional economic activities. Complex computer algorithms monitor financial markets to take advantage of, and possibly amplify, fluctuations for the purpose of speculation. When successful, the claims on the total stock of economic goods and services backed by money are increased, with the result that another way of disproportionately extracting wealth from everyone is created. As noted, the scale of this speculative activity now accounts for some 97 per cent of the financial flows around the world each and every day! It dwarfs the scale of the traditional three economic sectors to the point that much of the wealth today is purely extractive, thereby converting economies into anti-economies. This new sector helps to explain the unprecedented and growing gap between rich and poor: fewer than one hundred people own the wealth of half of humanity. This kind of speculation could have been stopped by a Tobin-like tax, which could have paid for a variety of things, including lifting the poorest one-third of humanity out of its misery while making the global economy much more stable.[76] The fact that such an important and necessary measure has still not been taken shows that democracy is powerless in the face of the influence of the wealthy and the secular idolatry of technique.

Ever since the Second World War, a subcategory of economic phenomena should have been treated as an economic sector in its own right, given its growing dominance in the industrially advanced nations. President Eisenhower, in his outgoing address to the American nation, drew attention to what he as a former general regarded as a disquieting development of the military-industrial complex.[77] John Kenneth Galbraith further articulated some of these concerns.[78] The tremendous accomplishment of Keynesian economics following the Second World War in stabilizing the economies of the industrially advanced nations for the first time in their histories had an unintended and unforeseen consequence. In managing the large public sectors with which governments could offset the fluctuations in the private economies, and in making these ventures politically acceptable, these governments had difficulty allocating the spending of surpluses gathered during years of economic strength to anything directly affecting the daily lives of their citizens: roads, parks, schools, hospitals, and so on. The reason is obvious: it is managerially and politically impossible to stop-start such undertakings, not because of need by the public but because of economic necessities generated by the private sectors of the economy. The surpluses had to be spent on things that did not directly affect the population. After all, it does not really affect a society whether or not it gets to the moon a little earlier or later, whether some of its citizens get to fly in a supersonic transport in two years or three, whether the power and efficiency of certain armaments rise by 5 or 10 per cent in the course of a few budgets, and so on. Moreover, these kinds of projects were much more manageable, since they could be contracted out to a defence industry on a cost-plus basis. Following the collapse of the Soviet Union, the so-called peace dividend did not materialize, and the military-industrial complex continued to thrive as if it had a life of its own.

The ongoing development of the military-industrial complex has had a profound negative impact on our sense of peace and security.[79] The application of technique to improve the efficiency and power of our weapon systems has been so spectacularly successful that these systems could no longer defend anyone but only threaten all life on the planet, if they were ever used in another world war, for example, or were triggered as a result of human or system errors. We ought to have recognized that our approach to peace and security needed to be fundamentally changed. It had to become preventive in its orientation. Policies for peace and security ought to have been guided by identifying (and, as much as possible, preventing) issues that would necessitate military interventions. For example, industrial civilization has deep umbilical cords connecting it to the biosphere. Any threats to these lifelines constitute a matter of economic life and death and would likely trigger wars. It would be more fruitful to redirect funding for the military-industrial complex to preventing these kinds of issues. Moreover, the existence of a vast sector in the economy organized on a cost-plus basis has had a huge detrimental effect on the so-called civilian economies. Many of the important debates that raged over these questions for decades appear to have been almost forgotten, and this continues to have significant economic implications.

Following the Second World War, knowing and doing continued to separate themselves from experience and culture as they adopted discipline-based approaches and organizations. This highly specialized knowledge became not only the most important factor of production but also the knowledge base on which the so-called industrial advanced nations evolved their ways of life. Hence, the "production," diffusion, and application of this knowledge began to constitute a category of phenomena with vast economic implications. At first, mass education systems scrambled to fill all the economic and social "niches" that needed to be occupied to complete the shift from a primary reliance on culture to a primary reliance on technique. Once this was accomplished, our mass education systems gradually became "manufacturers" of underemployment and unemployment. Many public universities and colleges (and especially community colleges) are increasingly run on a kind of business model. Enrolment levels are thus determined by departmental and faculty budgets, with little regard being paid to admitted students having a reasonable prospect of future employment in the areas they are going to be studying. A growing mismatch is created between the "outputs" of our education systems and the "occupational niches" that must be filled to maintain the technical order. Nor is there any correlation between the techniques used to endlessly reorganize our educational systems and the needs of our civilization to address its deep structural crises by means of resymbolization that will require fundamental educational reforms. In our present faculties and institutes of education, any attempt at true education by means of resymbolization is as unthinkable as it is in other professional faculties.

One of the other principal sources of the manufacture of underemployment and unemployment is the practice of medium-sized and large corporations (and their franchises) of converting as many of their full-time positions as possible to part-time and contract work with minimal or no benefits and very little job security. It is estimated that three-quarters of humanity has already been forced into the untenable position of having to pay bills as regularly as clockwork while employment is made more and more irregular. This constitutes one of the major attacks on civil society. It is depriving much of humanity of living normal lives and of meaningfully contributing to society. It is not very difficult to reverse this trend of driving everything down to the lowest common denominator. For example, educational institutions should be required to provide prospective students with information that will enable them to make informed decisions as to their chances of becoming underemployed or unemployed, and medium and large-sized corporations should be compelled to have no more than 10 per cent of their workforce be part-time or contract employees and to restrict their use of temporary employment agencies to less than 2 per cent.

In a contemporary mass society, there is possibly no more effective way of breaking someone's spirit than through underemployment and unemployment because all full and meaningful participation in a community depends on employment. How many job interviews can a person have and still be able to be upbeat and act as if they are the best person for the job? How many people laid off close to their expected retirement can bounce back from this betrayal of having loyally worked for an organization for their best years? How will we deal with the growing numbers of people who appear to have decided that the only way to survive is to "rip off" the "system" to the best of their ability? How many more people will decide that the only way to get some money out of the system is to litigate anything and everything whenever possible? In sum, there is little question that the "manufacture" of underemployment and unemployment may be regarded as an increasingly significant sphere of economic activities with devastating effects on human lives and communities. It constitutes yet another example of how a technical order increases performance by disordering everything else.

All this is further complicated by human resource departments being under tremendous pressure to invent creative ways of externalizing as many of the labour costs as possible. Employment in these departments may depend on driving down the wages of others, reducing benefits, and further externalizing these costs when labour contracts are renegotiated. The presence of these people on hiring committees can seriously distort job interviews because they tend to favour those people who are good at verbalizing the work, as opposed to doing it. There is little doubt that choosing between hundreds of applicants in a somewhat objective fashion with a very limited budget is next to impossible; but when it comes to counting the candidate's use of the latest fashionable concepts (in teachers' interviews, for example), things have really gotten out of hand.

We ought to get the message: we can no longer count on people who are displaced in agriculture finding work in industry, and people who are displaced in industry finding work in the service sector. Enterprise integration and the new intellectual assembly lines are automating this last sector, and now workers have nowhere to go except to join the ranks of the job seekers, the underemployed, the unemployed, and the working poor. This is essentially creating a new kind of economic sector, that of a growing underclass. This sector is now so pervasive that very few people are not directly or indirectly affected by it in their immediate circle. Some middle-class families may continue to be able to assist their children financially, but this potential is shrinking along with the middle class itself. Many employees feel that what is happening to them is unjust. Employers, however, tend to interpret all of this as these people being unwilling or unable to put in an honest day's work, and thus feel justified in taking the actions they do. It completes a downward spiral in which we are all losers. The system is so gridlocked that working conditions are steadily sliding back to the kind that we were once confident would never return. When employment statistics improve, governments take credit, and when they deteriorate, opposition parties rally, but all this political grandstanding does not in the least alter the situation. It can only be addressed through preventive approaches that can restructure the economic system so that unemployment, underemployment, and poverty will not be a large and permanent component. To politicize this structural issue is to ensure that no genuine solutions will be forthcoming. It is an issue to which neither the right nor the left has any answers whatsoever.

Structural unemployment, underemployment, and poverty, closely associated with technique and economic globalization, are making societies more dependent on their voluntary sectors, without which the full brunt of our economic policies would be even greater. Increasing numbers of people will also have to rely on a growing informal economy in which people exchange goods and services – an economy not mediated by money. In the next chapter, we will discuss the importance of delimiting social inequality. It is a policy capable of delivering substantial economic and social benefits, as well as the creation of alternative currencies to facilitate the informal economies as much as the regular currencies facilitate the formal economies. Such initiatives will become increasingly necessary as a growing portion of the population retires with little or no pension or other sources of income. Without these kinds of interventions, poverty among the elderly is bound to increase dramatically. As governments continue to cut back on social safety nets, as politicians go on promising unaffordable tax cuts, as corporations pay a fraction of their fair share of the costs of running a society on which they depend, as many pension funds are completely unsustainable (having been turned into little more than intergenerational Ponzi schemes), it is clear that drastic policy reorientations are urgently required. If we accept the current economic situation as the only realistic one, there appear to be no solutions; but this is precisely what we aim to challenge.

The interdependence between economies and the biosphere has had a direct effect on underemployment and unemployment. The overemployment of the biosphere that has resulted in the environmental crisis is directly related to the underemployment of people. As noted, during the late eighteenth and nineteenth centuries, economists assumed that, since the scale of human economies was small relative to the scale of the biosphere, "natural capital" would always be available to future generations and therefore did not have to be priced. Consequently, the costs of the inputs of matter and energy required to sustain the way of life of industrializing societies were essentially limited to those of their extraction, processing, and transportation. Such costs were artificially low relative to those of labour, which reflected the rapidly growing costs of operating and living in an industrial society. Entrepreneurs operating in highly competitive environments had no choice but to minimize their costs, and this was more easily accomplished by economizing labour than by economizing matter and energy. Once discipline-based approaches were applied to increasing the productivity of labour, spectacular results were obtained, with the result that the current global economy over-consumes nature because it "under-consumes" people. Hence, the environmental crisis and the deep structural unemployment and underemployment are closely related to one another. A study produced for the American Academy of Engineering showed that only 7 per cent of what we extract from the biosphere ends up as saleable products.[80] The remaining 93 per cent ends up as pollutants, which we fail to regard as products the system produces like all others but which cannot be sold. Hence, the overall productivity of materials and energy is extremely low, especially when we bear in mind that all this throughput is generally associated with a single use and very little value is recovered at the end of that use period.

Economics as Ignoring Interconnectedness

The limitations of this economic perspective can be summed up by comparing how everything is related to everything else according to it, and what lies beyond according to what we know about human lives, our world, and our history.[81] We are dealing with a thin layer of life surrounding our planet, in which everything is related to everything else. Significant variations in the density of these interrelationships generally correlate with relatively distinct but interrelated symbolically or biologically enfolded wholes.[82] If we focus on any portion of all of this, we will be looking at evolving situations, which can be analysed in terms of interacting categories of phenomena. Only a small minority of these situations will be dominated by a single category of phenomena, with the others making only small contributions that for the purpose of this analysis may be neglected. However, in the vast majority of situations several categories of phenomena are present, and this constitutes the fundamental challenge to our primary reliance on discipline-based approaches to knowing and doing. Ever since the beginning of industrialization, societies have had to recognize the critical importance of technological and economic phenomena in order to create a new dynamic equilibrium of their technology-based connectedness. Gradually this recognition has turned into our acting as if it is simply unrealistic to regard economic phenomena as anything but the key to everything important to human life and society. A secular religious commitment to these phenomena is implicit: we believe them to possess the capacity to bring forth everything we desire and hope for, or at least almost everything. Putting this into practice gradually rearranged how everything was related to everything else in this thin layer of life around our planet.

Of course, each and every civilization did the same thing because "acting as if" turns into "living as if." This occurred by means of a set of myths created as the deepest metaconscious knowledge, which meant that all ultimate meanings were alienated meanings. All of human history bears this out: what was self-evident to a particular civilization was not so to any of the others. The major difference between our civilization and those that preceded ours is that all our means are vastly more powerful, precisely because we sacrificed the integrality of everything else through the quest for power by the endless improvement of efficiency.

From the beginning of industrialization, the technological and economic development of industry in the economy necessarily meant the underdevelopment of everything else. Desymbolization has made our humanity anorexic; our societies have lost much of their capacity to sustain human lives; and the biosphere is so strained by the burdens we place on it that we have agreed that the situation is unsustainable. In other words, acting as if economic phenomena are close to omnipotent so that economies can deliver good work, prosperity, and "development" has had exactly the opposite result – that of disordering everything around us. Good work has been turned into unhealthy work, prosperity has become paying off the debt (which cannot be done through an anti-economy), and development has become the meaningless increases of GDP to the detriment of everything else. Everything that really matters to people and their communities is being turned into scarcities under the sign of money.

Contemporary ways of life are pushing everything to its limits, but it is not these limits that most people would question. We tend to disagree about how close we are to these limits and how much further we can go. Will it be possible to create a consensus that we have gradually backed ourselves into an unliveable and unsustainable corner during the last two centuries?

In order to uncover what drives us towards pushing the limits of human lives, societies, and the biosphere, the concepts of commoditization and reification may prove helpful. They reveal how we act as if, and consequently live as if, the idea that everything being related to everything else is non-essential for human history and natural evolution. For example, the technical division of labour requires the commoditization of human work, thereby turning it into labour. The commodity of labour is treated as being detached from a person's life, and labour markets risk becoming slave markets. People who hire themselves out for a wage and those who employ them need to understand that if the organization of their work influences them more than they can influence it, they will be alienated by that work; and if this alienation spills over into the remainder of their lives, they will be enslaved by that work. Under the influences of Judaism and Christianity, Western civilization at one point clearly understood what was involved in hiring someone for a wage and thus entrusting human work to a labour market. It was not forbidden, because the organization of societies frequently made it necessary, but it had to be done under strict limitations. Essentially, earning a wage is selling a "chunk" of a person's life. The ways in which this can be done fall between the following two extremes. On one end of the spectrum of possibilities, work is organized in such a way as to allow this "chunk" to remain as much as possible an integral part of a person's life. Doing so is hardly abstract. The findings of contemporary social epidemiology related to human work show that the primary indicator of how healthy that work is (mentally, socially, and spiritually) relates to the control people have over that work, thus permitting them to use their education, experience, motivation, creativity, imagination, and other personality traits as they see fit.[83] The opposite end of the spectrum of possibilities makes this virtually impossible, causing the "chunk" of labour to represent a distortion of a person's life, the effects of which spill over into the remainder of that life and the way the person participates in his or her community.[84] We have also demonstrated that most of the preventively oriented work redesign projects of many modern corporations involved the shifting of the organization of work towards the healthy portion of the above spectrum of possibilities, and that this tended to be good for both parties.[85] We will return to all of this later in this chapter. The point is that how work is organized deeply affects the level of commoditization of that work into labour, which also increases the externalities of any transaction on labour markets. The discussions on this point have become so ideologically polarized that we have all but forgotten that central planning is not the alternative to markets, since it did not work at all. Labour markets are a human creation and thus, like any means to an end, will do certain things very well, others not so well, and still others with a great deal of damage. For example, when the work of children was first regulated by labour markets, the results were so disastrous that we quickly realized that, if there were to be any future society, this means ought not to be used for that end, and so we forbade young people under a certain age to work for a wage. Labour markets were thus restricted to people over a certain age. Wherever these kinds of restrictions are relied on, the design of healthy workplaces can vastly reduce the negative effects of the commoditization of human work into labour.

The introduction of the technical division of labour, followed by mechanization, transformed the commoditization of human work into a reified function of a hand or a brain. The result was that scientific management began to apply discipline-based approaches to making these reified functions as efficient as possible, with spectacular success. We also know that this had such a detrimental effect on the workforce that, prior to manufacturing jobs being exported overseas under free trade, almost every large corporation was quietly experimenting with work redesign in an attempt to bring the vast negative consequences under control. Excellent solutions that were highly profitable were discovered, but much of this was lost when it became much easier to externalize as much of the labour cost as possible overseas. The technical organization of work leading to commoditization and reification would not be able to compete with healthy work, if labour markets were designed to operate within strict standards that would not permit an almost unlimited externalization of the costs of this commoditization and reification in labour market transactions. In other words, if the technical division of labour was not to make a great number of people as stupid as they could possibly become, its commoditizing and reifying effects would have to be delimited by means of labour standards. Work organizations would have to be "designed" on the level of symbolization, experience, and culture, which would guide and delimit the analytical exemplars taught in engineering and business schools.[86] If labour markets were to have goals determined by human values and thus remain as integral as possible to the interconnectedness of human life and society, labour standards would be the articulation of such ends. In contrast, if we act as if the interrelatedness of human lives and societies is irrelevant because of the omnipotence of technical and economic phenomena, the goals of labour markets can be surrendered to output-input criteria, which have no connection whatsoever (except by myth) to any human ends.

In the same vein, before land (representing the life-sustaining powers of the biosphere) can be divided into pieces of terrain and be traded in real estate markets, it must be commoditized. Today we do not act, and therefore do not live, as if the pieces of terrain we individually or corporately own remain nevertheless integral to a local ecosystem. No life but our own can respect the artificial boundaries of these terrains. As a result, if the life-sustaining capacities of the biosphere are not to be diminished, some limitations to the ownership of such terrains will have to be recognized by buyers, sellers, and society at large; these in turn necessitate the placing of restrictions on real estate markets. The environmental crisis is in part a reflection of real estate markets having created a powerful negative Market force with respect to the biosphere. To suggest that ownership really means stewardship is well intended, but unless its theoretical and practical significance is worked out concretely, it is nothing but a blowing in the wind.

None of this thinking is anything new, but we practise it to an inadequate extent. For example, when in a great many cities people were obliged to disconnect the downspouts from the storm sewers, we modified part of the property rights in the recognition that groundwater had to be more directly replenished and that treatment of contaminated runoff water during storms would be difficult to manage otherwise in a cost-effective way. This recognition could have prevented the building of a great many unnecessary storm sewers. When it comes to fracking, however, we appear not to get the obvious message, by acting as if particular functions such as those of groundwater can be reified. The recognition that this is impossible would again affect private property rights. It is not feasible to separate our conceptions regarding the ownership of terrain from our obligation to maintain the life-supporting capacities of the biosphere, that is by setting clear standards to ensure that pieces of terrain remain integral to the interconnectedness of everything designated by the concept of land.

There are a great many ways in which contemporary societies commoditize and reify cultures. For example: advertising, public relations, and the mass media utilize words as tools to create particular effects rather than to communicate meanings. By means of symbolization, languages reflect how everything is related to everything else, and a blatant disregard for meanings commoditizes and reifies all dialectically enfolded cultural wholes. Many current debates regarding intellectual property rights completely disregard the cultural and intellectual interconnectedness of human life. Everything we create builds on professional and cultural heritages, and hopefully will contribute to the heritages we leave for future generations. To accomplish this will require that limits on commoditization and reification be imposed as much as possible, so that the making of human history is not subjected to a mythical omnipotence of economic phenomena. For example, is the highly specialized scientific and technical knowledge generated by our public universities to be a public good (since its use by someone does not diminish its value for everyone else), or is it to become intellectual property and thus unavailable to most people, with enormous detrimental effects on the economy and human life?

It now becomes clear how the commoditization and reification of human lives, societies, and the biosphere were directly translated into technical "externalities" in applying the technical division of labour and subsequent mechanization.[87] These externalities were in turn translated into market externalities when goods were traded, and thus contributed to Market forces, of which the negative ones led to the undermining of how everything was related to everything else in the nineteenth and first half of the twentieth centuries, and later into technique-based equivalents in the planning system during the second half of the twentieth and beginning of the twenty-first centuries. We can act as if, and subsequently live as if, everything is not related to everything else, but this will simply distort this interrelatedness and disorder human lives, societies, and the biosphere. The economic reduction of human development to mere growth has achieved the exact opposite of what we intended, which is the inevitable consequence of the secular myth we now serve (much as the gods were once served). From this perspective, all consumer goods are a commoditization and reification of the culture-based and biology-based connectedness. In addition, they represent a certain commoditization and reification of the technique-based connectedness as they participate in human activities related to production, distribution, sales, use, repair, reuse, and eventual disposal. Each and every consumer good modifies this technique-based connectedness in some fashion, but this is reflected in neither its market price nor those established by the planning system. Nor does this price reflect the effects technique-based connectedness has on the culture-based connectedness or on the natural connectedness of the biosphere.

Hopefully, these kinds of arguments will contribute to a consensus that we have backed ourselves into a corner during the last two hundred years. Acting as if, and subsequently living as if, the interrelatedness of our world matters little in comparison to what economic phenomena can deliver has had the opposite effect to what we intended. We have created a civilization to fulfil our wishes and desires, and in doing so we have undermined how everything is related to, and thus depends on, everything else. We can now begin to see that the strategy of maximizing power, efficiency, and profits is indissociably linked to the maximization of risks to all life. Economic phenomena have turned out to be not as omnipotent as we believed them to be. They have brought us the great separation of everything human and natural, first from the economy and later from technique. A great deal of our societal and natural abundance, which used to be a free gift dominated by symbolization, experience, and culture, now became dominated by money and efficiency, and this fundamentally changed everything. Western civilization ought to have been forewarned, because at one point it did accept the difference between grace and mammon. The discipline of theology, however, made this distinction so divorced from life as to become impractical. We have discovered the consequences to our detriment.

From a historical perspective, there has been considerable variation in the ways cultures dealt with how everything was related to everything else. There has also been a substantial historical variability in the consequences that flow from respecting this interdependence of human life and the world. We will seek to show that if we are prepared to take this interconnectedness as our point of departure for our technical and economic thinking and doing, preventive approaches can be developed that will delimit the effects of commoditization and reification as much as possible and thus delimit the negative consequences of what began as Market forces and then became forces of technique. We will then have a better theoretical and practical grip on what we are externalizing in order to provide us with the ability to assess alternative courses of action in terms of what they mean for human life, societies, and the biosphere. What we began in chapter 1 with regard to the cult of the fact, and continued in chapter 2 with regard to the cult of efficiency, we will now pursue in relation to the cult of growth. Discipline-based approaches need to be complemented by efforts to resymbolize economies in order to reintegrate them into everything else.

Towards a Diagnosis

Fortunately, economists have generally recognized that market prices for goods and services are a poor measure of their value for individual and collective human life. The reasons for this are now clear. Meaning and values are established by means of symbolization, experience, and culture, which respect the interrelatedness of everything as much as possible. In contrast, market prices are established in the very narrow context of their supply and demand in individual markets, determined by *homo economicus* maximizing the utility derived from a wage or the profits derived from investments. Even if these markets could be perfectly efficient and all the actors could have perfect information, the results would be more or less as catastrophic as they are now. With the emergence of technique along with mass societies, technical planning all but replaced the Market, thus leaving markets to take care of any remaining gap between planning and consumer responses. Therefore, market prices continue to be a poor measure of value. Consequently, we need to challenge the cult of growth based on the dominance of economic phenomena by symbolizing the latter. Economic theories and practice can thus be dominated by economic design exemplars worked out in detail by means of economic analytical exemplars. This will open the door to more adequately taking into account how everything is related to everything else according to human values and aspirations, thereby internalizing many externalities now produced by technical and economic decision making.

There is no point using the traditional end-of-pipe approaches when seeking to internalize externalities into market mechanisms or in technical decision processes. What needs to be done is to design goods and services and their distribution by means of technical and economic design exemplars arrived at by means of preventive approaches. In this way, goods and services will continue to require commoditization and reification, but to a far lesser degree because they will be a more adequate embodiment of the culture-based connectedness and biology-based connectedness of our world.

There are three prerequisites for developing this potential. The first was described in chapter 1 as a rebuilding of our knowledge bases, which will make effective use of approaches based both on symbolization and on disciplines according to their opposite and unique limitations. The second was described in chapter 2, and involves the domination of individual techniques by the equivalent of design based on symbolization, experience, and culture. The third is the domination of economic phenomena in human life and society by symbolizing them according to their participation in the way everything is related to everything else in our world. Although this third prerequisite is an essential part of university reform, it takes on a special significance, given that our civilization treats economic phenomena the way the gods were venerated by earlier civilizations. It requires destroying the current economic world view based on the omnipotence of the economy. As such, this prerequisite goes well beyond university reform.

Preventive approaches can accomplish what most engineers, business managers, and economists are likely to dismiss out of hand, given the contrary evidence all around us. People brought up in these discipline-based approaches are convinced that making contemporary ways of life based on science and technique more economically competitive by raising social and environmental standards is at best well-intended wishful thinking. They are entirely correct insofar as doing so on a disciplinary basis leaves no alternative other than adopting end-of-pipe approaches. Consequently, increasing social and environmental standards would require adding on even more compensatory social, health, environmental, and other services as well as more devices and systems for trapping and containing pollutants. They are entirely incorrect if this can be accomplished by means of preventive approaches, which go beyond discipline-based ones to tap into the potential of symbolization, experience, and culture in order to improve compatibility and avoid collisions with human life, society, and biosphere. Although there are plenty of successful preventive approaches documented in the literature,[88] they were all piecemeal and thus sooner or later were overwhelmed by their surroundings, which of course had been organized and dominated by discipline-based approaches. From the outset, we must be very clear that re-equipping contemporary ways of life with a preventive orientation is not equivalent to adopting a piecemeal approach based on the precautionary and no-regrets principles. In the past, these principles were built into the cultures and ways of life of earlier societies as a consequence of symbolization. Even though university reforms and the introduction of a preventive orientation in our knowing and doing are to some extent an "all or nothing" affair, the implementation can be done in a carefully monitored iterative approach, as will soon become apparent.

Re-creating an Economic Dimension in Human Life and Society

When "science and technique changing people" created a new secular myth as the deepest metaconscious knowledge in people's brain-minds, the road was blocked to evolving contemporary ways of life other than discipline-based approaches. All human creativity on this front was blocked by these myths, turning science and technique into a kind of human destiny. Discipline-based approaches were used as if they had no limits, with the result that other approaches became unthinkable and unimaginable. Hence, what I am about to describe is entirely feasible in the absence of our secular myths, but it is highly improbable as long as we remain occupied with serving our new masters. Consequently, my point of departure is an admission of the improbability of our waking up in the near future with our secular myths having been unmasked in our individual collective lives, thus making us take another look at our lives in the world only to find them to be entirely different from what we supposed according to the cult of the fact, the cult of efficiency, and the cult of growth. If this improbable event were to occur, we would have to evolve a new collective intelligence by means of which we would apprehend ourselves and our world. (This is the true meaning of conversion.) Such a new intelligence would have to involve a new economic dimension of human life and society as well as of the biosphere.

However, implementing the kinds of strategies described below can itself help to generate a kind of practical wisdom for a more liveable and sustainable future.

Insisting on an economic dimension presupposes other dimensions as well as their differentiation from each other to reveal their relative meanings and values for human life and society, thus replacing the other dimensions in their contexts. This is very important. It is the blueprint for challenging and reforming our current scientific and technical division of labour, the organization of our universities, and all the other institutions that depend on them. These dimensions of how everything is related to, and evolves in relation to, everything else must be dealt with in a scientific manner by considering other categories of phenomena which make non-negligible contributions to the kinds of situations being analysed. Elsewhere,[89] I have shown how this may be accomplished for a variety of disciplines including economics and sociology, as well as disciplines related to professions such as engineering, business administration and management, medicine, and law. Implicit in this process is an inquiry into how the knowledge generated by any scientific or technical discipline is used in maintaining and evolving contemporary ways of life, including the desired and undesired consequences of such usage. It also implies the need to act on this understanding by enlarging the context of such knowledge in order to enhance the desired effects and prevent or greatly minimize the undesired ones. Especially mathematics in all its forms must not be excluded from these efforts on the mistaken grounds that our concerns should be limited to applied mathematics. I will leave a discussion of the non-existence of pure mathematics to others.[90] I will simply note that there are mathematicians who devote five to ten minutes of each lecture to this task and others who devote entire courses to reviewing, for example, the inappropriate use of statistics in a great many areas of application. It helps to make their students more aware of the implications of what they are learning and encourages them to take responsibility for their future decisions.

The improbability of what I am about to propose, and the power of our secular myths responsible for blocking it, may be illustrated from the following events. As noted, in 1995 an earlier version of the proposal being developed here caught the attention of the former Premier's Council of Ontario in Canada. This led to the creation of a round table to investigate the potential of preventive approaches for restructuring economies to deliver goods and services with a fraction of the current social and environmental burdens and how such a potential could best be tapped into. A similar proposal was spearheaded in 2003 by the presidents of the Natural Science and Engineering Research Council and the Social Science and Humanities Research Council (both of Canada), called the Society, Technology and Science 21 (STS21) proposal. It was also recognized as a leading Canadian innovation by the Canada Foundation for Innovation in 2002. Although all these initiatives became the victims of our economic alienation, it is clear that there was high-level recognition of the merits of what is being described here.

A more limited initiative was begun much earlier with a bold move by one of Canada's leading engineering schools based on the detailed study of undergraduate engineering education discussed earlier.[91] As a result of an elective sequence of courses in preventive engineering that were described by a significant number of students as life-changing, the chair of one of the largest departments invited a proposal to describe how two advanced courses in fluid mechanics and heat transfer would change if they were to take on a preventive orientation. Following a presentation backed by a nearly one-hundred-page document, the chair put the proposal to a vote. It split the faculty roughly down the middle, with the senior members voting against and the junior members voting in favour. The chair correctly concluded that he could not move forward under such conditions, especially because the junior colleagues who would commit to such a preventively oriented stream would have greater difficulties obtaining promotion and tenure, given that their senior colleagues would likely constitute a majority on the committees deciding these matters. Even though every dean in the faculty (except the two most recent ones) had acknowledged that such a venture was an essential part of the profession's obligation to society, the need to make this practical by creating a new stream within the curriculum with its own appropriate standards for tenure and promotion in this new intellectual and professional "culture" was not acted upon. This happened despite the fact that quantitative research instruments to score and monitor preventively oriented developments in any given course had been developed and tested. Nevertheless, the faculty took the bold step of making the introductory course on preventive approaches in the above-mentioned elective sequence into a compulsory course in the first year.

The elective sequence in preventive engineering continued for many years and undoubtedly constituted the most successful interdisciplinary program the faculty had ever offered. The students had no difficulty recognizing the potential of preventive approaches, but the departments refused offers of seminars to explain why preventive engineering could operate cost-effectively to higher social and environmental standards. When the director of the Centre for Technology and Social Development overseeing all these efforts retired, the last position in the faculty dealing with the abilities of engineers to protect the public interest was mutated into something entirely different. Because of student pressure, this director continued to offer some of the courses after his retirement, but the centre was shut down. Since the director had moved out of town, the department took the unprecedented step of having him driven to and from his classes. What this demonstrates is that the proposed scientific and technical reforms are intellectually and practically feasible, but the inertia in the system and the existence of all manner of pre-judgments are too great at this time. It would appear that nothing short of a major crisis will be required to turn the situation around. At that point, the plan outlined below could be implemented.

The principal challenge to developing a strategy to break with the cult of growth is that the kinds of issues and problems it seeks to address are interrelated and therefore interdependent. We act and live as if we can concentrate on what we regard as one of our most serious problems, such as global warming, and then when we have had some success we can turn to others. Such an approach is a non-starter. It must be remembered that because of the way we have organized contemporary ways of life, they produce and deliver goods and services but also "produce" all the issues and problems we face. For example, the way we are currently dealing with global warming is a manifestation of how our minds are possessed by technique. How everything is related to everything else in the biosphere is examined purely from the perspective of how this is affected by the release of greenhouse gases and how this translates into a warming effect on our planet. Attempts are made at reaching a consensus as to how far this warming can go, with the result that we are essentially attempting to establish the "efficiency" with which the earth's atmosphere can process greenhouse gases. All this is based on very complex and difficult to test computer simulations. Once consensus can be reached on what is likely the most reliable estimate, we think business as usual can more or less be continued within these limits. Everything that really matters to our lives and to all other life is a mere technical externality in this decision process. The only non-negotiable element in all of this is that economic phenomena must be touched as little as possible so that they can continue to right themselves, regain their omnipotence, and bestow their many blessings. When we superimpose on this picture all the other issues we face and how interrelated they all are, we see that the current approach proceeds by treating each and every issue in isolation, while assuming that everything else remains constant. Doing so is not a reflection of what we know about life in the world but a necessity imposed by discipline-based approaches.

Since a great many other serious issues and problems are related to the ways we have undermined how everything is related to everything else in human lives, society, and the biosphere, more comprehensive approaches will be required. Such approaches, however, can give rise to totalitarian ideologies and regimes, especially when a necessity is made liveable by means of secular religious attitudes that transform it into an ultimate good. Such a risk can and must be avoided: it is a struggle against our enslavement resulting from endowing science, technique, and the economy with omnipotence in their own domains, and thus is a struggle against alienation, not against alienated people. No political elite can bring this freedom.

It is also clear that such a comprehensive strategy cannot be an economic one. There is no separate economy, hence there can be no distinct economic strategy for reintegrating it into everything else – thereby turning anti-economies back into economies designed to serve human values and aspirations. It must seek to delimit all anti-economic activities, especially the making of money with money without any intervening economic activities. As noted, this can be accomplished by imposing a Tobin-like tax, of which the revenues could be used to finance the kind of strategy being proposed as well as redirecting our economic focus towards genuinely economic activities.

As noted, the proposed strategy is an attempt at creating some freedom in relation to what alienates and reifies us, namely, the technical and economic externalities resulting from our present enslavement to discipline-based approaches. The accompanying effects of commoditization and reification can be minimized by strengthening the way everything is related to everything else by integrating discipline-based approaches into human lives and societies, integrating the results into all our organizations and institutions, and incorporating these results into the reintegration of the economy into society and of society into the biosphere. Doing so involves four strategies associated with our dependence on knowing and doing, our dependence on work, our dependence on matter, and our dependence on energy.

Our Dependence on Knowing and Doing

Our current dependence on discipline-based knowing and doing is created through our systems of mass education, which ultimately depend on our universities. Hence, university reform must be the first focus of our strategy. This task will turn out to be not nearly as daunting as it may appear, provided that our secular myths no longer block the way. All the parallel efforts in individual disciplines, specialties, and professions will be fundamentally synergistic. For example, the restructuring of the discipline of economics into a specialty examining the economic dimension of human life, society, and the biosphere would involve an identification of all the other categories of phenomena that play a significant role in the economic situations being examined. As these categories are being considered, parallel efforts in other disciplines will be identifying situations in which economic phenomena play non-negligible roles. In this way, meaningful patterns of collaboration can readily be identified, while, at the same time, such parallel efforts provide intellectual checks on all the others. As new patterns of collaboration are being developed, disciplines on the way to becoming genuine scientific specialties will create new cognate areas for teaching and research, which will eventually help to create a new organization for the public university. This organization can once again be harnessed to the public good. Every emerging scientific specialty will begin to have a focus on one category of phenomena but will include its interaction with all other significant categories of phenomena that cannot be neglected. Gradually each such specialty will be turned into a dimension of the interrelated character of human life and the world. The specific positive effects on contemporary ways of life will gradually become more synergistic, while the negative effects will become more preventable. The overall effect of these synergistic efforts can create forms of science that are a great deal more scientific than most of our current disciplines; this will pave the way towards gradually harnessing technique to human values, and only secondarily, within this context, to efficiency and output-input ratios.

What this part of the proposed strategy will not be able to accomplish is to directly loosen the grip science and technique have on people's minds and cultures in order to ensure that their values are genuine and not a consequence of "science and technique changing people." Since I have already devoted the major portion of a previous volume to university and college reforms, the reader is referred to it for full details.[92] I will simply note that all these efforts depend on the integration of a knowing and doing embedded in experience and culture with knowing and doing separated from experience and culture. Since the strengths and weaknesses of these approaches are diametrically opposite, they are potentially complementary, provided that our current secular myths no longer stand in the way. Every new area of specialization must take on the equivalent of design, one that brings its practices and applications under the guidance of symbolization, experience, and culture. It is only on this level that different categories of phenomena interacting in situations being examined can be reintegrated into a more comprehensive understanding of the relevant aspects of how everything is related to everything else.

Our Dependence on Work

Our dependence on work in relation to the economic dimension of human life will be discussed here, and a discussion of the social dimension will be deferred to the next chapter. Much of human work is in the grip of discipline-based approaches to knowing and doing. The organization of this work inevitably involves a fault line, with one side primarily depending on knowing and doing separated from experience and culture (and thus largely disdaining knowing and doing embedded in experience and culture), while on the other side the opposite is the case. The consequences are far-reaching. The intellectual and professional division of labour distorts how everything associated with work is related to its context; it excludes a great deal of human talent; it renders effective equivalents of negative feedback loops impossible; and it divides the organization against itself because of the difficulties of communicating across the fault line. For example: in a corporation the office is essentially the locus of discipline-based approaches separated from experience and culture, while the plants are the locus of approaches embedded in experience and culture. Management essentially constitutes a "collective brain" which has difficulty understanding and using its "collective hand," with the result that the latter atrophies. Labour has a minimally developed "collective brain" as far as discipline-based approaches separated from experience and culture are concerned, and an overdeveloped "collective hand" embedded in experience and culture, which lacks effective guidance. Both management and labour face their unique deficiencies, which are the mirror image of one another but are both rooted in their inability to synergistically utilize the two modes of knowing and doing. This issue is also found in corporations running retail stores or franchises. It is somewhat less applicable to small entrepreneurial firms.

A second fault line results from the lack of complementarity between the work organization and the "infrastructure" composed of its machines (be they of a classical or information character) as well as the enterprise integration built up with them. As noted, the strengths and weaknesses of human beings and machines are diametrical opposites. The former live by non-repetitive adaptation, while the latter function by means of non-adaptive repetition. Thus far we have designed work organizations that undermine the resources and skills of people based on knowing and doing embedded in experience and culture. As noted, we have also turned these organizations into primary sources of physical and mental illness. From social epidemiology we know that healthier workplaces require the coupling of high demands to high levels of control because this permits learning, enhances motivation, increases job satisfaction, and reduces stress.[93] It reintegrates the hand with the brain and reduces the crippling "us versus them" attitudes associated with the fault lines and lack of complementarity between the organization and its technical infrastructure.

Finally, since these fault lines run through the entire work organization, any attempts at weakening and potentially eliminating them must be consistent throughout the organization and not limited to a few departments or sections. Localized efforts within the work organization would simply move these fault lines around, thus displacing the problems from one part to another. There must be no "islands" where the two modes of knowing and doing function synergistically within an ocean where they do not. Nor must there be any "islands" of healthy work in danger of being swamped by an ocean of unhealthy work. These kinds of situations have been the downfall of almost every preventively oriented and initially highly successful attempt documented in the literature.[94]

Similar kinds of fault lines exist between these work organizations and their broader contexts made up of the economy, society, and the biosphere. The fault line with society will be dealt with more fully in the next chapter.

All this can be further understood in terms of the reciprocal interactions that exist between employees and their work organizations. The latter place a variety of demands on the former, which they must meet with their own resources derived from their education, experience, creativity, motivation, persistence, and other personality traits.

If the demands swamp the resources, the work is crushing. If the reverse is the case, the work will be trivial and boring. Somewhere in between these two extremes, we find the healthy situations in which the resources of the employees are somewhat challenged by the demands placed on them, with the result that the latter can be met only by creatively expanding the former. Initially, this can be very stressful when an employee has no idea how to tackle the task at hand. After a frustrating period of mulling it over, however, the employee's intuitions as to how to proceed may lead to success. The completion of the task provides a great deal of satisfaction and presumably recognition for a job well done. The expanded resources of the employee eventually constitute a basis for advancement and promotion.

In a great many situations, the resources of employees are essentially paralysed by the work organization. For example, many job descriptions largely exclude either the hand or the brain. In the case of factory work, there is an unrecognized involvement of the brain to bridge the gap between how a machine, procedure, or system functions according to a knowing and doing separate from experience and culture and how it actually functions in an office or factory. The lean production system gave assembly-line workers additional responsibilities, to the point of actually stopping the line. This possibility had previously been written out of their job descriptions, but when the line almost never stops, there really is very little effective change.

When employees go home at night, they must replenish their resources by sustaining and being sustained by others, eating, relaxing, and sleeping. If these resources are not replenished, they are gradually "mined," and depleted. This leads to chronic fatigue and exhaustion, thereby reducing the availability of these resources. Such a reduction has substantial consequences for the work organization.

The reciprocal interdependence between employees and the work organization is healthy for the former when demands are coupled to high levels of control, and unhealthy when this is not the case. As a result, the introduction of the technical division of labour, mechanization, automation, and computerization has transformed relatively healthy jobs into mostly unhealthy ones during the last two hundred years of industrialization.

With this basic framework, we return to two previously mentioned preventively oriented initiatives. The Semco organization was built on the values of personal initiative, individualism, and freedom within limits.[95] These limits control greed by sharing power and making all operations as flexible and self-managing as possible. It includes a great deal of integration between both approaches to knowing and doing by incorporating into every job the mundane tasks associated with each job in order to eliminate dead-end jobs as much as possible. It allows roughly one-quarter of the employees to set their own salaries, and to decide whether to augment their resources by taking a short sabbatical or by formal courses. The leadership is evaluated by those whom they lead, and the results are publicly posted. Employees help to decide on the products they make and the way they are made and marketed. Time clocks are only used by employees for the purpose of self-management. Profit sharing is carried out on a democratic basis by giving employees an effective voice, because they receive training in reading and interpreting the financial statements of the company, which are available to everyone. They can vote when decisions are to be made regarding the acquisition of another business unit. In every respect, everyone is made as self-reliant as possible, thus shortening the equivalent of negative feedback loops to the greatest extent possible. The conventional wisdom is that such an organization cannot work because people cannot be trusted, and yet this company grew its profits and sales many times over as its reorganization advanced. Throughout this reorganization there were no management fads and no consultants! It would appear that even further advances might have been possible on the basis of a clearer understanding of the potential complementarity between the two modes of knowing and doing.

For example, additional advances might have resulted by also integrating the assembly methods pioneered in Volvo's Uddevalla plant, where autonomous working teams composed of members, each possessing all necessary skills for any task faced by the team, cooperated in assembling vehicles.[96] As previously noted, repetition was enormously reduced as each team member cycled through a series of tasks. These lasted eighty minutes in one shop organization and one hundred minutes in another. This kind of organization created unprecedented flexibility, as vehicles could be built directly to customers' orders with short delivery times. It also greatly facilitated retooling these workshops for building new models. All this could have been advanced even further by incorporating the just-in-time supply chains and the design for easy assembly procedures of lean production. It would have complemented and reinforced the hand-brain reintegration of a Semco-style organization. By reviewing the literature on our experiences with preventively oriented design exemplars for offices and factories, it is possible to synthesize all this by creating a comprehensive design exemplar capable of eliminating all the above fault lines by having human beings do what they do best and machines do what they excel at. It is at this point where we will go beyond the discussions in chapter 2 in order to examine these kinds of design exemplars in terms of how economic phenomena intermingle with others. Doing so will also reveal which masters we serve.

One of the most astonishing contradictions of our contemporary societies is the complete opposition that exists between the values we profess and the values implied in our workplaces. The above design exemplar for a work organization challenges the prediction of Adam Smith that the technical division of labour would necessarily make people as stupid as they could possibly become. It also challenges the widespread beliefs that authoritarian work organizations are required because people are irresponsible and thus must be enslaved, and that the responsibilities of management to shareholders trump all limits on commoditization and reification. Preventively designed work organizations eliminate all excuses not to practise our values on the grounds that discipline-based approaches are the only way in which nations can "realistically" organize everything. It is this "realism" that masks our enslavement to these approaches. By recognizing their limits, it becomes possible to practise our human and democratic values by rejecting the idea that only economic phenomena can "save" our world and deliver a future with jobs, prosperity, and growth.

To get to the heart of the matter, let us take the above exemplar for the design of a work organization and let us assume the highly unlikely possibility that its productivity and profitability are somewhat below those of a computer-based Taylorist organization using lean production. What conclusion should be drawn from this situation? The economic "realists" are likely to point out that this design exemplar for a new organization will not be competitive and therefore is without a future. To be consistent, this means that human freedom and democratic values have no role to play in the most influential and significant part of our life related to work, making all our political rhetoric a blowing in the wind.

If we are willing to contemplate that life is more than work, the above economic "realism" is absurd. As we have seen, there is overwhelming evidence that a lean production organization managed by an office based on enterprise integration is a significant source of physical and mental illness, with the result that such an organization imposes on society needlessly high levels of disease care, social support, coping mechanisms based on drugs, alcohol and violence, and family and community dislocations. In other words, this kind of work organization helps to produce anti-economies that extract rather than produce wealth. Designing a work organization that contributes to making us poor is not only economically irrational, it is socially irresponsible and, by any religious standards, immoral as well. It makes no sense whatsoever other than in a world dominated by secular religious commitment to the omnipotence of economic phenomena.

So what can be done practically if the unlikely situation arises in which a preventively oriented design exemplar for work organization is not quite as competitive as its current counterpart? One obvious solution is to give the corporations using this kind of design exemplar a tax credit commensurate with the reductions in health and social costs imposed on society. More importantly, we must draw the conclusion that clearly we are improperly using some markets for goods and services. Too much is entrusted to them, with the result that their operations must be curbed by raising labour, health, and social standards to create a more level playing field within which such markets can operate. In other words, preventively oriented design can help us discover in a completely practical manner how we can utilize technical planning and the remaining dependence on markets in a way that will help transform anti-economies into genuine ones. Technical planning and the highly diminished role of the Market are, as we have argued, tools with strengths, weaknesses, and harmful effects that are inseparable from one another. This means that we must push our preventive efforts to currently feasible limits to increase the results we desire and simultaneously decrease harm. This latter course of action will open up if we can transcend our secular religious attitudes to economic phenomena.

From a strategic perspective, the above kind of design exemplar for a work organization is likely to be pioneered first by smaller entrepreneurial firms or local cooperatives. They will have substantial advantages in their ability to adapt to scientific and technical innovations, preventive approaches, any turbulence of an economic, social, and environmental nature, and other shocks. Moreover, it is almost certain that the future belongs to them. There are many developments pointing in that direction. First, there is the issue of fossil fuels and global warming. Energy prices will almost certainly have to rise to the point that shipping costs will begin to offset the advantages obtained by externalizing production costs to the greatest extent possible.[97] In our desperate attempts to maintain the status quo, we appear to be willing to incur ever greater risks in gaining energy from fracking or from tar sands production. This kind of behaviour is all too typical when it is driven by secular religious commitments. Second, the countries in which global production is now being concentrated, especially in China, will not be able to withstand the devastation caused by all the externalities being absorbed by society and local ecosystems. They will have no choice but to increase protective standards, and they will be unable to do so under the WTO regime – unless another continent is willing to repeat their experiences. Third, even a limited decentralization of global production will involve countries with standards that are far beyond the Chinas and Indias of our world, with the result that this repatriation of production will have to involve preventive approaches to make them acceptable and economically viable.

These kinds of developments will hurt the transnational corporations because they are one of the primary loci of technique and thus a summit of resistance, born from excessive power, to having restrictions placed on their ability to externalize costs of all kinds to present and future generations as well as the planet. Some have pointed out that since corporations were granted the status of a "legal person," their behaviour as such is so psychopathic that we will not be able to tolerate it in the long term.[98] They are faithfully served by engineering and business schools which, under the WTO, see no need of teaching preventive practices of any kind.

In the end, it all comes down to our willingness to serve the cult of economic growth requiring that economic phenomena are to be treated as omnipotent. It is not a lack of practical technical and economic alternatives that stands in the way. It is ultimately our willingness to serve our present masters as faithful economic slaves. Once in a while there is a glimpse of hope, as when the Germans voted with their feet, making it impossible for Walmart to operate in Germany. If most people really understood what was happening in their communities and the world, they would do much the same thing, and change could come quickly, provided we are ready with the above kinds of reforms. It is the cult of economic growth that needs to be broken.

Within this general design exemplar, particular attention must be paid to our human resource policies, which, in a great many cases, externalize employment costs to the point of commoditizing people's lives, thus leading to high levels of dehumanization. For example, if we were at all serious about delimiting employment costs, we would immediately pass laws compelling large corporations to create as many real jobs as possible by permitting only a small fraction to be part-time workers with commensurate benefits. Why should our box stores have mostly part-time employees, thereby imposing many costs on communities as a result of disordering people's lives and families? We have created a large group of people who must juggle three part-time jobs to make ends meet, while each employer expects these people to be available. Worse, when two parents face these kinds of situations, the chances of getting a day off together are slim indeed. While this contributes to the GDP as a consequence of marriage counselling, divorce-related legal services, drugs, or alcohol, it makes a normal human life with a partner and children virtually impossible. It is unbelievable that we, who regard ourselves as so highly developed and humane, tolerate this level of human slavery.

These practices are hardly limited to the private sector. Local bus services externalize employment costs into the community to equally high levels by requiring newly hired drivers to be available for work six days per week without any benefits. This may last from three to seven years, when they finally may become official full-time employees. Again, the so-called part-time workers (who work full-time most weeks) have no input into when their days off occur, with the result that for years and years many of them share very few days off with their partners and children.

All this could readily be resolved if we were prepared to challenge the World Trade Organization by placing restrictions on labour markets to delimit commoditization and reification. At present, human resource departments encounter very few limitations in externalizing labour costs, and their members are forced to behave as psychopaths in imitation of their employers.

What is equally disturbing is that many contemporary unions are not really defending their members' interests in comprehensive ways. While wages and benefits are important issues, the enslavement of human beings by their work surely is even more so, especially since humanity has agreed that slavery is not an acceptable form of human life.

For some of us, the most embarrassing thing is the extensive support some sectors in the Jewish and Christian communities give to political parties who, more than others, serve the secular equivalent of the gods of the past. These are the Market (privatizing and commoditizing abundance and grace) and mammon (enslaving human life to wealth).

It amounts to a very convenient interpretation of the second great commandment. It is surprising that so many people believing in humane values support political parties that propose, enact, or enforce so-called right to work legislation, which essentially makes employees defenceless against employers by eliminating the possibility of forming unions. The greatest outrage against people who have to earn a living are the so-called zero hour contracts in England. Hundreds of thousands of people signed these contracts, which require them to be available for work at any time at the call of the employer without any guarantee of paid work. How is it possible that a civilized society can make this kind of work arrangement legal? How can we respect corporations like McDonald's, which in England has the majority of its employees on such contracts? In Canada, this company recently attracted a great deal of public anger because it was believed to be abusing current legislation that allowed for the importation of foreign workers while many Canadians go unemployed. We are slowly creating a return of slave-like conditions.

Once again, the creation of design exemplars for healthy work must transcend the usual political left-right arguments. We are confronted with a structural situation in which desymbolization has practically eliminated all pushback the cultures of communities used to exercise on employers, once a reasonable balance of power had been established following the first phase of industrialization. The complaints of employers and employees have to be addressed simultaneously, and much of this has to do with the kinds of jobs we have created. For example, it was not long ago when becoming a car mechanic involved the rendering of a useful service to the community. Under technical planning, efficiency ruled everything, with the result that cars were more poorly made and difficult to repair, and the availability of parts declined over time. Car mechanics lost pride in what they did, and eventually there was little to do but join the system and reduce everything to the making of money. We all know about having repairs done that were not necessary following faulty diagnoses. Mechanics internalized the idea that they were part of the corruption, and being proud of your work was a thing of the past. This situation is characteristic of almost every kind of technically organized work, from the manufacture of consumer goods to the construction of buildings, the running of public services, and more. If we do things that are useful to others, we feel needed, and there is a reward in this. Design exemplars for healthier work could reverse the current trends and restore a measure of dignity and pride in human work. The benefits to employers and employees would be incalculable, with a great many spillover effects into families and communities. People would once again come home from a day of useful, valuable work. Achieving this will contribute significantly to a certain resymbolization effort. However, if we continue to politicize the situation, we will get absolutely nowhere.

Our Dependence on Matter

As noted, no economy can create or destroy the matter and energy which it requires. It temporarily borrows them from the biosphere and returns them in ways that cannot be reabsorbed into natural cycles, while low-temperature heat must be radiated out into space. As a result, the network of flows of matter and the network of flows of energy of individual societies are integrated into the corresponding universal networks of an emerging global civilization. In turn, these universal networks are embedded into those of the biosphere.

An understanding of this interdependence permits us to derive three fundamental principles for developing superior design exemplars for the way we utilize matter and energy. For matter, these can be deduced from the following three questions: how do matter in general, and materials in particular, flow through the networks of flows of matter? What kinds of materials flow through these networks? How can these networks be redesigned to improve the productivity of matter in delivering the goods and services needed and desired by societies?

We have noted that almost all the materials we use flow through the networks of flows of matter in a linear pattern corresponding to a single use. Materials in the biosphere, however, flow through the corresponding network in cyclical patterns because the waste products of one transformation become the resources for another. For example, the waste gases we exhale contain carbon dioxide, which plants, by means of photosynthesis and water, convert back into oxygen and plant sugars. Initially, industrial ecology took this simple observation as the basis for proposing equivalent technical systems that would, as much as possible, transform wastes into resources.[99] This idea quickly became the victim of the cult of economic growth by failing to recognize that only by means of an all-out preventively oriented strategy is there any chance of making it somewhat feasible and manageable. As a result, industrial ecology gradually became reoriented towards end-of-pipe solutions for turning wastes into resources.

Concerning the question as to what kinds of materials flow through the network of flows of matter, quantitative estimates vary, but generally range around one hundred thousand kinds for a modern economy. For the overwhelming majority of these materials, we have little understanding of their effects on everything directly and indirectly related to human health and ecosystem viability. As a result, systematic design strategies of substituting less harmful materials for more harmful ones wherever technical and economic constraints permitted this are not yet feasible, but this will have to change.

Concerning the productivity of the matter flowing through the network of flows of matter (including its composites), we have noted that it has been estimated to be as low as 7 per cent, and this is for the delivery of mostly single uses. This is embarrassingly low; thus it should not be difficult to create a superior design exemplar for interrelating the network of flows of matter and its composites to the corresponding network in the biosphere.

The creation of a much more preventively oriented design exemplar for the way everything is related everything else via the use of materials can have many benefits. It involves the transformation of linear throughput patterns into open or closed loops in order to derive multiple services from a single resource extraction. In this way, goods and services can comprehensively be dematerialized, thereby reducing future resource crises and the risks of conflicts these would certainly trigger. It is therefore also a question of peace and security. Such a new design exemplar could substantially reduce the burden imposed on the biosphere, because, generally speaking, the most damaging part of our materials use is related to the extraction, processing, and refining of the resources. These could now become a kind of environmental "overhead" for as many uses as are technically and economically feasible. Doing so will also substantially diminish the rate at which wastes are returned to the biosphere, for the following reasons. Preventive approaches can improve the ratio of desired to undesired outputs of a great many processes, the wastes per use will decline, and the recovery of value from end-of-life products whenever this is technically and economically feasible can further reduce this waste stream. It will also be possible to more closely align the interests of producers and consumers by creating a service economy based on selling services, as opposed to selling the goods which deliver these services, wherever this is feasible. Finally, the required energy inputs can be substantially reduced, since the reprocessing of materials and the re-engineering of components require a fraction of the energy needed to make them from virgin resources. We will see that, by reintegrating the use of materials more effectively into ways of life to deliver as many services as possible from a single extraction of resources and by exploiting the full potential of preventive approaches while realigning the interests of producers and consumers as much as possible, the network of flows of matter and its composites associated with our contemporary ways of life will be substantially modified, as will be its connections to the biosphere. It is a matter of relating everything to everything else in a way that makes sense, serves human needs and wants, and can be sustained by the biosphere. It is not nearly as tall an order as it may appear, provided that we are willing to challenge our current commitments and prejudgments.

Achieving a more preventively oriented design exemplar for the use of matter is based on two hundred years of experience and carefully examining what works well and what does not. The three primary components are a modification to the business plans of corporations, the creation of a supporting level playing field by altering tax policies, and using state-of-the-art practices to transform so-called "free" trade. Beginning with the first, few if any business plans examine the technical and economic implications of relying as little as possible on virgin resources and as much as possible on value recovered from end-of-life products, or of doing the reverse. This observation is hardly surprising, given that engineering and business schools have ignored the experiments with preventive approaches and continue to rely almost exclusively on end-of-pipe "solutions" to labour and environmental problems. Our current recycling and recovery efforts are well intended and better than nothing, but they have absolutely nothing in common with a well-engineered and economically thought out design exemplar for the use of matter associated with the meeting of human needs and wants. The criteria for the design and production of materials, for example, would be completely altered. What, in the context of the current design exemplar, may be the best materials from an engineering and economic perspective may well turn out to be inferior in the context of the proposed design exemplar. It will become a trade-off between engineering characteristics and costs for a single use and those for multiple uses. A material whose properties can be restored or nearly restored following its recovery from an end-of-life product may well be superior overall, even though it is inferior to others in terms of a single use. In the latter situation, the costs of extraction, processing, and refining virgin resources, including their unpriced burdens, become a kind of environmental overhead for multiple uses. The new design exemplar may fundamentally alter what are deemed to be the most efficient or most economic alternatives in materials engineering. The same considerations must be given to the design and manufacture of each part, subsystem, process, and overall system. Based on the limited evidence where one or more details of this kind of proposal have been attempted (as in the remanufacturing of components), demonstrable economic advantages frequently occurred along with major reductions in environmental burdens. In other words, by developing a new design exemplar for the use of matter, it is possible that some aspects may become less efficient and economic than they currently are; but in the context of how everything is related to everything else by the use of matter, there is little doubt that all parties will be better served from an economic, social, and environmental perspective.

There is another way in which the kinds of materials borrowed from the biosphere are likely to change by means of this new kind of design exemplar for the use of matter. There will probably be a steady drop in materials that have a low compatibility with human life and the biosphere as a result of attempts to systematically substitute others that are more context-compatible. Such substitutions can reduce the need for compensatory end-of-pipe disease care and social support.

A third fundamental change that may take place as our use of materials is being re-engineered by taking into account a much broader context is the creation of a new structure of the network of flows of matter and everything built up with it. This restructuring will happen from the bottom up by means of the systematic application of preventive approaches – not by means of a top-down approach. We would expect a gradual but complete overhaul of the structure of the network of flows of matter, for three reasons. The first is that corporations will have to make a careful technical and economic assessment as to whether to continue to primarily rely on virgin resources, or whether to transform their practices by the comprehensive application of preventive approaches to recover value from end-of-life products and wastes as a source of materials and remanufactured components and systems. The second is a change in the tax system that balances the productivity of matter and energy with that of labour. The third reason is that together these changes will generate competitive pressures of an entirely different kind, now deliberately designed to internalize as many costs as possible in a manner that is beneficial to all parties. This will also help to create a new level playing field that will bring into action the potential of preventive approaches capable of cost-effectively improving labour, social, and environmental standards by moving away from end-of-pipe approaches. It amounts to a bottom-up approach to holding the World Trade Organization accountable to people instead of to technique.

As part of their new business plans, corporations can exercise the following hierarchy of strategies for improving the productivity of materials. As noted, pollutants represent products of the system, like all the others we utilize, with the only differences being that the former cannot be sold and that their disposal will become increasingly expensive.

The first, and likely the most beneficial, strategy in the hierarchy is to sell the services a product renders as opposed to the product itself. Consider the competitive advantages that can be derived by producing a degreasing agent, leasing it, taking it back, and restoring its original properties for a tiny fraction of the costs of producing it, leasing it again, and so on up to a total of ten uses. Such an approach realigns the interests of producers and consumers because competition is now based on providing services with the least throughput of matter and energy, which is in everyone's economic, social, and environmental interests. This design exemplar for degreasing agents has been tried out, with great success. This option excludes products whose use is dissipative, products whose irresponsible use is difficult to avoid, products whose monitored use may be a serious invasion of privacy, and so on. There is a non-trivial core of strategies where this may work extremely well, especially for products with a few subcomponents that tend to quickly outdate. For example, buying computing services instead of computers would allow manufacturers to make designs more modular in order to facilitate upgrades, while components that are less rapidly outmoded or whose performance improvements are marginally relevant to the functioning of the overall system may be retained for much longer periods. As noted, it is but a small step to move from the leasing of cars to the purchase of transportation services with varying levels of fuel efficiency, comfort, safety, and luxury. Nevertheless, from the perspective of the network of flows of matter, the potential benefits of an economic, social, and environmental nature are considerable.

A second strategy, with somewhat lesser expected overall benefits, involves the manufacturer having to relinquish control over the use-phase of a product. Doing so must not be compared to so-called take-back approaches. Although these have significant potential advantages, the full potential can only be realized if everything in the first strategy is carried out, with the exception of the product being sold with a prearranged take-back agreement. Such current agreements based on discipline-based approaches have a very mediocre performance record.[100] To use this as evidence to demonstrate that what is being proposed here will not bring the suggested benefits is to compare apples with oranges. No current take-back system comes anywhere close to what is being proposed here.

A third strategy, with still lower expected benefits, may have to be considered if the above two strategies are not feasible, as is the case where collection costs may be too great. This may be partly due to an excessive concentration of global production. Consequently, in such cases the only phase of a product's cycle under the control of a producer is its design and manufacture. A good example is Design For Environment (DFE), which is rarely taught in engineering faculties, who continue to concentrate on completely ineffective Life Cycle Analysis (LCA).[101] The latter has not been able to decisively resolve the simplest of issues, such as the use of washable ceramic cups in cafeterias versus disposable styrofoam or paper cups, or the use of washable cloth diapers versus disposable ones. The number of assumptions demanded by a comprehensive LCA is such that no firm conclusions have ever been reached, and yet we continue to use input-output LCA to generate all kinds of meaningless numbers that serve little purpose other than the lobbying of government, the moulding of public opinion by public relations, and so on. It is simply impossible to use the input-output models of a national economy for the purpose of LCA because of the highly integrated global materials and products system. Doing so will create crippling externalities to the model that make the conclusions even more uncertain than those reached by the conventional LCA approaches. In contrast, the DFE approaches pioneered by T.E. Graedel and B.R. Allenby[102] map portions of the network of flows of matter and of the network of flows of energy relevant to a particular product's cycle to identify potential environmental "hot spots," in order to focus preventively oriented design efforts in a cost-effective manner. Intended as a scientific approach, however, LCA is incapable of functioning as an effective engineering approach.

With the above three strategies, the potential of preventive approaches will have been exhausted. Nevertheless, four additional strategies can effectively deal with problems that cannot be prevented. In decreasing order of expected benefits, these include the reprocessing of any scraps of material occurring during production, turning pollutants into resources by creating real or virtual industrial eco-parks, preventing the mixing of any waste streams that can be "mined" for valuable resources, and preparing whatever wastes remain for safe disposal in landfills, where they must deteriorate sufficiently before the liners of these landfills begin to fail.[103]

Elsewhere,[104] we have shown that the above hierarchy of strategies should be complemented by other categories of preventive approaches aimed at energy use, the organization of work, and the development of the urban habitat, because there will be additional synergistic interactions that reinforce one another.

These developments can be encouraged and reinforced by the gradual creation of a fairer tax system to rebalance the productivities of matter and energy with those of labour, in order to slow the overconsumption of nature in favour of creating more work. Our current taxation systems encourage entrepreneurs and corporations to do the exact opposite by taxing labour and dealing with matter and energy as if these were abundantly available and thus without a need to be priced. It is surprising how vested interests and market ideologies have managed to distort all attempts at redressing this situation. When so-called green taxes were an issue in a Canadian federal election, scare tactics of all kinds were used, and the cult of growth triumphed easily. It was embarrassing how poorly the proponents of a green tax were able to defend their proposals. It will probably be a long time before a political party dares to incorporate this issue again into its party platform. This is another example of the politicization of a serious issue leading nowhere.[105] The necessary scientific, technical, and economic reforms are structural in character and thus cannot be politicized in a mass society dominated by mass media, where public opinion renders democracy impotent in the face of these kinds of issues.

The proposed tax reforms are simply a question of giving the correct signals to entrepreneurs, corporations, universities, and governments. The present end-of-pipe approaches, including cap and trade market approaches to pollution, cannot accomplish anything close to what we are proposing. Shifting the tax system to rebalance the productivity of matter and energy with that of labour will have vast implications of all kinds. It must therefore be carried out slowly, iteratively, and preventively but with a clear goal in mind. For example, taxes based on the consumption of matter and energy raise issues of social justice, since the poor would be paying a disproportionate tax burden compared to the wealthy. There are also practical difficulties in making the system manageable because the tracking of each and every flow of matter or energy is clearly not feasible. A great deal of practical experience has shown that a very good indicator of environmental damage resulting from overconsuming the biosphere is the Gross Energy Requirement (GER), which may well make it possible to use indicators of this kind for taxation purposes. Taxes could then be levied on exchanges of materials, parts, subassemblies, products, fuels, and so on. If we are really serious about this issue, the practical difficulties can surely be overcome, but everything must be done in such a way that the equivalent of negative feedback loops will provide adequate opportunities to make adjustments as the new tax system emerges. It should also be noted that this shift has generally been regarded by its proponents as revenue neutral. Labour taxes will be reduced to the extent that taxes on material and energy flows begin to generate equivalent revenues. The bookkeeping of flows of matter and energy will have to become mainstream, and here the resources of public universities oriented towards preventive approaches will be an essential resource.

This brief summary of what has been more extensively dealt with in the cited earlier works shows that it is possible to arrive at a design exemplar for the use of matter to satisfy human needs and wants in a manner that takes into account how everything is related to everything else. The design of each material, process, and technical system will embody something of this interrelatedness, and this will inevitably reduce the negative consequences of commoditization and reification. The issues we currently face are thus addressed at their very roots, with the result that the technical externalities in all technical planning necessitated by the use of discipline-based approaches can be reduced. Doing so will diminish the externalities created in markets for the goods and services produced by means of these kinds of design exemplars. This in turn will contribute to lowering the magnitude of the negative Market forces or the equivalent selective pressures on human history resulting from the use of technique. As noted, a growing use of preventively oriented design exemplars can serve as a practical guide to bringing the World Trade Organization into some kind of human control by restricting the roles of markets and dominating technical planning in order to ensure that their negative effects will not undermine their positive roles. It is but a small step in eroding the cult of the fact, the cult of efficiency, and the cult of economic growth. By means of these design exemplars, however, it will be possible to reimpose some human values and thus to diminish our enslavement.

Our Dependence on Energy

Although most so-called industrially advanced nations have no clear materials strategy, they generally do have an energy strategy. Unfortunately, with a few exceptions these are an unmitigated disaster compared to what is possible with a preventively oriented strategy. Since not a single human activity can either create or destroy the energy on which it depends, getting a strategy wrong affects almost everything in a society. Developing such strategies by means of discipline-based approaches has led us inevitably into a trap. Achieving an energy strategy that is as compatible as possible with human life, society, and the biosphere can change a great many things for the better. In fact, for decades I have argued that the best economic strategy we could adopt is a preventively oriented energy strategy, because, from an economic perspective alone, it would outperform most current economic policies. In addition, it would have substantial social and environmental advantages.[106] It could make a significant contribution in turning our anti-economies into genuine economies.

Our current energy strategies have had major negative impacts on the availability of investment capital because they are needlessly capital-intensive and often account for a significant portion of the debt of a community. For example, in my jurisdiction, we will still be paying down the debt created by nuclear energy long after many of its components have reached the end of their operations, and we have not even begun to pay for long-term storage of radioactive waste. Worse, no matter how expensive the solutions to waste storage may turn out to be, no safety can be assured for the hundreds of thousands of years they will have to last. If nuclear plants had been obliged to carry adequate liability insurance, they would never have been put into operation. Massive subsidies into photovoltaic systems were made when, in some countries, the systems were almost already competitive without subsidies. Contracts with the operators of wind farms have led to enormous controversy because of the health effects of such farms and their proponents' heavy-handed tactics, arguably similar to those of the tobacco industry. There is now a virtual paralysis on this front. On the demand side, most government programs for the improvement of energy end-use efficiency have not had the success to warrant the investments.

All this massive investment has significantly contributed to unemployment and underemployment, since these energy strategies created relatively few jobs compared to some alternatives.

These strategies also contributed to a needlessly rapid depletion of fossil fuels. Since parallel distribution systems are technically and economically not feasible, gas and electric utilities were generally given a monopoly in their jurisdictions. Doing so made a great deal of sense except that selling energy as a commodity instead of a service opposed the interests of utilities to those of their customers by resulting in very low energy end-use efficiencies. In some jurisdictions where utilities were mandated to act as energy service providers, one kilowatt-hour could often be saved at half the cost of producing it, with the result that energy end-use efficiency improvements became the best way of negatively generating power. In other words, had energy systems paid as much technical and economic attention to the production and distribution of energy as to its use, enormous energy savings could have been realized. The result has been a needlessly rapid drawing down of fossil fuel reserves as well as avoidably high environmental burdens, leading to global warming and other issues.

Energy markets have been distorted for decades as a result of the enormous subsidies provided by governments to force utilities into adopting nuclear power, to encourage oil companies to develop exploration and infrastructure, to provide military protection for shipping oil out of the Middle East (which could have doubled the cost of a barrel of oil), and to construct the interstate highway system in the US (which favoured automobile use and urban sprawl). Agricultural policies that substituted energy for land, inappropriate building codes, and other initiatives resulted in energy prices that did not reflect the real costs, especially in North America. This led to an energy addiction, with the result that when there was still significant manufacturing in North America, the production of a unit of GDP required twice as much energy as in Western Europe or Japan.

This energy addiction spread to the south. The oil shocks plunged many nations into severe debt to pay for oil imports, thus diverting financial resources away from essentials and, in many cases, leading to the imposition of draconian measures by the International Monetary Fund. This action almost without exception hurt the poorest and most vulnerable portions of their populations.

The lack of correlation between energy prices and costs also permitted the concentration of global production in a few countries. The scale of this concentration would have been far less if energy prices had been higher because that would have undermined the advantages gained by corporations from an almost unlimited externalization of all costs.

Finally, these energy strategies had considerable implications for peace and security. The use of nuclear power plants for the production of nuclear weapons is well known, and this is certainly not the only concern. Many of the above consequences directly and indirectly contribute to economic, social, political, and environmental instabilities, with far-reaching consequences for peace and security.

In sum, there is very little about our contemporary world that has not been directly or indirectly affected by the disastrous energy strategies of the last half-century. All this is well known, but since we are locked into discipline-based approaches, little change has been forthcoming, and even then the same kinds of patterns reoccurred. This is the bad news – which holds the key to the good news.

Turning the situation around begins by fully recognizing how the use of energy directly affects the way everything is related everything else in human life, society, and the biosphere. Since no human activity can create or destroy the energy on which it depends, all the activities of the way of life of a society are interrelated by means of a network of flows of energy. These networks, associated with the ways of life of the so-called industrially advanced nations, are suspended in the larger network corresponding to these flows in the biosphere. Once again, we can ask three questions about these networks in order to arrive at some basic principles for redesigning our energy strategies: How does energy flow through these networks? What kinds of energy flow through these networks? How can the networks be reorganized to make them more compatible with human life, society, and the biosphere? We need to improve the way everything is related to everything else through the use of energy.

Concerning the first question, the second law of thermodynamics implies that linear chains of successive energy transformations cannot be reorganized as closed loops. Hence, the best we can accomplish is to extract as many energy-based services as possible from these linear chains until the remaining low-temperature heat is returned to the biosphere and radiated out into space.

The kinds of energy flowing through these networks ought to be derived from renewable sources to the greatest extent possible. Until now, the availability of renewable energy has not yet been affected by human usage, but when it does it will have to be priced unless we wish to repeat the same kinds of mistakes we have made with regard to the extraction of material resources. It is also important not to confuse energy sources with energy carriers. There was a lot of silly talk regarding hydrogen economies, as though hydrogen could be extracted from the biosphere the way fossil fuels are. If we are not to contribute to the latter's depletion, the only sustainable source of hydrogen comes from the electrolysis of water, but this requires a long chain of transformations. Even if each one could be brought up to an efficiency of at least 80 per cent, the energy payback from primary sources powering this chain of energy transformations would be tiny (a great deal less than that from electricity).[107] Hence, the widespread use of hydrogen would translate into a massive increase in the use of primary fuels. There are also many mistaken expectations from a greater use of electricity, especially in transportation. It is as essential to undertake energy accounting as it is to do financial accounting in these matters. What concerns us is the net energy available from the network of flows of energy as derived from the inputs from primary sources, and not its gross energy production. If there are two alternative ways of powering energy to benefit human life and society, a comparison ought to be attempted of the total investment in primary fuels to deliver the service. We have not yet begun to do net energy accounting. It is the energy parallel of our misuse of the GDP.

As far as the third question regarding the networks of flows of energy is concerned, it has been noted that possibly the best economic strategy under our current conditions is a preventively oriented energy strategy. The reasons are rooted in the massive and widespread effects any energy strategy will have on almost every economic activity as well as its negative effects on society and the biosphere. Since our current strategies have very poor ratios of desired to undesired effects, their redesign by means of preventive approaches can improve the net energy available from these networks of flows of energy. This would make economies more competitive, largely by greatly diminishing undesired social and environmental effects. No current energy strategy of which I am aware even comes close to having a preventive orientation. Increasing the energy derived from renewable resources is but a tiny portion of such a strategy. Efforts on the supply side must be matched by means of a comprehensive approach on the demand side. José Goldemberg and others[108] have shown how this can be accomplished in a systematic fashion. This approach begins by tabulating all the services required to sustain a way of life which involve the transformation of energy and grouping them into convenient categories, such as domestic, commercial, industrial, agricultural, resource extraction, and transportation. In this way, the equivalent of a map is produced showing how everything is related to everything else via the network of flows of energy associated with a way of life. This approach presents the results in tabular form, beginning with a first column listing the energy-based services required to sustain a way of life. The second column tabulates the required technologies to deliver these services. It is subdivided into three columns listing the energy technologies in predominant use, the most efficient technologies that are (under the current circumstances) still economic, and state-of-the art technologies that are not yet economic. The third column tabulates the energy inputs required by these technologies. The fourth column is like the third except that any energy carriers, such as electricity and hydrogen, are replaced by their primary fuel requirements. In the fifth column, this information is organized to show the total energy inputs into the network of flows of energy as derived from the different primary sources.

The energy inputs from primary sources should be compared in terms of the effects they have on the net energy availability derived from the network of energy flows. For example, at one time there was a great deal of controversy over a claim that the building, running, and decommissioning of nuclear power plants required an investment of primary fuels that was nearly equal to the energy these plants delivered during their time of operation, which would mean that building nuclear plants would not have increased our net energy availability. All this controversy did not lead to the obvious conclusion: it is important to systematically check the net energy availability derived from all primary sources. In the case of renewable sources, the energy invested in their production, operation, and decommissioning must be deducted from the energy generated during operation.

Once such a map of how a way of life depends on various primary energy inputs has been constructed, different policy scenarios can be systematically investigated. Some issues are immediately obvious. For example, in a great many industrially advanced societies, the energy leaking out of its building stock both in winter and summer is vast but can easily be reduced by improving building skins. Goldemberg et al.[109] have made many suggestions, such as comparing a technology's thermodynamic first law efficiency with its maximal attainable second law efficiency, in order to assess the potential for improvement and explore alternative processes based on different energy transformations. We have barely begun to scratch the surface of what is possible in creating superior design exemplars for the use of energy by exploiting the potential of preventive approaches. Such design exemplars can be optimized by means of discipline-based approaches, although they cannot be derived from them. They can substantially improve the net energy availability while lowering the negative impact on human lives, society, and the biosphere.

Once again, these preventively oriented energy strategies have many potential synergies with other such strategies. For example, the above hierarchy of seven strategies by which corporations can improve the productivity of the use of their material resources and reduce their environmental burdens has considerable implications for the network of flows of energy. Similarly, the design of more liveable sustainable urban habitats appears to go hand in hand with substantial reductions in their energy requirements. Finally, the redesign of work organizations to tap the full potential of preventive approaches also has significant energy implications. Jointly, they can considerably increase the productivity of our energy resources by making the network of energy flows more compatible with the economy, society, and the biosphere. The concerns for efficiency are now embedded into a design exemplar, which ensures that any gains are not achieved at the expense of the broader context.

This design exemplar stands in sharp contrast with the policy of deregulation. As noted, regulated public utilities were created to ensure that there would always be adequate supplies for customers, since they had no other supplier to turn to. Deregulation amounts to little more than the following: turning over the profitable portions of the networks of flows of energy to private enterprise, while the non-profitable portions (including nuclear plants) remain a public responsibility; ensuring adequate supplies; and opening up newly created energy markets to speculation. Deregulation simply did not work. In most, if not all, jurisdictions, governments had to step in and stabilize energy prices. Before this occurred, the Enrons of this world all but destroyed the gains that had been made with integrated resource planning and turning regulated utilities into energy service providers. There is a fundamental incompatibility between a network of energy flows regarded as a commons and as a sector that can be exploited for private gain. For example, a company that is running three power plants will have every incentive to find a way in which, during peak hours, it can significantly reduce the output of one of them in order to provoke a spike in energy prices while reducing its fuel costs in that plant. In this way, operating only two plants would make it possible to make a greater return on investments than operating three. We have also witnessed the endless speculation engaged in by energy companies who own no power plant, no distribution grid, and no fuel resources, and thus make their returns purely from speculation, to the detriment of everyone but themselves. The deregulation of the network of flows of energy has had the same effects as it had on all commons.

Our Dependence on Food

In many jurisdictions, the portion of networks of flows of energy associated with agribusiness will have to be examined carefully if they are not to undermine the above kinds of efforts. Agricultural reforms will have to be undertaken in parallel with significant changes to energy policy. The current agribusiness system was first developed in a North American context, where family farms had plenty of land (the life-sustaining capabilities of local ecosystems) but a shortage of farm labour. These conditions were the exact opposite of what is encountered today, which has resulted in the overexploitation and the underutilization of people. Initially, mechanization was supposed to solve the labour shortage problem, but it transformed highly diversified and comprehensively efficient farms into increasingly highly specialized, capital-intensive, and vulnerable businesses.[110] The dilemma may be simply put as follows. It was financially impossible for family farms to purchase equipment to facilitate the growing of a diversity of crops on a small scale. The overhead of this ever more efficient and larger equipment was cost-effective only if ever more land was devoted to the corresponding crops. It pushed all family farms towards monoculture-like operations. As these monoculture practices were imposed on local ecosystems, the food supply of some insects and animals became very large while for all others it was greatly diminished, and this produced imbalances in their populations. Natural population control mechanisms were either greatly weakened or destroyed, which necessitated that farmers create their own by means of pesticides and herbicides. As predator populations evolved and adapted to the new conditions, ever larger quantities of pest controls were required. It was not long before the crops themselves became affected. This eventually led to the necessity of genetic engineering, with its corresponding risks. Much of our global food supply has been reduced to a handful of species, often genetically engineered and thus without any track record of long-term viability. We are thus gambling with food security. Moreover, monoculture mines the soils for the same set of nutrients, which is only partly offset by adding fertilizers. Consequently, soil cycles are disturbed, thus causing them to be impoverished, and this is frequently accompanied by soil erosion. In addition, the global food supply has been greatly impoverished in terms of certain nutrients as mining of the soils continues. Any increase in the quantity of food produced is thus undermined by a decrease in its nutritional quality. Moreover, there is growing evidence that the latest herbicides and pesticides are negatively affecting the bacteria in our digestive systems.

The mechanization of the farm required ever larger inputs of fertilizers, pesticides, herbicides, genetically re-engineered species, and terminator seeds (to protect manufacturers' investments), as well as fuel and equipment. The result was that in essence energy was substituted for land in a manner that is entirely unsustainable in the long term. The ever-growing capital intensity of these monoculture operations led to an enormous concentration of land ownership, as small-scale operations could not survive. It compelled many societies to depopulate their rural areas and absorb the displaced people in their urban centres. The capital required to create one agricultural job soon began to exceed that of many industries. All this required techniques of all kinds, thus creating what we refer to as agribusiness. Efficiency had to be imposed everywhere.[111] From a historical perspective, it launched one of the largest unregulated experiments on the food supply of humanity. Seed companies took control of much of the food supply, since only genetically modified organisms (GMOs) can be efficient in this newly created hostile environment. It is another example of the fact that technique expands by means of positive feedback, because an intervention by one group of techniques creates a variety of context incompatibilities. Rather than addressing these at their roots, we use end-of-pipe approaches, resulting in the addition of a layer of techniques which in turn cause further incompatibilities, and so on. In many nations, the situation is so dire that citizens are not permitted to know whether they are eating genetically modified organisms or not. The defenders of agribusiness argue that without it we will not be able to feed a growing global population, but others have shown that the root problems are the enormous waste of food and the decline of nutritional value increasingly coupled to threats to human health. In any case, the current agribusiness system is "mining" soils and exposing humanity to enormous risks if the new genetically modified crops turn out to be vulnerable as a result of their polluting the DNA pool. Agribusiness is contributing significantly to fossil fuel depletion, which is increasing urbanization, making cities less liveable and sustainable, adding to unemployment and underemployment, and placing much else on the altar of technical efficiency. New preventively oriented design exemplars for energy use will have to restructure agribusiness so as to produce healthy food, and at the same time undo or greatly reduce the many negative consequences imposed by this sector of technique and embodied in the new McTimLobWallMon seeds of this world.

In the network of energy flows corresponding to agribusiness, technique has succeeded brilliantly in improving the performance of everything by disordering everything it touches. Economically, it represents a system of non-sense because the long-term costs and risks far outweigh the gains. What will we do if, a few years from now, it turns out that the polluting of the DNA pool threatens the globally over-represented efficient crops? What will happen of the health effects of GMOs are not only confirmed but found to grow over time? How long can we export nutrients from the south to the north? What will we do with all the people being displaced by this system who have lost their livelihood? What will we do with the corresponding acceleration of the overemployment of nature and the underemployment of people? Traditional agricultural approaches worked with nature, taking advantage of the local context, while the discipline-based approaches of agribusiness work against nature. Agribusiness is all about efficiency and not nutrition: everything must be processed to achieve a long shelf life for global distribution, have the best possible appearance for packaging and advertising, and aid in creating a kind of addiction to foods that artificially stimulate taste and smell but compensate for the loss of nutrients in ways that are frequently difficult or impossible for the human body to absorb. Traditional approaches selected varieties of plants that were ideally suited to local soils and climate conditions. In the course of generations these were then further adapted, with the result that plants thrived, and the best possible crops were produced with the least burden on ecosystems. Now that we know a great deal more about the interconnectedness of agribusiness with everything else, it will not be difficult to come up with design exemplars that work for people and with nature, including energy usage. The World Trade Organization must be held responsible for the effects it is having on the poorest farmers and our long-term food security. The productivities of matter and energy will have to be balanced with that of labour in agribusiness, as everywhere else.

Our Dependence on an Urban Habitat

The urbanization of a growing portion of humanity began with industrialization. With the continuing depopulation of the countryside by agribusiness and a growing global population, it will almost certainly continue in the near future. Since this urban habitat can neither create nor destroy the matter and energy on which it depends, it must temporarily borrow them from the biosphere, with the result that the preventively oriented design exemplars discussed earlier must be applied here also. In addition, these urban habitats must be made a great deal more liveable than they are now. Fortunately, the needs of reducing the material and energy intensity of these urban habitats and improving their liveability converge extremely well. However, rarely are these two considerations integrated into a comprehensive overall urban design exemplar. Initiatives to make the urban habitat more liveable, such as the new urbanist movement, urban villages, transit-oriented design, pedestrian pockets, and smart growth, also need to carefully consider their material and energy intensities. This can usually be accomplished without much difficulty. Elsewhere,[112] we have shown that the pioneering work of Jane Jacobs (which led to design principles that underlie most of the above initiatives) implies a synergy between knowing and doing embedded in experience and culture and knowing and doing separated from experience and culture. Making this explicit constitutes a blueprint for advancing these kinds of approaches. Since some of this was already discussed in chapter 2, I will take the liberty of referring the reader to it, including the referenced earlier works.

There is little doubt that, with more than half of humanity now living in urban habitats, unless we make these habitats more liveable and sustainable, all the previously mentioned initiatives will not bring the critical mass of change that is required for our future life on this planet.

Developing the Design Exemplars

Once new design exemplars have been developed for our dependence on knowledge and skills, work, matter, energy, food, and an urban habitat, they will need to be worked out by means of a detailed analysis, using specialized knowledge of the kind that will emerge from university reforms. This knowledge will also have to contribute to making every aspect as optimal as it can be within the context of the whole. At first, discipline-based approaches will continue to play a role, but as university reforms advance, the new scientific and technical specialties will permit context-appropriate efficiency.

It is these kinds of strategies, based on carefully analysed and optimized design exemplars, which jointly can deliver far more than current economic policies. All this is very different from our attempts at balancing budgets, creating jobs, growing the economy through free trade, and paying down the debt. What we have tried to demonstrate is that such an economic agenda cannot deliver what it promises. Anti-economies, when stimulated by economic measures, will produce more of the same effects, and all the political illusions in the world will not change this. It is simply impossible to behave as if economic phenomena are inherently capable of delivering the kind of future towards which we aspire. The implicit devaluation of the contributions made by many other categories of phenomena and the way all these participate in the way everything is related to everything else will continue to strengthen our anti-economies, with everything this involves. It is the summit of economic non-sense. Our creation of a discipline-based science and technique is disordering human life and the world, making it impossible to create wealth and ensure a liveable and sustainable future. Only if we go to the root of the problem will we be able to have budget surpluses, create real jobs, turn growth into development, convert forced trade into free trade with its mutual advantages, and end our borrowing from future generations. As has always been the case in human history, one day people will look back at our era and wonder how it was possible that almost everyone believed in our secular economic theology. The answer is the same as always: our creations have taken hold of our minds and cultures. We will turn to this enslavement in the next chapters.

4 The Cult of Disembodied Communal Life: The Anti-Society

Divided Human Life

This and the following chapter will deal with "technology changing people" as well as the later "technique changing people," as designating the influences of the previously described developments during the first and second phases of industrialization on communities and their members. Briefly put: ways of life that strive for efficiency give rise to forms of human life that are very different from those guided by symbolic cultures. In the first chapter, we attempted to show that daily-life knowing is the opposite of discipline-based knowing. The former is related to individual collective human life in the world, while the latter is focused on the domains of disciplines that have been separated from that life and the world as a consequence of a process of abstraction. Daily-life knowing aims to understand everything in relation to everything else, but this relative character is made absolute before adulthood by the onset of myths. Scientific knowing implies the transcendence of these limitations by means of an intellectual division of labour that assigns one category of phenomena to each corresponding discipline, transposing that category of phenomena into a domain organized as what we have referred to as reality. It thus separates that category from all others assigned to their own disciplinary domains, mathematically models everything in that domain as much as possible, and gathers as many "facts" as the models permit. The absolute character of this scientific knowing is based on proceeding as if the unknown will simply deliver refinements and extensions of what is scientifically known, which, once again, can only be assured through myth. Discipline-based science as a whole proceeds as if human life and the world are built up from distinct domains each dominated by a single category of phenomena.

However, the discoveries of the last two hundred years, particularly with regard to our growing use of technology and technique, appear to imply that domain-based structures are rarely encountered in any living entities (natural or cultural), with the result that discipline-based science appears to treat life as if it were organized in terms of what we have referred to as reality. In this way, a knowledge of the living is represented in terms of what is non-living, including technical processes as information machines. There have been a variety of attempts to loosen the restrictions imposed by our current scientific division of labour by the creation of hybrid disciplines (such as biochemistry or social psychology), systems thinking, and interdisciplinary programs (such as cultural studies, urban studies, and criminology); but these and other efforts do not appear to be able to shed the fundamental limitations of discipline-based approaches and the intellectual division of labour based on them. As such, scientific knowing is the exact opposite of daily-life knowing, in the sense that the former all but ignores how everything is related to, and evolves in relation to, everything else, while the latter accounts for it as much as possible. Each approach to knowing could be better protected from its own inherent limitations by regarding the other approach as its complement, but this possibility is currently ruled out by the myths of our civilization. We have unleashed an unprecedented destruction of our life as a consequence of treating it as non-life as well as having created an exponential growth of knowledge about non-life. In sum, relative to daily-life knowing, scientific knowing is an anti-knowing, while in relation to scientific knowing, daily-life knowing is an anti-knowing, in the sense of each one being the opposite of the other in almost every respect.

In the same vein, chapter 2 attempted to show that daily-life doing rooted in symbolization, experience, and culture constitutes an anti-doing relative to its discipline-based counterpart rooted in technique, while technique constitutes an anti-doing relative to its cultural counterpart. Daily-life doing is guided by the context in which it occurs and is understood by means of symbolizing how everything relates to everything else. In contrast, the technical approach to doing severs all connections with human life and the world by means of a triple abstraction, leaving a single category of phenomena occupying the domain of a technical discipline. It thereby imposes the goal of improving the desired output obtained from the required inputs by means of an endlessly repeated transformation accomplished by a particular member of its category of phenomena. Consequently, technical doing can have but one goal, that of performance and efficiency, which measure something on its own terms separated from everything else. It introduces a reorganization of life based on the principle of repetition, as opposed to ongoing adaption and evolution. In contrast, daily-life doing seeks to improve the contribution something makes to how everything relates to, and evolves in relation to, everything else, in accordance with necessity or human desires expressed by means of cultural values. This contribution is excluded from consideration in the pursuit of power and efficiency. The result is that technical doing dominates over any context and, paradoxically, undermines power and efficiency by disordering this very context, on which everything depends. Daily-life doing is absolutized by the myths of a culture, while discipline-based doing is absolutized by the facts emerging from a triple abstraction and the accompanying objectivity of performance measures. Daily-life doing contributes to the creation and evolution of a symbolic universe, while discipline-based doing transforms this universe into a reality divided into separate and distinct domains. These are open to measurement, quantification, and mathematical expression, thereby simplifying their complexity by excluding all dialectical and enfolded relationships characteristic of cultural and natural life respectively.

In chapter 3, these analyses were extended to understanding economic phenomena as intermingled with everything else in human life and society, or alternatively, as having a kind of a priori importance over all other categories of phenomena, which in effect encloses economic phenomena into a domain from which they rule over all other phenomena. In the former case, the economy is a dimension of the way a culture mediates between human life and an ultimately unknowable universe; in the latter, the economy becomes an enslaving domain in the reality of a divided human life and the world. Political economy once regarded economic phenomena as evolving within the culture, while economics as a discipline examines the economy as if it were inert to all non-economic phenomena or as if economic phenomena can be understood in their own right because all other phenomena remain essentially static. Economics thus examines the interplay of economic phenomena in a static world, fully conforming to what we have already discovered regarding discipline-based approaches to knowing and doing. The practical consequences are vast and deeply significant. Economies tend to be appropriate to a culture as long as they can enfold the technology-based connectedness of human life and society into their culture-based connectedness. In this way they can greatly constrain the roles assigned to markets, so that a culture and not these markets determines the value of everything. Economies become relatively distinct domains when this is no longer the case, at which point they produce growth by disordering human life, society, and the biosphere. Such growth is entirely different from development that seeks to serve the values of a community. In order to tolerate this rupture between growth and development, economies must be treated as good in themselves and societies must abandon their own cultural good. Economies organized as domains rapidly turn into anti-economies, which extract rather than produce wealth. Such economies are clearly inappropriate to a culture and equally inappropriate to the biosphere on which they ultimately depend. They are no longer an economic expression of how everything is related to everything else. Instead, they have become the negation of this interdependence, of which economic phenomena are no longer a particular expression.

In terms of what has happened during the last two centuries, economic phenomena ceased to be one manifestation of the cultural approach, but instead became the expression of an economic approach. It in turn quickly became one kind of technical approach as human life, society, and the biosphere became understood and dealt with as a reality of non-life. Politicians now began to consult economists as to what was "realistic" in the domain of economics to which everything must bow, just as the rulers of the past consulted the religious authorities. Before economies became distinct from societies, human beings were alienated by the myths of their culture and thus enslaved. When economies became treated as if they were built up with domains, this alienation was complemented by unprecedented levels of commoditization, reification, and externalization. Any concept of a public good has now all but disappeared, and political differences have been dwarfed by the necessity of making economies as efficient as possible lest we perish for lack of growth. In sum, economies largely guided by cultures are the opposite of anti-economies based on technique in almost every respect.

What will become of human life and society when knowing and anti-knowing, doing and anti-doing, and the daily-life economy and the anti-economy coexist in a way that disorders all life? For another perspective on this issue, consider the irrationality of rationality and what embodies either one. Rationality cannot respect the interrelatedness of everything because it involves imposing a goal, which then divides the interrelatedness according to what is important, marginally relevant, or irrelevant with respect to the goal, thereby distorting the interrelatedness. Technique is perfectly rational in that it pays careful attention to everything in order to improve efficiency while discounting or neglecting everything else. The irrationality of this rationality has become plainly evident in the disordering of everything living. Cultures are non-rational in the sense that they do not belong to the domain of rationality. Culture-based reasoning attempts to impose sense when dealing with the interrelatedness of everything by specific interventions responding to necessities or human aspirations. What makes sense in this way cannot be transformed into the non-sense of reality. The way cultures violate sense is rooted in the way the deepest metaconscious knowledge symbolizes the unknown as more of the known, thereby creating a symbolic universe of false meanings that distorts the interrelatedness of human life and the world. The "rationality" of sense expresses itself whenever people live lives in the world by means of a shared way of life and culture. The rationality of technique expresses itself whenever people impose the goal of efficiency as a criterion of non-sense, because everything is judged on its own terms as opposed to its meaning and value for human life. Cultures alienate human life, while technique reifies it. Cultures, however, respect the integrality of communities and ecosystems, while technique undermines them through commoditization and reification. Cultures evolve an uninterrelatedness based on false meanings, while technique moves forward by treating this interrelatedness as a reality of which every distinct element can be perfected on its own terms. From a cultural perspective, technique is an anti-culture of non-sense, while from the perspective of technique a culture is irrational. A complementary relationship can be established by accepting a culture as the sphere of sense and thus of non-rationality, and technique as the sphere of rationality and thus of non-sense, provided that technique and culture each remain within their limits. As a result, culture and technique have diametrically opposite strengths and weaknesses, but as a symbolic species we cannot live by technique. Of course, most of us will be much more comfortable with the kind of rationality we have grown up with and lived, because it is supported by living myths and not by the mythologies and ideologies of the past. Nevertheless, it is not our task to judge history but to live responsible lives in this world.

The Erosion of Social and Cultural Support

As noted, the concept of culture emerged precisely when its ability to sustain individual and collective human life was weakening. By evaluating everything in relation to everything else, a culture acted as a control structure that integrated all aspects of a society's way of life according to its myths, thereby sustaining the lives of its members. A parallel development occurred with the emergence of the concept of an environment, when the life-sustaining functions of the biosphere could no longer be taken for granted. Prior to industrialization, traditional societies sustained their members by cultures working in the background through their lives as symbolized by the organizations of their brain-minds. As a society assigns more and more of its organization of the activities of its way of life to discipline-based approaches, it enters into a gradual transition, creating what we commonly refer to as a mass society, whose members minimally depend on cultural support. In order to examine this in greater detail, it is essential to expand the model of culture (briefly summarized in the Introduction) to include what is commonly referred to as our psychological, social, and spiritual life.

If we place brain plasticity at the very centre of our understanding of what it is to be a symbolic species, the pioneering works of Freud and his followers (who founded psychoanalysis) must be reinterpreted in this context. Many levels of metaconscious knowledge mediate between the lives we live and our physical and biological selves. As a result, explanations of these lives and selves do not have to be exclusively related to our physical and biological selves. It is highly probable that a great many such explanations will need to refer to people's metaconscious organization of their brain-minds, which has resulted from differentiating and integrating the experiences of their lives. For example, a great deal less of our lives will need to be attributed to physical and sexual drives; wherever such explanations are appropriate, they will need to be interpreted in the context of how such drives are experienced and lived in the culture under study. With a great deal of hindsight, our current understanding of being a symbolic species makes some of the psychoanalytic explanations of human behaviour appear somewhat amusing. This interpretation converges with one of the earliest such interpretations, that of Karen Horney,[1] and later ones such as Robert J. Lifton's.[2] All this is easily understood when we recognize that, prior to the separation of the technology-based connectedness and the culture-based connectedness and the ordering of the former by the Market and the latter by what remained of culture, it was as if something of people's symbolic humanity was being spun off, causing a divided life that was reintegrated by the myths of capital, progress, work, and happiness. From this perspective, psychoanalysis was symptomatic of what was happening to human life in the course of industrialization. The latter opened a gap between cultures (working largely unnoticed in the background) and technique (taking over much of their former role). Hence, the explanations of psychoanalysis were believable because the role of culture had been greatly weakened.

Much the same kind of reinterpretation is required in relation to social biology, which became plausible as the process of desymbolizing cultures was highly advanced. It is also worth noting that the development of the so-called mechanistic world view in Western European cultures implied that much of life could be understood as non-life, with little or no need for symbolization and culture. This became particularly evident during the second generation of mechanistic world views, which replaced the classical machine with the information machine. We will return to all of this in the next chapter.

The metaconscious knowledge implied in the organization of the brain-mind roughly corresponds to what (in a non-symbolic and non-cultural perspective on human life) is commonly referred to as the unconscious or subconscious.[3] These terms will now be reserved for the innate organizations of the brain-minds of babies prior to their being modified by experience. Metaconscious knowledge may be associated with our personal lives, our being members of a society, and our collective ability to get on with our lives in an ultimately unknowable universe. This knowledge is the cultural equivalent of what is commonly referred to as the collective unconscious. It is fundamentally influenced by the life-milieu of the group or society in relation to which it exists because without it life would be unthinkable. As such it is the creator and sustainer of the group or society. It symbolizes vitality relative to everything that threatens it.

The summary of the model of culture presented in the Introduction, from the perspective of babies and children growing up in a community, could just as well have been presented from the perspective of separation from it, of which the final separation is death. We have noted that communication between human beings dialectically enfolds their lives into each other. A life lived with a spouse, children, relatives, and friends is enfolded into their lives and vice versa. When someone moves away and no further contact takes place, it is as if a portion of our life atrophies and dies. It is not simply a question of our memories of that person becoming less vivid, but of that part of our life no longer evolving and thus stagnating and eventually "dying." The same is true for people's active participation in adapting and evolving the way of life of their community. Some aspect may atrophy when alternative ways of approaching the relevant situations become part of a working culture, while others may be "born" as new situations arise and are dealt with in new and unique ways. When a significant other person dies, the portion of our life shared with this person essentially dies as well. It is for these and other reasons that depth psychology has paid a great deal of attention to "mental images" of vitality and life on the one hand and of separation and death on the other. The concept of a mental image appears to correspond to what, in this study, is referred to as metaconscious knowledge, except that the latter is not made up from images.

When the process of industrialization is interpreted from a cultural perspective, it becomes evident that it generated a great deal of metaconscious knowledge of separation and death, and created a new kind of vitality and life that eventually became dealt with as non-life. Prior to industrialization, a community interpreted the interrelatedness of everything, beginning with its local context, while industrialization increasingly imposed a universal perspective that was necessarily inadapted to the local context. For example, the growing dominance of the technology-based connectedness over the culture-based connectedness meant that the former was no longer primarily mediated by a culture. As a result, many of the necessities of life became directly expressible in terms of the laws of thermodynamics, thus eliminating the "softer" cultural mediation. The need to organize this technology-based connectedness led to what was possibly the first universal institution, namely the Market (which can be described in terms of "market mechanisms"). When knowing and doing separated from experience and culture, universal discipline-based science and technology emerged, which were no longer dimensions of the way any local culture mediated its relationships with everything. Gradually, what appeared to be missing from human life began to be conceptualized in terms of symbolization, experience, and culture. Ernst Cassirer began to speak of the cultural animal.[4] Much later, Robert Lifton began to speak of the symbol-hungry person.[5] These and many other developments had a profound influence on disciplines such as psychology. They led to a widespread acceptance that human activities depended on mental representation.[6] Some now regarded humanity as a symbolic species.[7] Before we can fully appreciate the influence industrialization had on human life, it is essential to briefly explore the psychic dimension of civilization in general and the dependence of human life on psychic metaconscious knowledge in particular. It provides the key to our understanding of how industrialization and everything that followed weakened the abilities of societies to sustain human lives.

We have noted that human embryos develop by internally creating human organs and tissues in building the physical self. One of these organs is the brain, which is equipped with a very limited organization at the time of birth. It implies a subconscious knowledge with which it is genetically endowed, which "knows" how to launch the organism in the direction towards becoming a member of a symbolic species. For example, the structure of the throat is incomprehensible without the anticipation of language. The limited organization of the brain is ready to receive the further developments resulting from each and every experience being symbolized by neural and synaptic changes to it. The senses and their locations on the body represent unique vantage points from which everything will be experienced. The position of the thumb opposite to the fingers anticipates a complex and symbolic manipulation of what will be encountered in the world. The attachment of the head to the spinal cord and vertebrae anticipates the emergence of an erect position. As such, the organism implies all the subconscious knowledge required for the unique symbolic mediation of its relationships with the world by means of the culture of a community. These developments orient the organism towards not taking for granted what is perceived but to go beyond it to become a member of a symbolic species, which mediates all its relationships with the world and itself by means of a culture received from a community. Although this kind of explanation supersedes those based on instincts and archetypes, it raises many other questions which discipline-based science is ill equipped to answer. For example, does each cell "live" the equivalent of a dialectical tension between the whole it enfolds and the specific function it expresses? How are we to interpret its being internally and externally connected to all the other cells? We are thus encountering similar questions to the ones raised by some physicists.[8]

In the same way that subconscious knowledge genetically created in an organism implies a transcendence and an orientation, so also the organization of our brain-mind orients individual experiences towards the living of a life. This simple phrase, "living a life," is next to impossible to understand by means of a discipline-based science. It is as if the organizations of our brain-minds symbolically organize all our experiences into a life (metaphorically compared to processes of interpolation and extrapolation). At the same time, these organizations work in the background, forming each and every experience. It also means that our lives are inseparable from the lives of our communities. If a community is to remain alive and well it must maintain a dialectical tension between individual differences and its cultural unity as the context of all dialectically enfolded social relationships and groups formed by its members. The cultural unity is created by the deepest metaconscious knowledge (or myths), with the central myth (or the sacred) anchoring the hierarchy of values of a culture.

The creation of an inner symbolic representation of our lives in the world eliminates the need for a mechanistic concept such as negative feedback. On the basis of this inner representation, we know exactly where we are and what to expect. If the resulting experience is "aligned" with, or "centred" on, this expectation, we can proceed in confidence. If it is not, it must be fitted into the inner representation of our lives in the world in a different manner. We may have to pause, look over the situation, and think about it before continuing to respond in order to achieve a better fit. The closest we come to a "set-point" is our life in the world, into which each and every experience must be fitted as a moment and expression of that life. Without a great deal of metaconscious knowledge implied in the organization of our brain-minds, all this would be impossible. This metaconscious knowledge is acquired by virtue of the fact that all our experiences are transcended by their differentiation and integration into the organizations of our brain-minds.

It is not necessary for this metaconscious knowledge to constitute mental images of any kind. It is possible to show that by systematically integrating and differentiating the experiences of babies (somewhat before but mostly after birth), a great deal of metaconscious knowledge develops which is never explicitly transmitted and yet is fully evident in the lives of these babies, beginning shortly after birth.[9] The process of differentiation orders all experiences according to their being perceived as more or less similar to the ones they resemble, or as something new that must be distinguished from previous experiences. For example, by differentiating and integrating all the experiences of face-to-face conversations, what may be conceptualized as a cluster of these experiences implies that if we stand a little too close or too far from other people we do not receive the anticipated responses. Eventually, this "cluster," enfolded into the organizations of our brain-minds, begins to imply an objective conversation distance characteristic of the culture of our community. All other elements of the working culture of the community necessary for making sense of and living in the world are acquired in the same way and at the same time. No representation of the human brain-mind by means of a computer or neural net can simulate this kind of dialectical enfoldedness.

The unknown as the "in between" and the "beyond" confirms everything known and lived. Anything radically different cannot find a place in our lives. The same is true for all metaconscious values and the moralities built on them. The central myth (the sacred) equally excludes all alternatives. In this context, mental images must be understood as a way of imagining and talking about all this metaconscious knowledge implied in the organizations of our brain-minds. The concept of a mental image is somewhat inappropriate, given the fundamental differences between what is seen and heard, between the image and the word, and between "reality" and a symbolic universe.[10]

Human skill acquisition is thus an integral part of culture acquisition: a kind of macro-level skill of making sense of, and living in, the world. The previously noted model of human skill acquisition developed by Stuart and Hubert Dreyfus shows that the first three stages (novice, advanced beginner, and competence) are characterized by applying rules, learning exceptions to these rules, and solving problems.[11] The transition to the next two stages (proficiency and expertise) is marked by an increasingly effortless recognition of the significance of a situation, and later, of how to respond to it. It is based on what in this study we call metaconscious knowledge, implicit in a highly differentiated and integrated body of experience. The traditions of pre-industrial societies were much more than a simple catalogue of successful prior responses to situations that were taught as skills to the next generation. By being differentiated and integrated into the organizations of people's brain-minds, they became symbolically enfolded into a culture, thereby expanding its metaconscious knowledge. Moreover, in their fifth stage these skills enfolded psychic metaconscious knowledge.[12] Although initially there is no personal investment in applying rules invented and taught by someone else, as people move through the five stages of skill acquisition this completely changes. Especially in the fourth and fifth stages, people must commit to, and take responsibility for, the body of experience they are acquiring. In order to push the limits of their skills they run the risk of making mistakes and damaging their reputations or professional prestige, and they will not be willing to do this without a strong commitment to becoming the best in their field. There is a component of artfulness in the use of any body of knowledge.

The above kinds of situations can be a source of worry, stress, and anxiety. Our lives are full of experiences that do not align with, or centre on, what we expected according to our prior experiences. What are the results when the metaconscious knowledge of our psychic life exhibits tensions with the metaconscious knowledge of our social life, or when both of them place stresses on our spiritual life? What kind of stresses will result when, over a prolonged period of time, the demands placed on our lives cannot be easily dealt with by the unique balance we all develop between perceiving, feeling, intuiting, and thinking as we deal with them in an extroverted or introverted mode (to use the topology developed by Carl Jung)?[13] What happens when we experience internal tensions between the metaconscious knowledge of our daily affairs and the deeper metaconscious knowledge of myths? What happens when some experiences are so threatening to our lives that they must be repressed in order for these lives to continue? What happens when the door to anomie, relativism, and nihilism is cracked open? This can happen in a time that will in the future be recognized as the end of a historical epoch, when a culture and its way of life are less and less able to give meaning, direction, and purpose to the lives of the members of a community. Such questions draw attention to some of the sources of tensions that can occur in individual and collective human life. They can also illuminate why certain sociocultural conditions during a particular phase in the historical journey of a community tend to make some kinds of tensions very common and others much less common, thus contributing to what may be considered normal or abnormal behaviour. They could also help to explain the differences between one social stratum and another. In any case, many of these kinds of tensions cannot be understood apart from their cultural context. Once again, exclusive reliance on discipline-based approaches will lead to an inadequate understanding of our lives.

We now have a conceptual framework with which we can further explore what happens to human lives and communities when they are divided between lives lived in relation to the technology-based or technique-based connectedness and the culture-based connectedness of human life, between the economy and society, and between an economic or technical approach and a cultural approach. These tensions developed most strongly in the United States, a society whose members were much less well supported by the social fabric than was the case in West European societies, from which many of the immigrants came. Karen Horney identified three contradictions marking American life just prior to the Second World War.[14] First, there was a contradiction between the celebration and approval of competition, performance, and success and what remained of the Jewish and Christian practices of humility and love for one's neighbour, which meant dealing with this neighbour in exactly the opposite manner. The Christian community largely accommodated this contradiction by splitting itself into two streams, which each separated the two commandments to love God and to love one's neighbour in order to emphasize one at the expense of the other. Doing so divided a conservative community emphasizing the vertical dimension from a liberal community emphasizing the horizontal dimension of Christianity. The second contradiction was between the celebration and approval of an unlimited stimulation of consumption in order to keep up with one's neighbours and the accompanying frustrations in satisfying these needs as a way of belonging, fitting in, and receiving the approval of these neighbours. People became acutely aware of the inadequacy of their lives, including their incomes, social ties, influence, access to good schools, health and social security, and a great deal more. There were the additional frustrations of all this material consumption not meeting their non-material emotional and psychic needs. The third contradiction was between the freedom, rights, and dignity American society claimed to offer equally to all its citizens and the daily-life experiences of how they were restricted. Most people could not lead their lives as they saw fit, nor did they succeed in getting what they desired even if they were educated and worked hard. It was a conflict between the desire to determine one's own fate and severe social constraints, creating a sense of helplessness and powerlessness. It should be noted that these three contradictions rigorously correspond to the necessities imposed on human life and society by the first generation of secular myths, which give priority to the technology-based connectedness of human life over the culture-based connectedness. As noted, it was very difficult to make these necessities liveable, especially with an increasingly desymbolized culture. These three contradictions were about to be intensified by the influence of technique following the Second World War. The cult of efficiency required the improvement of everything on its own terms without any constraining influence of context, thereby destroying the fabric of relationships of human lives, communities, and ecosystems. The building of the technical order could not be achieved without disordering all life. With hindsight, it is clear that Karen Horney anticipated the growing battle between the cultural order sustaining life and the technical order sustaining efficiency.

The American people found themselves in a predicament. For each one of the three contradictions, they would have to choose to live according to one pull or the other. Regardless of their choices, the pulls opposite to the ones they chose would result in a great deal of tension and, depending on their personalities, would lead to anxiety and even depression. Horney also explained the three strategies Americans commonly rely on to cope with these contradictions in their lives.[15] The first seeks to obtain the affection of others to ensure that they will not hurt you, but doing so can come at a great cost to your own life. Complying with the wishes of others to avoid negative reactions from them frequently requires submissive behaviour and thus the suppression of the self. It may include a submission to institutions that demand certain moral and religious behaviour because failure to do so may cause additional tensions with others. In the extreme, people may become submissive to the point of tolerating being abused by others, refusing to defend themselves, and being indiscriminately helpful to others. It is also an extreme avoidance of competing with others. This behaviour can cause a great deal of anxiety, of which the people themselves may be unaware.

The second strategy is the exact opposite, based on competing with others in order to gain power over them and thus reduce their ability to hurt you. It may be achieved by surrounding yourself with symbols of power and status, by seeking superiority from an advantage you may have over others because of beauty, intelligence, charm, etc., or by outperforming others to achieve the success that translates into other advantages. Having power over others undoubtedly greatly reduces their ability to hurt you, but it comes at a great cost to your own life because power is destructive of all values and everything else cultural. In the extreme, it can translate into a great deal of anxiety caused by internalizing these conflicts with others.

The third strategy avoids contact with others as much as possible because they cannot hurt you when you withdraw from them. A distance and independence from others can be created by meeting internal and external needs with possessions of all kinds, but their usage is often so riddled with anxiety that they bring little enjoyment because these needs are not really satisfied. What is being consumed constitutes a barrier against others, and this requires the choking off of internal emotional needs. Another way of withdrawing from others and from one's self is to not take anything seriously by means of a cynical attitude, not uncommon in intellectual circles.

Because each of these three strategies has its own unique blend of advantages and disadvantages associated with the tensions between opposite poles, most people rely on a mix of them. When one of them gains the upper hand, the internalized conflicts associated with it may lead to neurosis as a way of coping with a deeper underlying anxiety.

Horney has persuasively argued that the above three contradictions create a powerful tendency towards the formation of a neurotic character.[16] Neurotic behaviour tends to deal with a wide range of situations in the same way, thus making it inappropriate in a great many cases. A "normal" response is guided by a culture to fit a situation. For example, in a particular culture at a particular point in time, certain situations may arouse suspicion in most members, and this would be regarded as "normal" by them. If some of the members are suspicious in a great many other situations, their suspicions may not be warranted and do not fit the context. In other words, it is not a question of the culture's not sufficiently differentiating between the situations that receive a normal response from most people and those that are treated by others as if they are essentially the same. It would appear that the explanation must include the metaconscious social self overriding the culture, as it were, in order to avoid a deep underlying anxiety that causes a great deal of pain; and this could be a sign of neurosis. The kinds of lives people live are often the result of circumstances beyond their control, but when this is not the case it may be the consequence of an underlying neurosis. For example, according to all appearances some people should be happy because they have what it takes to be successful in their careers and to attract others, and yet the opposite appears to be the case. Neurotic persons tend to develop fears along with behaviour patterns to protect themselves from such fears, well beyond the range of situations in which other members of the same culture would be afraid. These fears are therefore not the result of circumstances but of conditions in their own lives that have been metaconsciously symbolized and to which their metaconscious social selves respond more directly. Moreover, their culture has not developed any resources for coping with this kind of behaviour. This makes it very difficult for neurotic people to get on with their lives; they tend to suffer much more than most other members of their culture. Frequently, they are not aware of this.

Human life in America just prior to the Second World War was characterized by an overdependence on the approval and affection of others, which became excessive in neurotic persons. It was the extreme outcome of the processes of desymbolization that accompanied industrialization, for the simple reason that there was not a historically "mature" society in America to begin with. To compensate for a weakening social support, there was an indiscriminate hunger for appreciation and affection. It was internalized as an inner insecurity which, in neurotic people, created feelings of inferiority and self-doubt, or alternatively, a need for self-aggrandizement. For most people it became more difficult to say no to others, to express one's own wishes and feelings, to know exactly what a person wanted, and to assert one's own interests. It required a stronger self-assertion, often to the point of becoming aggressive and domineering; or when a person found this impossible, to become submissive or to withdraw. In the former case, this led to a metaconscious knowledge of the social self that became insensitive to others, and in the latter case, one that became over-sensitive to others. In other words, to make life liveable required either a struggle against everyone or a defensiveness as if the whole world was against you. In either case, this led to a great deal of anxiety as a disproportionate reaction to situations that were interpreted as being threatening to a person's life. A "cultural niche" had been created for the development of the neurotic character. Since their anxiety is largely metaconscious, people are often not aware of it, which makes them feel totally helpless because there no longer is a fit between situations and their responses to them. It is another consequence of "technology changing people," particularly their metaconscious social selves.

The situation was somewhat different in the industrializing societies of Western Europe, which were much more resilient because the process of desymbolization began with "mature" dialectically enfolded traditional societies. The metaconscious social selves of their members became divided as their lives and communities were unfolded into relatively distinct spheres related to work, family life, shopping for the necessities of life as wage earners, and relaxing. Each sphere of activities tended to be carried out with very different people, with the result that each person had one "life" with fellow workers, another with family, another with a crowd of shoppers, and another with fellow parishioners, club members, or a sports team. This division of human life (not unlike the technical division of labour on which it was superimposed) was further reinforced by the layout of the urban habitat, largely zoned according to these four spheres of activities. This division was internalized as multiple metaconscious social selves, each corresponding to one of these "lives"; these were often further subdivided as one "life" with one person, another with a second person, and so on. This also occurred in traditional societies, but the results were very different because the much lower levels of desymbolization allowed all these "lives" and metaconscious social selves to be integrated to a much greater degree. There was a much greater unity in everyone's personal life and, since these lives overlapped, in their communities. This centred and unitary self and a cohesive community were undermined by the desymbolization that resulted when what was taken to be real began to dominate what was taken to be true. There continued to be non-trivial differences between the mass societies in North America and those of Western Europe. Much of this can be explained by the very different conditions under which industrialization began in these parts of the world. The social forces at work were very similar, but the contexts into which they were launched were, and continue to be, very different. Later in this chapter, we will return to the ways technique affected the above three contradictions in the United States following the Second World War.

The possibility of a neurotic character provides further evidence that the metaconscious processes of differentiation and integration are not mechanistic in nature. When the metaconscious psychic and social selves become profoundly damaged as a consequence of "technology changing people," they can transform these processes to the point that they no longer reveal the meanings and values required for normal human life, but only the meanings and values profoundly affected by a deep underlying anxiety stemming from a profound loneliness, which occurs when the social fabric of a community is weakened by the desymbolization accompanying the process of industrialization. This helps to explain the well-known phenomenon of the loneliness of individuals who conduct much of their lives in crowds instead of a sustaining social fabric.

Neurosis also demonstrates the inadequacy of a discipline-based approach to mental illness. How can a psychologist or a psychiatrist understand neurosis without an understanding of the sociocultural conditions under which neurosis is created? What is required is a kind of psychosocial history into which the understandings of the relevant disciplines are enfolded.

Desymbolization and Divided Lives

The integrity of our symbolic life may be threatened as a consequence of desymbolization. This term refers to a deficit in how a particular experience might have been symbolized if it had been fully lived in relation to how everything is related to everything else according to our time, place, and culture.[17] Symbolization may be constrained by limiting the vantage point established by our physical, social, and cultural embodiment in a situation.[18] It may also be constrained by limiting our participation in the relationships with whoever or whatever is the focus of our attention.[19] Additional constraints can occur when our commitment to do the best we can in a situation is weak or externally discouraged or prevented.[20] Finally, the level of freedom or enslavement of our life will affect the way our life works in the background in many situations, thereby influencing the symbolic connections we are able to make.[21] Low levels of desymbolization can be expected to result in richer and fuller lives because of the density and quality of our relationships. High levels of desymbolization will impoverish the web of relationships within which our lives are lived.[22]

Desymbolization is not a new phenomenon. It has occurred throughout human history. For example, in pre-industrial traditional societies, the social division of labour led to desymbolization. Their traditions were internalized into, and symbolized by, the organizations of the brain-minds of the members of these societies. They could be used as a kind of "mental map" for people to orient themselves within their lives in the world. These mental maps would be highly elaborated with high densities of symbolic connections in areas where people contributed to a way of life, and less well elaborated for other social roles according to the levels of contact people had with those occupying these roles.[23] For example, the mental map of a peasant belonging to a feudal estate may have been highly desymbolized compared to what would be possible by imagining a scenario in which all the mental maps of all the members of that society could be superimposed on each other in order to create the highest levels of symbolization possible in that society within the constraints that affected everyone. In contrast, the mental map of an aristocrat in the same society would be less desymbolized because he would have been dealing with the entire estate and all the people on it. Moreover, he would likely travel more than anyone else on the estate to further enrich the symbolization of the estate-related experiences, and this enrichment would receive a further boost if he had enjoyed a good education, complemented by a great deal of reading. A society's denying the same opportunities to his spouse helps to explain the often extreme gender differences in this regard.

The process of industrialization led to a great deal of desymbolization of the lives of the people involved, as well as of their societies.[24] For factory workers, this was in part related to very long working hours under harsh disciplines that left them with little energy to rebuild the kinds of communities they had left behind when they were forced off the land. After a long working day, they had little time or strength left to devote to their spouses, children, neighbours, and others, with the result that a kind of existential and social poverty was superimposed on their financial poverty.[25]

In addition, these factory workers could only participate in production by means of their hands, to the point that it was not long before nervous fatigue began to characterize work organized on the basis of the technical division of labour. Nervous fatigue occurs when a great deal of energy must be spent to mentally suppress the fact that these hands are connected to everything else in a person's life.

The factory owners and entrepreneurs experienced the diametrically opposite kind of desymbolization. They were so busy mentally organizing and reorganizing the local portion of the technology-based connectedness related to their business that they delegated what needed to be done to others, who by doing the work at least had some "hands-on" experience of what was happening. Consequently, these owners and entrepreneurs were robbed of the possibility of living their experiences in other ways.

As a result, the introduction of the technical division of labour, followed by the mechanization of human work, superimposed two new forms of desymbolization: one related to participating as a hand, brain, or intermediary between the two, and the other existentially impoverishing the fabric of social relationships that sustain everyone. The quality and strength of all the relationships established during work are negatively affected by the restriction of embodiment and participation as a hand, brain, or intermediary, the restriction of commitment to the utility of a wage or the profits earned, and the enslavement of work to the "world" of work organized in terms of domains of non-sense. As noted, social epidemiology has shown that the life lived at work and the life beyond remain somewhat integral within a person's life, since the effects of the former spill over into the latter.

Some of our own experiences can shed some further light on what this all means. Exposure to any kind of repetition, monotony, or sensory deprivation generally makes us withdraw to the greatest extent possible.[26] Such a withdrawal often begins on the physiological level by things vanishing from our field of vision if we cannot detect any changes over an extended period, by our being unable to use words or phrases that have been heard repeatedly from a closed loop recording, by our dozing off when in the middle of the night the road is empty, leaving only the monotony of street lights flashing by, by our daydreaming when the instrumentation on a control panel shows no significant change for a long time, and by our inability to endure sensory deprivation for any length of time, following which professional help is often required. Our nervous system has evolved in relation to a living social and natural life-milieu and is unable to cope very well with the "world" of non-life of classical or information machines. In the case of the former, repetition is built in, while in the case of the latter, it is programmable by means of rules or algorithms. Information machines and the systems built up with them cannot evolve and adapt other than by software changes.

Turning to the economic "life" of the first generation of industrial societies, the behaviour of *homo economicus* may well approximate that of wage earners enjoying a very low standard of living, but from a cultural perspective, it represents a very high level of desymbolization associated with an enslavement of human life to the economic mechanisms of *homo economicus* and the Market. Observers like Karl Marx essentially spun off everything that did not belong to industry and the economy to a "superstructure."[27]

As noted, before anything in human life, society, or the biosphere can enter the economy and be ordered by the Market, it must be commoditized. Such commoditization begins with desymbolizing, and we have become so accustomed to this that we pay little or no attention to human work being disconnected from life to be turned into labour, to land being disconnected from the biosphere to become pieces of terrain, to materials and energy being disconnected from the biosphere to become resources, and so on. In other words, we live and act as if this desymbolization is entirely normal and acceptable. Perhaps our own experiences can shed some further light on what this means. The closest we come to being commoditized is when someone stares at us. Staring is very different from looking at someone because the other person is treated as an object, from which no reciprocity is expected or wanted.

Being treated in this manner can be so upsetting that a forceful response is often required as a way of coping with this humiliation. The resulting experience is that of non-life, and it helps build the metaconscious knowledge of separation and death. In our contemporary mass societies, this experience of staring at someone is hardly limited to pornography; we are so used to high levels of desymbolization that we all have become numbed by it to varying degrees. When these kinds of experiences build a great deal of metaconscious knowledge of non-life, a level of psychic incapacity may become widespread, and anxiety and neurosis may follow.

The tensions within the symbolic universes of industrializing societies multiplied and grew as one element after another became unfolded from the traditional dialectically enfolded universes. The technical division of labour and its subsequent mechanization unfolded the technology-based connectedness from the culture-based connectedness and the economy from society, and necessitated a major realignment of the social structure, political framework, legal institutions, morality, religion, and aesthetic expression (soon regarded as the mere cultural embellishment of "real" life).[28] Industry and the economy were built up from domains that jointly took on the characteristics of what we have referred to as "reality," while the symbolic universe continued to be the locus of what was true for human life and society.[29] Human lives and societies were thus divided between what was "real" and what was "true," with everything this entailed for people's psychic, social, and spiritual metaconscious knowledge. The separation of what was real from what was true involved tensions between what was organized in the image of non-life (the machine) and symbolic life. Cultural values were in tension, with prices established by market mechanisms and thus not by human freedom but out of technological and economic necessities.

This division of human life and the tensions between what was lived as being real and what was lived as being true were multiplied further when knowing and doing separated themselves from experience and culture following the first phase of industrialization. This separation represented a creative solution to the limitations of knowing and doing embedded in experience and culture that had been uncovered by industrialization. To understand the predicament in which people found themselves, we will briefly revisit the four limitations of technological and economic knowing and doing embedded in experience and culture, which was begun in the Introduction.

The first limitation stemmed from the senses. As the technologies and processes of some industries advanced, they became increasingly inaccessible to the senses and thus to experience and culture. In the electrical and chemical industries, this inaccessibility to the senses was present from the beginning because there were no meaningful observable phenomena that correlated with what was really happening in the technologies or processes.

Second, the cultural approach quickly proved itself poorly suited to building up the new technology-based connectedness. Since the latter was built up with the technical division of labour, it became organized in terms of interconnected domains which did not depend on any context other than the inputs received and the intermediary outputs exported. Within each domain, an instance of one category of phenomena transformed the inputs received from an upstream domain into the outputs to be passed on to a downstream one, until the final desired output was achieved. Within each domain, what mattered was the most efficient and profitable way of transforming the inputs into the outputs. To ensure this, these domains had to be assessed on their own terms apart from the broader context. It did not take very long before the associated knowing and doing became separated from experience and culture, and eventually this led to discipline-based approaches that limited the context taken into account to a single category of phenomena. In other words, it was increasingly the technical significance of the inputs and outputs and their transformations relative to the local technology-based connectedness that needed to be considered, as opposed to the meaning and value it all had in relation to the broader context of the remainder of the entire technology-based connectedness, as well as its embeddedness in human life, society, and the biosphere. The cultural approach paid far too much attention to the broader context to be of much use. Symbolization had to be restricted to the context of the local technology-based connectedness. Doing so involved an abandoning of culture-based reasoning in favour of rational goal-directed approaches. Desymbolization made the living of a life more difficult by forcing people to move from goal to goal and to solve one problem after another, mostly related to building up the technology-based connectedness and everything associated with it. We continue to think of living our lives in terms of an endless series of problems to be solved.

A third limitation came from the technologies of the first generation of industrial societies being based on a technological tradition embedded within the larger tradition corresponding to a way of life of these societies. Their knowing and doing embedded in experience and culture encountered real limitations when these traditions became swamped by new situations to which prior experience could not be applied. Consequently, all technological skills were forced back to the third stage of the human skill acquisition model, which was characterized by problem solving. Initially this was accomplished by means of culture-based reasoning, but this reasoning quickly encountered related difficulties, as we shall see. It was gradually replaced by goal-directed behaviour associated with problem solving. For example, how could an entrepreneur rebalance the output of one domain with the required input to another, or how could a particular transformation in a domain be made more profitable? Such problems were dictated by the local technology-based connectedness, whose purpose and goals took the form of output-input ratios to the exclusion of all cultural values. The result was a gradual mutation of culture-based reasoning into the kinds of rational approaches that eventually led to the phenomenon of rationality as described by Max Weber.[30]

A fourth limitation came from intense competition, which made other culture-based approaches ineffective compared to the new emerging practices. As more nations began to industrialize, they had to compete with each other for many of the same markets for their goods. These competitive pressures did not in the least respect the above limitations imposed on technological and economic growth. It took ever more gifted people longer periods of time to arrive at ever more incremental technological improvements, and this could only be made cost-effective by spreading the fixed costs over a great many machines in order to gain economies of scale. These developments necessitated the growth of entrepreneurial firms, and this could be achieved in an effective manner only by diffusing goal-directed behaviour throughout the firms. Diminishing returns were encountered everywhere. Consequently, individually and jointly these firms approached a kind of asymptote to technological and economic growth, and some observers expected that the pressures within the economic systems of the time would soon cause them to self-destruct. The principal reason why this did not occur was an entirely new development pioneered in Germany. It involved the application of science in industry, and this led to the discovery that discipline-based approaches to knowing and doing were ideally suited to building up the domains of the technology-based connectedness of industry and the economy. This development essentially eliminated all four limitations to the application of knowing and doing embedded in experience and culture, when applied to things having the structure of what we have referred to as reality. In Germany, the spread of discipline-based approaches in industry, the economy, government, and the military was so rapid, and produced so many advantages, that Germany quickly became the leading industrial power.

The phenomenon of rationality began to dominate the second phase in the process of industrialization, when technological knowing and doing were no longer guided by the cultural approach based on symbolization, experience, and culture but on discipline-based approaches.[31] The rational approaches were goal-directed in the context of the technology-based connectedness, while cultural approaches were life-directed in the context of the culture-based connectedness of human life and society.

In the course of a little more than a century, the cultural approach was essentially eliminated from industry and the economy. Human life and society became characterized by an ever greater tension between their technology-based connectedness and their culture-based connectedness, to the detriment of the latter. Before this could lead to major dislocations, the effects of "technology changing people" permeated the deepest metaconscious knowledge, thereby creating the first generation of secular myths in human history. These myths implied that all earlier cultures had been wrong. The good life was not to be attained by following moral and religious precepts but by hard work in order to develop the technology-based connectedness, and this in turn would bring progress and happiness. Capital was the ultimate value, in relation to which all other values had to be repressed.

The shift from a primary reliance on the cultural approach to a primary reliance on the rational approach with only a secondary reliance on the cultural approach is the key to understanding many of today's tensions in human life and society. This shift does not mark a mutation in the way the human brain-mind functions, because the historical rapidity with which it came about could not have been the result of genetically transmitted adaptations. Nor could it have been the consequence of "technology changing people," because that always lags behind "people changing technology" and takes a long time to reach the deepest levels of metaconscious knowledge. We must therefore look for explanations in the way the context taken into account by symbolization could be altered, so that desymbolization could be seen as beneficial under certain conditions. It turns out that this is the case when we are dealing with a world recreated in the image of "reality."

When we deal with a daily-life situation by means of symbolization, experience, and culture, a foreground-background distinction is established according to how our attention is directed, but anything in the background remains available in case any shift in attention is desired or necessary. The integrity of how everything is related to everything else thus remains intact. In contrast, when we first learned physics or chemistry in high school, our attention was directed towards the domain of a discipline exclusively populated by physical or chemical phenomena. We could only enter this domain by means of our imagination and certainly not by our experience. As our learning progressed, we acquired the ability to compensate for friction, inertia, air resistance, and more, but even if we ever reach the frontier, the domain of physics will remain unlike the world in that it is exclusively populated by physical phenomena.

Our "experiences" of a domain of a discipline and those of daily life differ in three fundamental respects. In these "experiences," the foreground is now a domain, while the background remains the daily-life context in which we deal with the discipline. The foreground is now discontinuous with the background because the former has the structure of "reality" while the latter is integral to a dialectically enfolded symbolic universe. Finally, the metaconscious knowledge built up from domain-based "experiences" is discontinuous with the metaconscious knowledge built up from daily-life experiences. Simply put: a goal is substituted for living a life and keeping everything in context. In the case of a technical speciality, its domain has been separated from all of life and the world by means of a triple abstraction. It rules out every possible human activity except that of ensuring that any phenomenon operating in the domain is as good as it can be on its own terms, that is, in terms of how efficiently it transforms the inputs received into a desired output by means of one instance of one category of phenomena. Because of the triple abstraction, this domain has no relationship to our daily-life world other than by means of the inputs it receives from it and the output it returns to it. This domain has a technical significance in relation to all others that help achieve a final result, but all human meanings and values have been filtered out by means of the triple abstraction. Its efficiency can only be made absolute, that is, relative only to itself and on its own terms. It represents the most radical desymbolization in all of human history. It should hardly be surprising that Max Weber warned humanity that the phenomenon of rationality would shut us into an iron cage, that is, would enslave us to rationality.[32]

The only way out of this cage would be to spin off our symbolic nature, but what kind of freedom could this possibly bring?

For now, it is important to understand that when our attention is directed towards a domain, the process of symbolization, derived from experience and culture, becomes so highly desymbolized that it is almost unrecognizable. The rational approach essentially amounted to a highly desymbolized approach; many observers deduced that it was objective because all values had been eliminated from it. It was a soothing fantasy because the application of any scientific or technical-based knowledge does not remain within the domain of a discipline but is redirected towards a living world. This world is so different in character from a domain that the criteria that must be met before domain-based knowing can be applied rarely occur. In such cases, all the other categories of phenomena not found in the domain will be disordered. Hence, there can be no scientific understanding of riding a bicycle, for example. There can be an understanding of the physical aspects of doing so, but that hardly exhausts everything involved in bike riding. Similarly, there can be no technical discipline-based understanding of anything other than what we have reorganized in the image of "reality." In other words, to evolve a way of life on the basis of countless scientific and technical disciplines is to reorganize it in the image of non-life, and this is hardly objective. It is analogous to what we have already discussed in relation to the economy, where the application of the discipline of economics turned it into an anti-economy because of the effects of capital, market forces, and technical planning. In the same way, the application of discipline-based approaches to evolving and adapting a way of life transforms societies into anti-societies.

The rational approach could not possibly shut a symbolic species into an iron cage if it were not applicable to almost every area of human life. For example, a bureaucracy is a unique organization in which human beings work together on the basis of goal-directed behaviour instead of symbolization and experience. A mission statement is formulated, and it in turn is broken down into the goals of different departments. These in turn are divided into a variety of sub-goals that can be assigned to different sections within each department, and these sections are organized by means of job descriptions connected by reporting relationships. Such bureaucracies were later re-engineered in the image of the computer by means of enterprise integration, which, through large databases, coordinates flows of matter, energy, composites built up with them, labour, capital, and highly specialized discipline-based knowledge. Each flow is made as efficient as possible with respect to the local technique-based connectedness. The implications are that everything relevant to the goal is taken into account, while everything unrelated to it is ignored. This "unrelated" matter includes how everything is related to everything else and how everything will remain related to everything else and evolves in relation to everything else, despite the imposed goal and the subsequent intervention in this relatedness. The consequences are the previously discussed disordering of human life, society, and the biosphere. In other words, the desymbolization that occurs when a domain is used to understand or modify anything in the world that does not conform to what we have referred to as "reality" directly corresponds to the disorder it creates.

Superimposed on this extreme desymbolization resulting from discipline-based approaches in science and technique is the ongoing separation of knowers, doers, and intermediaries. The knowers now operate in terms of domains learned in schools and not by doing. These domains are separated from experience and culture and thus from the world of the doers. The intermediaries must now bridge the gap between the domains of the knowers and the doers' experiences of the consequences. As before, none of the three groups can fully participate in their activities as human subjects, with the difference that the brain function is now fully rationalized and technicized and is thus separated from experience and culture, while the hand continues to be involved on the basis of knowing and doing embedded in experience and culture. This separation superimposes an additional desymbolization, as before, but it is now qualitatively sharper and more divisive. Of course, the doers and intermediaries work in a setting that has been designed and built by means of discipline-based approaches, to which they continue to apply a knowing and doing embedded in experience and culture, with all the previously noted difficulties this involves.

As the above developments continued to grow and spread in societies, another kind of desymbolization superimposed itself on the others. In attempting to explain why the Industrial Revolution occurred when it did, David Landes[33] and Jacques Ellul[34] came to very similar conclusions, to the effect that the social and historical conditions at the time provided enough technological, economic, social, and cultural "resources" to proceed. These historical conditions were most developed in England and somewhat less on the Continent, but they existed nowhere else in the world of that time or during any previous times. All that was needed was a decisive opening move that would deeply disturb the technology-based connectedness, thus requiring that these conditions be turned into resources for industrialization. Today we appear to have little difficulty speaking of human resources or of our being an item on a budget line. We also appear to have little difficulty speaking of social, cultural, and natural "capital." Such sentiments continue to reflect a deep metaconscious knowledge of ourselves that we must all serve the present system of technique. We have metaconsciously come to recognize that "this is how the system works" as a consequence of its acquiring its own internal mechanism of development in which we can no longer interfere on the basis of human values, because we have metaconsciously endowed technique with the highest value as a consequence of "technique changing people."[35] We no longer symbolize technique in order to gain a better understanding of its meaning and value for human life and society as a first step towards weakening our enslavement to it and eventually mastering it as a creation we intended to serve us. A universal technique is necessarily inadapted to local conditions and thus to serving our local needs. Instead, we have reorganized human life in the image of non-life and have turned ourselves into a resource for technique.

I have attempted to show that the transformations of technology and society in the course of industrialization and technicization have amounted to a comprehensive undermining of our humanity as a symbolic species. The core of human life has become a kind of non-life increasingly dominated by the rational approach, which later mutated into the technical approach. We have conducted a war against ourselves as a symbolic species.[36] It marks a kind of separation from, and death of, our symbolic life, even though babies and children continue to depend on symbolization, experience, and culture in order to learn to make sense of and live in the world as they grow up. Nevertheless, their experiences are now full of desymbolizing influences.[37] Contemporary societies organize their affairs as if they can more or less dispense with symbolization, and yet without it the next generation cannot grow up to acquire the discipline-based approaches, and without all this there would be no technique.

The organizations of the brain-minds of the members of contemporary mass societies have been divided into two networks of "clusters" of "experiences." One of these relates to their participation in technique as knowers, doers, or intermediaries, differentiated from the other network, that of "clusters" of "regular" experiences with family members, friends, and neighbours. The metaconscious knowledge implied in the former network is very different from that implied in the latter. The one network represents an extreme level of desymbolization, while the other continues to be more or less carried on the basis of symbolization, experience, and culture. This development also significantly affects people's metaconscious social selves, which are now internally divided and thus minimalized. It may be concluded, therefore, that industrialization followed by technicization has been accompanied by a great deal of psychic, social, and spiritual metaconscious knowledge related to disconnection, separation, non-life, and death.

All this desymbolization was accompanied by the creation of vitalizing metaconscious knowledge in the form of the first two generations of secular myths. These myths were not only secular in character but also different from their traditional precursors in being affected by desymbolization. In order to clarify this apparent contradiction, a distinction must be made between the two roles performed by secular myths. Their first role is to symbolize the unknown as more of what is known and lived, thereby excluding anything that is radically different. It is the function that allows millions and millions of people who, on the surface, agree on very little to nevertheless share the same historical journey because their way of life has become the only way of life. The second function of secular myths is to anchor the hierarchy of values of the culture of a society in the greatest good encountered by its members. By symbolizing the unknown as more of what is known and lived, this greatest good becomes the absolute greatest good, without any limits. It follows that the first function is not affected by desymbolization because it "interpolates" and "extrapolates" human experience regardless of its level of desymbolization. In contrast, the second function, that of providing an ultimate reference point for cultural values, is deeply affected by desymbolization. This helps to explain the changes in advertising and art during the last fifty years.[38] The desymbolization of cultural values is compensated by the use of the organization of the brain-mind as a radar by the other-directed personality. Anyone bringing up teenagers will know that appealing to traditional values makes no sense to them when their "radar" tells them that what most of their peers do is normal. This "normal" has become normative in the presence of highly desymbolized values. We will return to this in this and the following chapters.

For now, I will simply point out that the first generation of secular myths metaconsciously created by the members of the first generation of industrializing societies was universal in character. Capital, work, progress, and happiness had no specific cultural content because they accompanied the building of a technology-based connectedness (separated from the local culture-based connectedness), were organized by a universal Market (which set prices to replace cultural values), and increasingly relied on universal output/input measures independent of all cultural values. These secular myths were thus ideally suited to provide vitalizing metaconscious knowledge for life being recreated in the image of non-life, because life has always been lived in relation to a local interrelatedness. This helped to spin off people's symbolic functions, but it did mark an age of ideology, which put some existential substance on the bare bones of non-life.

The second generation of secular myths was centred on an unlimited reliance on the discipline-based approaches to knowing and doing that jointly made up the phenomenon of technique (and later, the system of technique). Because of the extensive desymbolization of culture, the self-regulating character of local communities based on culture all but vanished, with the result that the state had to step in and constitute a major organizational locus. This had the effect of gradually making everything political, as if the nation-state had no limits.[39] The supporting myths to those of technique and the nation-state were those of science and history. The latter became a myth because it excluded all historical journeys other than our current universal one. We may think in terms of economic growth based on free trade supervised by the World Trade Organization, but its role amounts to little more than making the world safe for technique by destroying all barriers against it. The secular myths of technique, the nation-state, science, and history vitalized the recreation of human life in the image of non-life. No age of ideology was required any more, since these secular myths anchor what we will soon examine as integration propaganda, which amounts to a technical recreation, in the image of non-life, of what was accomplished by customs and traditions in earlier societies.[40]

We have previously suggested that as a consequence of "technology changing people," industrializing societies were first dominated by the tradition-directed personality, then by the inner-directed personality, followed by the other-directed personality of our current mass societies. These metaphors were well chosen because they imply a spinning off of the symbolic functions and the desymbolization of the metaconscious self. When people's lives work in the background via the organizations of their brain-minds functioning as a mental map, the human subject acts as a user and interpreter of that map, free to decide that, even though a certain situation has been responded to in a certain way for countless times, all of a sudden that response no longer appears to be appropriate. Moreover, that map is the accumulated experiences of one's life, with the result that a person can be fully committed to it. From the perspective of embodiment, participation, commitment, and freedom, the use of the organization of the brain-mind as a mental map involves little structural desymbolization. This personality type dominated traditional pre-industrial societies and could not endure very long once the effects of "technology changing people" became significant.

Around the end of the first phase and the beginning of the second phase of industrialization, people's lives worked in the background via the organizations of their brain-minds, functioning more like a gyroscope or a compass. Their deepest metaconscious knowledge was as yet not affected, with the result that it could perform its symbolic functions for a limited time even though the desymbolization of human life had increased substantially. Compared to the previous type, for the inner-directed personality embodiment, participation, commitment, and freedom had been significantly restricted; the metaphor of a gyroscope makes this clear. All a person could do was to note the direction it pointed and follow it or be guided by the deepest metaconscious values set in an earlier time, since there was little guidance to be had of any other kind. The further a society moves away from the time when this gyroscope was set, the less viable this personality type becomes.

Finally, the other-directed personality type has to make sense of, and live in, the world with a culture that is so highly desymbolized that it is of minimal use because "technique changing people" has now reached the deepest metaconscious knowledge, making it impossible for this knowledge to act as a gyroscope. Such people's inner "symbolic emptiness" can function only with externally provided myth-information, with the result that their lives only minimally work in the background. They have to compensate for the additional constraints on embodiment, participation, commitment, and freedom that have resulted from further desymbolization under the pressures of technique and the nation-state. They are now compelled to use what is left of their inner resources to scan what is happening around them as a kind of radar in order to "go with the flow."

The process of industrialization, culminating in the establishment of the system of technique, appears to have progressively spun off much of what was central to our being a symbolic species. The accompanying metaconscious knowledge of separation, non-life, and death that plays such a significant role can only be dealt with by the compensating technique of integration propaganda. We do well to remember how Germany solved the many difficulties associated with its being the leading industrial society during the early twentieth century, including massive unemployment, inflation, and high levels of desymbolization. A frenzy of Hitlerian violence was its attempt at rejuvenation and the reinvention of a glorious future.

Anti-Societies

The next chapter in the story of technique takes us to the United States and its involvement in the Second World War. The preparation for, and the conduct of, a mass war created the perfect conditions for pushing the potential of technique to its limits. After all, the conduct of war is the ultimate goal-directed rational approach. Nothing matters other than what can ensure victory. What is at stake is much more than killing or being killed. The nation must believe that it is engaged in an all-out life-or-death struggle for its existence and future. Everything can and must be sacrificed to ensure that future. As a result, power and efficiency must characterize everything being done, from the training of recruits to the mass production of all the equipment and weapons, the organization of the employment of the recruits and this technology, the organization of supply lines, and a host of other issues, including dealing with dead and wounded soldiers. It is hardly an accident that operations research, so widely applied to almost anything today, had its primary roots in the Second World War. As a discipline, operations research applies mathematical methods to create a model of a particular activity in order to enhance its performance in obtaining a desired result. It is a kind of technique of techniques because the activity is mathematically organized into a set of domain-like components, each represented by one or more equations. For example, the surveillance of the seas for the purpose of locating and destroying enemy submarines can be executed by a variety of possible strategies. Mathematical models of each of these strategies can be prepared and optimized, at which point it becomes possible to select the most efficient one. Operations research, when applied to warfare, can safely disregard anything that is not relevant to winning.

The United States was uniquely prepared for the fullest development of technique, for a variety of reasons. Because of the lack of farm labour, it had invented interchangeable parts used in farm equipment to facilitate repairs in remote locations. It must be recalled that the machinery of the earlier phases of industrialization was often built up from custom-fitted parts, not unlike the way the crafts built things in the past. Without the invention of interchangeable parts, Henry Ford could not have taken the next step of creating the assembly line. Scientific management was invented around the same time. The Fordist-Taylorist production system constituted the next step, and it made the country into the leading industrial nation in the world.[41]

The United States constituted a highly unique society very different from its European counterparts, even though many of its citizens had emigrated from there. Gradually, frontier communities helped to forge a society that had no history other than that of the Native people. Few of its members had come to America to recreate what they had left, because they desired to invent a better future for themselves. In other words, there was an extraordinary economic, social, political, and legal plasticity in the United States compared to its more established European counterparts: there was no long-established tradition, not even a highly desymbolized one.

Arguably one of the most significant contributions to the development of technique that came from the war effort was related to human techniques.[42] Imagine the task of putting together an army of recruits. People from all walks of life had to be trained to face life-and-death situations together as perfect strangers. To accomplish this, it was essential to classify these recruits according to their intelligence, aptitude, physical abilities, and so on. Psychological techniques of all kinds were developed to measure these human qualities. Based on this information, people had to be assigned to different kinds of training to prepare them to successfully carry out the many functions on which an army depends. Once this was accomplished, the recruits who were to work together on a particular function had to be made into a well-functioning group, whose members needed to cooperate despite being strangers to each other. Techniques of group dynamics and of organizational behaviour were developed to "engineer" social bonds as well as possible because there were no precedents based on experience and culture. In other words, the war effort identified an urgent need for an entirely new category of techniques to develop the "human resources" to ensure the greatest possible efficiency and power of all operations, and this effort continued during peacetime.[43]

Following the Second World War, American society faced a very similar challenge. A growing use was being made of discipline-based specialized knowledge to adapt and evolve the American way of life. There simply were not enough university and college graduates to staff the kinds of organizations making use of this highly specialized knowledge, which Galbraith has referred to as technostructures.[44] It became impossible to have most people entering universities and colleges coming from the higher social strata of society. There simply would not be enough graduates. Mass education had to be based on people's abilities to contribute to discipline-based knowing and doing (or on their lack of these abilities). To do this effectively created a need for the same kinds of testing techniques as those developed during the war effort. High schools had to identify those students who could excel at this new discipline-based knowing and doing separated from experience and culture and encourage them to go on to higher education. The nation needed them in order to succeed in the post-war world. It thus became necessary to set the traditional ethnic or religious prejudices aside. The female half of humanity also had to be drawn into the recruitment effort. Human rights legislation was drawn up to ensure that everyone with the urgently needed talents could go to university or college. In other words, it was not some new upwelling of humanism and a desire for equality that drove this development. All of a sudden people who had been on the margin of society in combating traditional prejudices and discrimination found themselves in the mainstream. The human techniques that had been developed during the war became essential to ensure America's success in the new age. This, however, set up new dialectical tensions and contradictions in American society: if everyone was equal going into the mass education system, they came out highly unequal, depending on their ability to participate in the new knowing and doing.[45] There was also a contradiction between the universal character of this new knowing and doing and the knowing and doing developed in a local context.

As in the war, the social bonds among the members of certain groups who were strangers to one another had to be engineered in order to facilitate the synthesis and application of the highly specialized knowledge held by their members. It was not long before we saw the same kind of engineering enter the schools as progressive education. The techniques developed during the war found further applications in group learning and the new classroom.[46]

As far as industry was concerned, although much of it was converted back to the service of a civilian economy, a substantial military-industrial complex remained. This development was greatly fuelled by the Cold War as well as by the widely practised Keynesian policies, which had the unintended effect of boosting the kind of scientific and technical developments that were not directly relevant to the daily lives of the American people. Nevertheless, it made no sense to create an economy that fundamentally depended on the development of weapons systems that technique had made so efficient that their all-out use could no longer defend anyone but only destroy everyone. Even when the Cold War ended, the military-industrial complex continued to thrive because it was the closest America came to having a technology policy. It constituted a frontier on which advanced technologies and techniques could be developed, almost regardless of their costs, in the name of the ultimate conceivable struggle for the security and future of the nation, and thus for its life or death as part of the combat between good and evil. In this way, advanced technologies and techniques could be developed until they became viable in the non-military sector of the economy. These efforts diverted enormous human talent and resources away from the civilian economy, with devastating consequences.

There are deep structural reasons why the economies of the most powerful nations continue to have large public sectors after Keynesian economic policies were abandoned. Governments could no longer accumulate surpluses during good years to assist them in getting through lean years, once technique turned economies into anti-economies that are incapable of generating such surpluses. These public sectors became safe havens for a significant part of the technical frontier.

The civilian economy inherited all the techniques associated with mass production developed during the Second World War. No longer was a great deal of this mass production destined for war and destruction. Consequently, peacetime mass production had to be matched by mass consumption, and this in turn required mass advertising to match consumption to production. The development of advertising also benefited from the human techniques developed during the war. The usual explanation that mass advertising simply creates informed consumers could not withstand the most elementary scrutiny when the information content in a set of advertisements was examined. Such advertisements contained almost no information about the products they represented. They operated on the deepest metaconscious levels by employing the symbols that emerged there and bonding them to products. These became symbolically endowed with non-material abilities they did not have in themselves – an echo of the myths of progress, work, and happiness. When the advertising agencies did their homework in probing for symbols that indeed corresponded to something deeply metaconscious in the lives of consumers, advertisements were very successful in ways of which people were unaware. To carry all this mass advertising necessitated the expansion of the mass media.

All these developments in turn resulted in a growing number of daily-life activities being carried out in crowds: in factories, offices, shopping malls, theatres, large schools, and universities, as well as the "virtual crowds" of television watchers. All this needs to be interpreted in the context of the ongoing desymbolization of society. Carrying out a growing number of daily-life activities in crowds was accompanied by unprecedented levels of desymbolization and thus a weakening of cultural support for guidance and direction. The considerable literature on crowd psychology suggests that people in crowds are vulnerable to external influences, which even cause them to engage in behaviour that normally would not come up in their heads.[47] The mass media in general and advertising in particular were successful because of the decline in cultural support and hence the growing need for a compensating equivalent.

Following the Second World War, there were essentially two explanations for what was happening to human life and American society. Both inadequately touched on the question of how Americans were being sustained by their way of life and society.[48] The most common interpretation regarded the processes of industrialization, modernization, and secularization as essentially liberating individual persons from the kinds of social constraints encountered in earlier societies. They had become free to be themselves, reach their true potential, and live richer and fuller lives of their own choosing. They had found wanting their traditional and religious points of reference, and this was made possible by their becoming rational and secular persons. These individuals now enjoyed a high standard of living accompanied by a multitude of possibilities from which they were free to choose whatever helped them fulfil their values and aspirations. All this was complemented by American democracy, which ensured that this individual freedom would translate into governments that followed the collective aspirations of these liberated individuals. Human rights legislation would ensure that everyone could participate in all these developments to the greatest extent possible. The widespread recognition that many people were being left behind moved some people to demand that government play a greater role because it alone could address this issue, while others insisted on government staying out of it because the problem was that these individuals were not behaving responsibly and were not taking charge of their lives through hard work. This common interpretation entirely ignored our being a symbolic species, with the result that all these liberated and so-called individuals nevertheless still required a culture (or its equivalent) as a shared way of making sense of, and living in, a community by means of a way of life in the setting of a local ecosystem. It is simply unimaginable that each and every liberated individual was now inventing for himself or herself the equivalent of a culture. Doing so would clearly be a crushing task; this burden had always been shared with others, jointly building on a rich cultural inheritance from previous generations. In other words, this explanation of what was happening to human lives and society was fatally incomplete because such a society should have gradually disintegrated for lack of an adequate social and cultural unity.

The alternative explanation essentially suggested that human lives and American society had lost their way and were in decline as a consequence of a widespread abandonment of the Christian morality and religion. As people lost an orientation and points of reference, their respect for law and order had also declined. It was therefore the responsibility of those who still had traditional moral and religious commitments to bring America back to its moral and religious roots. It had to be called back to God and the Bible. The moral and religious life of the nation had to be restored, and this gave certain groups their reason for being. Something equally fundamental is missing from this interpretation. It assumed that the only possible orientation and points of reference had to be the Christian morality and religion, and that abandoning it amounted to the creation of a moral and religious vacuum. Had there been such a vacuum, then the supposed decline of American society for some two hundred years should have led to its collapse. Without the emergence of two new generations of secular myths, such a collapse would have occurred, but it is precisely this development that at least in part filled the supposed vacuum. Did the hard-line democracy of the McCarthy era not accomplish the same kinds of secular religious functions that Communism and National Socialism had in other parts of the world – by building on the deep symbolic metaconscious developments of their adherents? Possibly the most serious error in this interpretation resulted from the fact that Christianity had made morality and religion a personal affair centred on individual salvation or social responsibility. All moralities and religions throughout human history have functioned on the level of a community, as is evident from such disciplines as cultural anthropology, the sociology and history of religion, and depth psychology.

From a cultural perspective, both explanations implied that the developments of the last two centuries had transformed Americans into cultural Robinson Crusoes, each person building his or her own "cultural island." They confirmed something that was deeply felt by many Americans, namely, that something fundamental had changed in the way American society sustained them. According to the first interpretation, they had to become secular and rational in order to be liberated from, and live without, the kind of cultural support a symbolic species had always required to be able to live in an ultimately unknowable universe. According to the second interpretation, they had to rely on a morality and a religion that could no longer be sustained by their deepest metaconscious knowledge, which included new secular myths. Both explanations ignored how the door to anomie, relativism, and nihilism had essentially remained shut. Somehow, people must have found new ways of orienting themselves.

From a sociological and cultural perspective, the question needs to be addressed as to how the emerging American mass society was sustaining its members in the post-war era. These members must have been sustained by something other than cultural support in order to find a meaningful orientation and reference points for living in the world, which could bind them together. If modern physics has taught us anything, it is how utterly complex our relationship is with an ultimately unknowable universe. I will not repeat what Niels Bohr and Albert Einstein had to say on this subject, and will refer to more recent works.[49]

One of the very few, if not the only, sociological studies of how the emerging American mass society was sustaining its members was undertaken by Jacques Ellul.[50] I have always been amazed by the reactions of my social science students in general, but especially the sociology students, to his description of human life in a mass society. It clearly resonated with them as a plausible explanation of what was happening in their lives. They simply had never encountered any description that came as close to helping them understand what they were experiencing. It is the only analysis I know which meets head-on the question of how human lives can be sustained in societies whose cultures have been desymbolized by industrialization for some two centuries. This analysis has been largely ignored because it goes against our current ideologies, by means of which we tell ourselves that, despite the decline of every living entity on the planet, we will be all right as long as we develop technique. Since some aspects of Ellul's interpretation were dealt with in the introductory chapter and since his work was extensively discussed in previous works,[51] it will be referred to only briefly in the following comparison of traditional societies to contemporary mass societies.

The first difference between contemporary mass societies and their traditional precursors is the way each new generation is socialized into a unique way of making sense of, and living in, the world. In traditional societies this involved becoming a member of a symbolic species and gradually entering the symbolic universe of a community, and learning to live within it by means of a way of life and a culture. In contemporary mass societies, each new generation must gradually be socialized into a technical order, even though doing so begins with entering what little remains of the culture-based connectedness. Very gradually and metaconsciously, children discover that they are involved in two streams of experiences: one related to the reality of this technical order and the other related to what is true in the lives of their significant others and their playmates. Initially, babies and toddlers will not differentiate these two streams of experiences. As their attention becomes increasingly drawn to electronic toys, television sets, computers, and all manner of screen-based devices, they will slowly and metaconsciously discover that, although similar in some ways, these experiences are very different from being told stories.[52] In the latter, the foreground may be discontinuous from the background, but both belong to what in a traditional society would have been the embryonic beginnings of a symbolic universe. As shown in their behaviour, children gradually and metaconsciously learn that their experiences associated with these technologies are highly desymbolized, as is reflected in a significant loss of embodiment, participation, commitment, and freedom. These developments will be further reinforced when they enter high school and learn discipline-based sciences. For example, their behaviour begins to show that they have learned the difference between intuitive physics and school physics. They learn to relate differently to what we have referred to as reality in the form of the domain of a discipline, and to a dialectically enfolded, culture-based connectedness in their personal lives with those of significant others. For example, there may be a temporary confusion about computers appearing to think somewhat like themselves but never expressing any emotions. Similarly, their behaviour shows that they learn to respond differently to persons on a screen from the way they respond to persons who are fully present.

As the kinds of metaconscious knowledge associated with the two streams of experiences become more and more differentiated from each other, children are gradually being socialized into the consequences of what began when the technology-based connectedness was unfolded from the culture-based connectedness: the introduction of the technical division of labour, its reinforcement when knowing and doing were separated from experience and culture, and the spinning off of almost everything related to a way of life in a dialectically enfolded symbolic universe, with only people's personal lives remaining as its residual "inhabitants." As teenagers they learn to join the other members of their mass society, whose organizations of their brain-minds are divided between the "lives" lived by participating in the evolution of the technical order and their personal lives lived in relation to the personal lives of others. People learn to defer to specialists for all knowing and doing, and to the state for everything related to a way of life in a sphere that has either been politicized or privatized. They learn when to rely on discipline-based approaches and when they can rely on whatever is left of symbolization, experience, and culture. The division of people's lives into these two streams of experiences was successively integrated by two generations of secular myths. These myths prioritized people's participation, first in the technology-based connectedness and then the technique-based connectedness, over everything else in life. The technical order comes first because its knowing, doing, and organizing are without limits and thus capable of bringing everyone whatever they desire.[53] Almost everything of any consequence to a mass society and its way of life is now separated from people's experience and culture. As noted, what was left of traditional moralities and religions had to become a personal matter because public life lived in relation to the technical order was guided by a statistical morality and new secular political religions. In the same vein, public opinions now govern public life, while private opinions are essentially limited to people's personal lives.

The above kinds of developments are greatly reinforced by the way the web relates everything to everything else, extensively, by way of hyperlinks. It represents an interrelatedness of non-sense; that is, hyperlinks have nothing to do with the meaning and value of the content as established by symbolization, experience, and culture. A technical substitute for the meaning and value of any information is created by means of sophisticated statistical techniques for analysing the networks formed by hyperlinks and their associated websites. One of the many significant factors of such analyses is the frequency with which such websites are consulted and updated as a web-based reflection of, and reinforcement of, public opinion and statistical morality in regard to this information. On a deeper level, all this represents a fundamental rearrangement of how everything is related to everything else by means of techniques. The technical order is extended into what, prior to these developments, was the order of sense. These kinds of developments represent another important step towards the emergence of *homo informaticus* associated with the anti-societies that have highly desymbolized cultures.

This raises the troubling issue of how far this desymbolization can go before babies and children will have enormous difficulties growing up, since they continue to depend on symbolization, experience, and culture. Already, it appears to affect language acquisition, creating a considerable diversity of learning disabilities. We may well discover that search engines represent the most powerful techniques invented thus far for the systematic desymbolization of our world of sense. The more successful they become, the more our reliance on the meaning and value of information established through symbolization, experience, and culture will be censured as being subjective and unreliable. This development is increasingly self-reinforcing as children and teenagers become more dependent on "googling" everything, causing them to leave behind what little they have acquired of a symbolic universe of sense in order to enter into the reality of the technical order built up without any reference to sense. The more we all become enclosed in this technical order, the more our escape by means of resymbolization will prove to be difficult. We have begun the re-engineering of all cultural meanings and values into their universal technical opposites by the most comprehensive and massive desymbolization of all cultures to the benefit of a global technical order, which in the end cannot sustain a symbolic species and which is unsustainable by the planet.

A second important difference between contemporary mass societies and traditional societies is the way culture no longer mediates between the metaconscious self and everything else in human life. In a previous work,[54] I have suggested that cultural mediation includes a scientific, technological, economic, social, political, legal, moral, religious, and aesthetic dimension. In a traditional society, each was a particular expression of the cultural spirit of an age (a kind of cultural DNA); collectively, they mediated all relationships with a local context in a way that was unique to a particular time, place, and culture. In contrast, in contemporary mass societies the scientific, technical, economic, and social dimensions have become universal in character and thus inadapted to a local context. We have a universal science, technology, and technique as well as a global economy. The latter is the result of a great deal of technical planning supplemented by what remains of the Market. We will soon show that the other dimensions of mediation are dominated by human techniques. As a result, the domains of the disciplines involved now mediate directly or indirectly between our metaconscious selves and everything else. This intermediary is very different from a culture, which respects how everything is related to everything else as much as possible. In other words, technical mediation tends to "filter out" everything unrelated to a domain, a process that includes but is not limited to the previously discussed triple abstraction. No domain-mediated interventions in our lives and the world respect their interrelated characters, as is evidenced in the human crisis (to be discussed in the next chapter), the social crisis resulting in the emergence of anti-societies, and the environmental crisis that has made contemporary ways of life unsustainable. Technical mediation is never neutral. For example, a telephone "filters out" the eye etiquette, facial expressions, and body rhythms from face-to-face conversations. Text messages and emails filter out even more, and the social media require a reorganized participation as compared to that of daily life. As we will see in the next chapter, this has a profound effect on many young people, even to the point of making some of them uncomfortable with face-to-face conversations because they are not technically mediated and must therefore be approached in a very different way.

A third important difference between contemporary mass societies and traditional societies is the way their social fabrics sustain the lives of their members. Generally speaking, traditional societies had limited social mobility, as most children followed in the footsteps of their parents. The importance placed on this lack of social mobility is indicated by its being attributed an ultimate religious significance and justification. For example, in the Christian Middle Ages it was commonly believed that God decided where to place everyone in the social hierarchy at birth, with the result that people could not question their social positions without challenging everything these societies stood for. Other societies fixed the social positions of their members in accordance with how the gods judged their previous incarnation: if a person had lived a good and virtuous life, they would merit a higher social position in their next life, but if they did not, their position would be a lower one. The universality of this approach to social mobility likely corresponds to something fundamental. This would appear to be the necessity of transmitting a way of life based on knowing and doing embedded in experience and culture from one generation to the next. When work was organized mostly within an extended family, children would receive a kind of informal apprenticeship by helping out their parents or grandparents with whatever needed doing. This informal apprenticeship could then be followed up by a formal one in which they would become the apprentices of a master in order to move to the final stages of human skill acquisition and reach the level of expertise. The historical records show that this almost guaranteed that the entire cultural inheritance would be passed on from one generation to the next.

In contrast, a low social mobility is disastrous for a contemporary mass society. If the need for discipline-based approaches related to the building and evolving of a technical order is greater than what can be met by the usual elite going on to higher education, social mobility must be increased as required. We have noted that, following the Second World War, the United States had to do everything possible to widely recruit people with talents in discipline-based approaches to knowing and doing in order to meet the growing demand created by a way of life that increasingly depended on such approaches and less and less on their culture-based counterparts. Today we need fewer and fewer people to evolve a way of life, with the result that for a great many people, higher education no longer guarantees an upward social mobility. It has become a recipe for underemployment and unemployment. The higher social strata of a mass society are now generally occupied by those who have considerable financial resources and social connections as well as a reasonable ability in discipline-based approaches to knowing and doing, and those whose abilities are exceptional. It is as if every member of a mass society is placed into a personal "social elevator" at birth for an individual ride to their eventual social stratum. These personal elevators are powered by the occupant's social influence, wealth, and abilities as well as the need for them as determined by the technical order. Accordingly, some people ride to the top or the bottom, but most arrive somewhere in between. The consequences for families, groups, friendships, and other social bonds are devastating: they become strained to capacity and, in many cases, break up because people move into very different spheres and have difficulty understanding one another. We are approaching a situation where it is nearly impossible to develop meaningful, lasting, and sustaining social bonds and relationships, and thus the social fabric of mass societies is becoming incapable of sustaining people's lives. In traditional societies, such conditions were encountered only when the societies were in an advanced state of disintegration and near collapse.[55] These developments result in a further desymbolization of the cultures of mass societies.

A fourth difference between mass societies and traditional societies is that the members of the former live in a reality-like "world" while the members of the latter live in a symbolic universe. This "world" is constituted by a bath of images diffused by the mass media and the social media, into which its members are immersed.[56] In addition to constituting a "world," the bath of images is also a technical substitute for the role custom and tradition used to play in traditional societies, but with several important differences. First, customs and traditions evolved according to what was meaningful and valuable for individual and collective human life and thus with what was true, while integration propaganda is of the order of what is real. Second, customs and traditions were experienced through symbolization and thus enfolded into people's conscious as well as their metaconscious selves. In the case of integration propaganda, the "experiences" limit embodiment, participation, commitment, and freedom to such an extent that people become spectators to their world, as we will see. The bath of images contains everything people need to know for participating in their society and living in the world: what to eat and drink, what to wear and own, how to relax and enjoy yourself, what to ask a doctor, the risks to be covered by insurance, and how the events around the globe affect human life.

Generally speaking, prior to industrialization, human life was very local. A great many events that were important to people's lives were local ones. Consequently, they were directly accessible to experience, or indirectly accessible via the accounts of family members, friends, and neighbours. There were events beyond the local group that occasionally had a significant influence, but this was rare. As a result, the symbolic universe of the local group was deeply shared, and even those of neighbouring villages were noticeably different. Strangers stood out in all kinds of ways. The local "scale" of human life was completely transformed as a consequence of industrialization, beginning with production no longer being geared to local needs but to markets around the globe.

In a mass society dominated by mass production, mass consumption, mass media, and now the social media, very few events important to people's lives are directly or indirectly accessible to experience the way that was possible in the past. The mass media inform us about events all around the globe that could potentially have a significant influence on our lives. Such events no longer contribute to the enrichment of what little remains of the culture-based connectedness and the symbolic universe of our personal lives. We can readily confirm this by making an inventory of all the events dealt with on the news during one week. It becomes immediately evident that one event may be covered for several days in a row for at best a minute or two, but we have no idea whatsoever what led up to this event and what happens to it after coverage ceases. The "world" of the mass media is about as discontinuous and fragmented as it can possibly be. Worse, the reports of any event have been taken by an editor from a diversity of sources, with snippets recontextualized into a news story that "works" for the audience's need for understanding and for the sponsoring advertisers targeting this audience. The dominance of images over the word further distorts what is being reported and how, and what is not reported. For example, long-term trends are not very amenable to the image-dominated media. Hence, it is next to impossible to have a genuine understanding of how our world is evolving and the way this affects our lives.

The "event-dots" of our world cannot be symbolized and thus become part of our lives and our world. Some external guidance is required as to how we are supposed to fit all these event-dots together. This is done by patterns of understanding that take the form of commonplaces and stereotypes rooted in our secular myths: we are reassured that life is progressing in a direction of technical advancement and economic growth; that this economic growth stimulated by free trade is the solution to most of our woes; that education can keep our society in the global race; that democracy divides the world into our friends and foes and that it will triumph over the latter; that medicine can deliver us from disease and aging; that technical goods can bring us the life we desire and keep us in touch with each other and the world; and that our history is moving in the right direction and improving human nature. These kinds of external collective patterns have been all but substituted for the metaconscious knowledge that used to integrate a person's experiences into a life as well as guide the collective experience of a community on its historical journey. The sense of everything is now created externally to people's lives, produced by public relations firms or public relations experts acting on behalf of corporations, governments, and organizations of all kinds. Much of what "works" on the media resonates with, and reinforces, the "myth-information" we have acquired about our lives and the world.

Making sense of the "event-dots" that the members of an anti-society encounter each and every day of their lives may be compared to the interpolation and extrapolation of highly desymbolized, weakly differentiated, and thus very blurry dots on a piece of graph paper that are spread far apart, with the result that a set of adjacent curves can be fitted through them. Whatever metaconscious knowledge may result from the symbolization of the event-dots of the media is likely to be rather blurred and vague, with corresponding effects on its meaning and value. This blurriness leads to a great deal of desymbolization, with similar effects on the use of language.[57] Accordingly, we may expect that the metaconscious knowledge that we gather from the mass media and social media will suffer the same fate. This is particularly important for the metaconscious self, and for the deepest metaconscious knowledge that constitutes the myths of a highly desymbolized culture. That this is the case has been largely confirmed by the evolution of advertising and art in the last fifty years.[58] Initially, advertisements bonded consumer objects to symbols of youth, sexuality, and death.[59] As the influence of "technique changing people" grew over several decades, attempts to replace weakening symbols with corporate logos became more common, and these were associated with the trusted providers of the "goodness" granted by technique. Such efforts were reinforced when corporations associated these logos with the life of the community by sponsoring its events and naming its meeting places, from sports centres to concert halls as well as university buildings.

A fifth difference between mass societies and their traditional counterparts is the way they rely on myths. This difference involves what may appear as a contradiction in our arguments thus far. If the levels of desymbolization were as high as I have claimed, would the development of metaconscious knowledge not be so impaired as to lose its ability to generate myth? If this were the case, could integration propaganda replace myths, given what we know about the limits of the much more intense totalitarian propaganda?[60] To answer these kinds of questions it is important to recognize that the meanings and values of the experiences related to the technical order have been desymbolized to the point that the use of language has been substantially impoverished.[61] It may be expected, therefore, that the role our secular myths play ought to have suffered the same fate. We have suggested that the evolution of advertising and art during the last half-century would point in that direction. All this is true. What may appear as a contradiction can be explained in terms of our theory of culture. The organizations of our brain-minds appear to evolve on the basis of two distinct but independent processes: differentiation and integration. Desymbolization appears to have a much greater effect on the former than on the latter. The process of differentiation seeks to uncover the significance of an experience for our life by symbolically placing it into that life and differentiating it from all other experiences. When embodiment, participation, commitment, and freedom are significantly limited, this process is undermined and the meaning and value of that experience are desymbolized, with the result that our ability to fully live this experience is similarly diminished. All this has profound repercussions for our entire lives because each moment of these lives is what all the others are not, since the process of differentiation is based on a symbolic dialectical enfolding of each experience into our lives. Consequently, the presence of a stream of highly desymbolized experiences in the lives of the members of mass societies weakens all meanings and values established by the process of differentiation, right up to and including the central myth, which continues to anchor the hierarchy of the desymbolized values of the culture.

While the process of differentiation creates unique and distinct moments of our lives, the process of integration creates and evolves these lives by the organizations of our brain-minds symbolically interpolating and extrapolating all experiences as moments of one life. As such, all experiences contribute to the metaconscious knowledge that sustains our lives. This extends all the way to the deepest metaconscious knowledge by which we symbolize the unknown as the myths of our culture, which make life in an ultimately unknowable universe possible. This metaconscious knowledge also sustains the identity of our individual and collective human lives despite ongoing adaptation and evolution. We remain recognizable despite constant change. It would appear that the deeper the metaconscious knowledge lies (its depth correlating with how much of our lives is integrated by particular metaconscious knowledge), the less it is influenced by any single experience, no matter how desymbolized the experience may be. This is especially true when we approach the collective metaconscious. Nevertheless, the interpolation and extrapolation of highly desymbolized experiences will lead to similarly desymbolized metaconscious knowledge, previously explained in terms of a broad curve fitted through fuzzy data points. In other words, the unknown will be symbolized as more of what is highly desymbolized; anything radically other will be unthinkable and unliveable, thereby locking us into the downward spiral of desymbolization. Nevertheless, no matter how highly desymbolized this metaconscious knowledge is, it will continue to rule out anything that is radically other than what we know and live. In other words, even when all meanings and values and thus all metaconscious symbols are highly desymbolized, we continue on the path of what we know and live. Everyone is placed into the following dilemma: the fact that the deepest metaconscious knowledge takes the form of our current generation of secular myths delimits what can be known and lived, but within these limits the myths provide little guidance because of the desymbolization of all meanings and values and thus of the ultimate meanings and values they represent. As a result, contemporary mass societies would disintegrate and collapse if some external guidance were not provided for the other-directed members.

Other-directed people depend on integration propaganda to complement their highly desymbolized culture. It would appear, then, that integration propaganda in conjunction with metaconscious secular myths has kept the door shut to anomie, relativism, and nihilism. The organizations of the brain-minds of the members of contemporary mass societies metaconsciously rule out any knowing other than science, any doing other than technique, and any organizing other than what is done politically by the nation-state. We metaconsciously acknowledge this when we say that "time will tell" or "history will judge," which makes sense only if the single remaining historical journey for humanity is exclusively based on this universal knowing, doing, and political organizing. It is no longer essential for the social fabric of mass societies to integrate their members into a dialectically enfolded society capable of adapting and evolving its culture, other than to allow babies and toddlers to be launched in the direction of a symbolic species and to sustain of what is left of people's personal lives and their joint culture-based connectedness.

A sixth important difference between mass societies and traditional societies is that the former undermine our being a symbolic species, while the latter sustain it. Similarly, the difference between a tradition-directed and an other-directed person is that the latter corresponds to a minimal and divided self "living" in a residual culture-based connectedness surrounded and dominated by the technique-based connectedness. This is the subject for the next chapter.

An illustration of the above can frequently be found in situations in which the old and the new coexist in people's lives. When children growing up in a relatively traditional family reach their teenage years, their metaconscious selves have become clearly other-directed. For example, when asked for permission to go and see a movie, a parent may reply that she is not keen on the idea because the reviews describe it as a gratuitous celebration of violence for the sake of violence, and Hollywood should not be rewarded for making it. The teenager objects: almost everyone in the class has seen it, so why is he not allowed to see it? For a more traditional parent, such an argument is incomprehensible because what is normal does not become normative other than through the intervention of traditional values that no longer emerge in the metaconscious of teenagers. Reciprocally, the argument of the parent is incomprehensible to the other-directed teenager, for whom whatever everyone else does is normal, and what is normal is normative in the absence of effective cultural values. Such values cannot be inculcated by the convictions of parents because they are no longer culturally supported. Desymbolization has shut out the way cultures would have sustained such situations in traditional societies. It is making the raising of teenagers by traditional parents a next to impossible task. Most parents know this very well, but they may not understand the reasons why things have changed so dramatically in such a short time.

In the same way, the other-directed teenager picks up public opinions, which differ fundamentally from private opinions formed on the basis of direct experience, reading up on a subject and thinking about it, or (ideally) both. The other-directed teenagers live by the public opinions gathered from their peers and social media as well as by a kind of statistical morality that makes what everyone does normative in a highly desymbolized culture. All this is further reinforced by the schools, which divide the students' learning about themselves and the world into relatively distinct subjects that do not build on any corresponding knowledge they may have gathered from experience. The learning of mathematics, physics, and chemistry leads to an even deeper division of the world into the one of daily life and the one of domains that cannot be lived in the same way.[62] In other words, it does not take very long before teenagers begin to live as if they have few, if any, cultural resources and therefore must rely on others, supplemented by a myriad of facts and opinions available from their mobile computers (euphemistically referred to as cell phones). These kinds of developments are constantly reinforced as they encounter activities and events structured by means of technique.

A seventh difference between mass societies and their traditional precursors is that the former have become entirely dominated by a new kind of citizen. It is the outcome of a long process that began when corporations were granted the status of legal persons.[63] This situation in mass societies is unique because, as a result of high levels of desymbolization, these legal "persons" no longer face significant resistance. Compared to human citizens, they are what a highly desymbolized symbolic species can only approach: technical entities without any cultural resources to constrain anti-social and psychopathic behaviour. These corporate "citizens" thus have all the rights and the power to get around almost any obstacle or obligation. To make this liveable for their executives, they have to become addicted to greed, as evidenced by skyrocketing salaries and bonuses for CEOs that have little or no correlation with corporate performance. Because of desymbolization, the influence and power of these corporate citizens have grown to a point where not a single government is willing or able to redress the power imbalance between citizens to rescue some democracy. Instead, governments continue to increase the imbalance by negotiating free trade agreements that further strengthen corporate rights by undermining the ability of communities to run their own affairs. Step by step, the world is made safe for technique and unsafe for human beings, communities, and ecosystems. In the face of these trade agreements, all political differences are entirely trivial and meaningless. How can we claim to have a functioning democracy when a corporate citizen can exercise a level of influence that cannot be matched by millions of human citizens? The corporate citizens do not need to vote, nor do they have to unveil a political platform. Their collective influence on society can readily ensure that what is technically required will be supplied by society and the biosphere. Who will protect our drinking water from the effects of fracking? Who will protect the liveability of rural areas and communities from wind farms, which generate energy as well as acting as one of the most complex noise generators we have devised? Who will protect our food supplies from the experiments with GMOs and the latest generation of pesticides and herbicides? Who will protect our public sector institutions from undue corporate influences? Who will protect our right to "buy local" and prevent our jobs from being exported? Who will stop the jobs of our children and grandchildren being exported, thus depriving them of a meaningful future? No one across the entire political spectrum has any idea how to deal with technique and the myriad of collisions it creates with everything that matters to us. There appear to be few, if any, politicians left that even mention the public good. This is another reason why contemporary mass societies with their corporate legal "persons" have been turned into anti-societies and anti-democracies.

A final important difference between mass societies and their traditional precursors is the organizational role cultures play in traditional societies. Metaconsciously, symbolization gives everything a place and a value relative to everything else in individual and collective human life, thereby helping to organize and evolve societies. This made local communities highly self-regulating, leaving the state a very limited organizational role that was largely political in character. With industrialization, the economy was separated from society by the technology-based connectedness dominating the culture-based connectedness in mass societies, although the reverse was still the case in traditional ones. This created unprecedented levels of desymbolization: the economy now had to be organized by means other than a culture, while the members of society had to use the organizations of their brain-minds first as a gyroscope and then as a radar, which soon had to be complemented by integration propaganda. The state had to play an ever-greater organizational role everywhere, regardless of the political regime, causing the political sphere to expand and to take on a more organizational character. Everything became regarded as political, which implied an almost unlimited role for the state. The nation-state thus became one of the two central myths of contemporary societies. Three secular political religions have thus far been anchored in this sacred: communism, socialism, and hard-line democracy. In the twentieth century, these three secular political religions performed exactly the same functions as did their traditional precursors, which were now relegated to people's personal lives. The growing power and role of the state made it all in all – a role that was previously played by culture.

What is emerging as a result of these eight differences may be called anti-societies, in the sense that almost everything in these societies has been turned into its opposite: culture-based knowing into domain-based anti-knowing; culture-based doing into domain-based anti-doing; economic mediation guided by cultural values and institutions into anti-economies organized by technical planning and what little remains of the Market; and everything else, both social and human, into resources for evolving a technical order with a residual dependence of people's personal lives on what remains of highly desymbolized culture-based connectedness.

In other words, symbolization, experience, and culture as humanity's approaches for understanding and living in a world in which everything is related to everything else have been turned into their opposites, namely ignoring this interrelatedness so as to understand and deal with everything on its own by separating it from its context or by comparing it to other forms of the same thing. Domain-based knowing and doing and the organization of society by the state proceed as if the culture-based connectedness of human life and society has nothing to contribute. The knowers can exclusively rely on their discipline-related domains, and in their personal lives on integration propaganda. Those who are on the receiving end of this knowing and doing have to rely on integration propaganda and other human techniques, including human factors, public relations, group dynamics, and pedagogy, as we will soon consider in greater detail. In the past, people lived in symbolic universes based on the meaning and value of everything within them, which enabled them to live according to what was true for them, while people in contemporary mass societies live in a universe created by the mass media and the social media, which is delimited by our secular myths and has the character, or near character, of what we have referred to as reality, as opposed to a dialectically enfolded symbolic universe. It becomes difficult for people to *live* many of their experiences, in the sense that they cannot make them meaningful moments of their lives because of a lack of embodiment, participation, commitment, and freedom. For example, people cannot *live* the myth-information of the "world" created by the media, although they could live in the symbolic universe of daily life in past cultures. The former provides a bundle of "signals" to the "radars" of the other-directed personalities, while the latter created metaconscious "maps" for people based on their experience and participation in a way of life. What people "experience" in the "world" created by integration propaganda has become more "real" and "important" than what is left of our daily lives. It has become much more "real" to "live" on the media than to live our daily lives.

The eight differences between anti-societies and traditional societies also help to explain an extraordinary diversity of metaconscious knowledge of separation, disconnection, non-life, and death in the members of mass societies. As a result, many people experience a great deal of anxiety and depression spurred by vague intuitions rooted in the fact that the desymbolized metaconscious knowledge of their lives has been misaligned and decentred in relation to what (as other-directed persons) they have learned their lives ought to be, according to the mass media. There is a vague sense that, as members of a highly desymbolized symbolic species, they have become a resource for technique. They are powerless against the "system," and this is very stressful. The experiment of making humanity in the image of non-life appears to have a much greater impact on our mental health than most people are willing to admit.

Human Techniques and Anti-Societies

We have thus far argued that the desymbolization of human lives and societies begins with the way science, technique, and the nation-state constrain embodiment, participation, commitment, and freedom. Although we have suggested that becoming an expert in a field requires a commitment to the related body of experience so that the risks associated with learning and doing things can be faced, we have far from plumbed the depths of what has happened to human commitment. This commitment is no longer sustained by the deepest metaconscious knowledge acting as a kind of ultimate meaning and value symbolized by secular myths. These myths are limited to ruling out any radical alternatives, and thus genuine solutions based on attempts to resymbolize our situation and take charge of it are also excluded. In other words, some kind of technical complement to what little commitment is possible must be created in highly desymbolized cultures. A new spiritual commitment is engineered by means of human techniques. In order to understand this, we must revisit what we have discussed thus far in relation to technique from the perspective of how it first undermined our commitment and then re-engineered it by means of human techniques.

Prior to the introduction of the technical division of labour and mechanization, people could produce something that was their own as an expression of their experience, skill, creativity, and imagination – something of which they could be proud. They might be able to leave a mark on their world and possibly live on in their accomplishments and gain a measure of immortality from the recognition of others who outlived them. In other words, human work could contribute to a metaconscious knowledge of vitality and a degree of immortality. These possibilities were out of reach for most members in later societies because embodiment, participation, commitment, and freedom were severely constrained, be it in very different ways. Consequently, the desymbolization of work was widespread, and most people simply did what was necessary in order to get on with their lives to the greatest extent possible. Work was not a virtue but a necessity.

With the introduction of the technical division of labour, mechanization, and industrialization, all this changed. People now worked as a hand, a brain, or an intermediary between the two. A person working as a hand would endlessly repeat a production step in a domain-like setting, which was connected to other such settings that jointly structured the work organization and the factory that housed it into what we have referred to as reality. In turn, all this was integrated into the technology-based connectedness of an industrializing society that excluded, as much as possible, the culture-based connectedness of human life. There was no meaning or value in this work, but only a technical significance for the local technology-based connectedness. It was the performance of the task and not the contribution to human life and society that mattered. It was therefore very difficult to derive any human satisfaction from that work, and only draconian factory discipline could maintain this emerging order.

For the people working as the brain or an intermediary between it and the hand, there was little meaning and value other than an increasingly well developed theoretical technical significance relative to the local technology-based connectedness and its performance. For all three groups of participants in industrial production, there was therefore very little human satisfaction such as that derived from making a meaningful and valuable contribution to the community. Ultimately wages, wealth, and power were the only substitutes for whatever human satisfaction people may have derived from their work in pre-industrial days. Factory work and its organization and supervision led to metaconscious knowledge of working in the image of the machine: a non-life characterized by different forms of constraints rigorously eliminating any real embodiment, participation, commitment, and freedom.

During the early phases of these developments, there were intense power struggles as to what kind of society would emerge. Would technology have to be constrained to fit into what remained of a traditional society and thus serve traditional cultural values and aspirations, or would an entirely new kind of society emerge, with incredible prospects for wealth and power? As "technology changed people," the former choice essentially disappeared as the gradually emerging first generation of secular myths made it increasingly unthinkable and unliveable. All human freedom on this point disappeared, and the struggle between management and labour was reduced to a fundamental implicit agreement as to where society was heading, and an explicit disagreement over who would have the power to make what kinds of decisions and who would receive what share of the benefits. There was no longer a fundamental choice to be made. People had to adapt themselves to what was and bury what might have been. Culturally, this was justified by the new secular myths. Two centuries have passed without any real preventive approaches having been applied to the redesign of human work. This appears incomprehensible unless we analyse in some detail what changes in commitment took place.

Once human freedom had been decisively limited, adapting to the work organization gave people the considerable benefit of fitting in, belonging, and becoming happy by contributing everything industrial symbolization had to offer.[64] Stubbornly contesting the situation would have achieved the exact opposite: it would be discouraging to see yourself as a cog in the work organization, and working out the intuitions of metaconscious knowledge of non-life and death would be enough to make anyone depressed. Such people would have been frustrated and unhappy, since they did not fit in and contributed to nothing of value, other than a critique that made the opposite choice; and most people did not want to hear this. After all, it amounted to a waste of a person's life and a likely sinking into despair. Consequently, there was a growing convergence between the requirements of the emerging technology-based connectedness and the associated work organizations on the one hand, and people wanting and even needing to belong to it on the other hand. Especially the knowers began to so closely identify themselves with the work organization that they served it wholeheartedly and became the "company men" (still almost exclusively males in those days). Eventually, a considerable literature emerged about the organization persons, the climbers of the corporate ladders, and the business cultures that held it all together. The unions that had at first defended people's lives now simply defended their wages and benefits. Much of industry became the expression of the convergence between the necessities imposed by technology and the economy on the one hand and the need for people to belong, to be useful and respected, and to commit to what they are doing on the other. This brings us to the heart of our misconception of freedom, which is the opposite of this convergence. To choose not to obey necessity and to struggle for freedom in relation to what alienates us is much harder and much more challenging than to go with the flow.[65]

The kinds of choices people made in relation to their work must not be isolated from the many other choices they had to make in every other sphere in their lives. Such choices tended to be contradictory, and these contradictions in their lives had to be resolved in one way or another. Making these choices and deciding how to live with the contradictions amounted to a kind of spiritual act, which affected people's being and lives to their very depth. It is here that the cult of the fact gradually began to play an important role in the development of technique, a role that continues to this day in our own lives. As noted, facts are seen as being independent from the intellectual framework by which they are gathered and on which they depend. As such, these facts become absolute and thus asocial and ahistorical – a kind of eternally valid set of "bits" of what we have referred to as the reality that supposedly lies beyond or underneath the symbolic universe in which we live. Any possibility of interpreting, assessing, and judging these facts becomes impossible because it would require access to the intellectual framework from which they ought not to have been dissociated. In daily life, we constantly acknowledge this when we say that we must be realistic, that is, we must accept what is as real and not confuse it with what ought to be, according to our values. There is no role for symbolization, experience, and culture by which *what is* is transformed into its meaning and value for human life and society. The supremacy of *what is* also imposes the necessity of extrapolating this forward in time, as opposed to intervening in the situation according to the perceived meaning and value as a basis of orienting the evolution of *what is* by means of a culture. There is thus no room left for our participation as a symbolic species: the cult of the fact spins off our symbolic participation. In other words, to acknowledge that something is real and to accept this as if it were normative amounts to an enslavement to necessity. If humanity had treated what it experienced at face value, we would have never become a symbolic species. The cult of the fact makes us accept what is necessary as real, but more than a highly desymbolized culture is required if we are to be convinced of the excellence of what the cult of the fact compels us to do.

Being realistic about our work means that we might as well adapt to it. Besides, if we manage to accomplish this, less of the negativity of our work will spill over into our families and daily lives. We might also be better equipped to face the many contradictions in our lives. We are surrounded with time-saving gadgets, but we never appear to have time for anything. Our needs are constantly stimulated through advertising, and yet we do not have enough income to translate all these newly created needs into market demands. We are bombarded with information, but no matter how much education we have, it always appears to be insufficient to understand it all. We are told the virtues of unrestricted global competition, but we feel a lingering sense of having some responsibilities for our children, grandchildren, and others. We are told that we are all equal, but we are regularly confronted with numbing statistics of how anti-economies are steadily amplifying the levels of inequality (fewer than one hundred people are said to already own one-half of humanity's wealth). Our upbringing and education emphasize the value of each and every individual person, and yet we can participate only as a hand, brain, or intermediary between the two, never as an entire subject. Our participation in a technical order excludes us from living our lives, and yet we experience the need to do so in our personal lives, in relation to what little culture-based connectedness is maintained by doing so. There are many more such contradictions.

In order to live with all these contradictions, a metaconscious knowledge of disconnection, separation, non-life, and death must not be permitted to dominate that of vitality and life. In order to achieve this, many of these contradictions must be rendered harmless by separating them from each other and then suppressing one of the elements. It is the well-known role of prejudices, commonplaces, and ideologies. During the war effort, which enormously intensified the contradictions in American society, these defences were insufficient, and human techniques had to be developed to boost group efforts and create collective patterns capable of integrating the elements of these contradictions deemed to be essential for a rapidly expanding technique-based connectedness. At the same time, these techniques had to degrade the contradictory elements as being trivial, irrelevant, or heretical to human lives. The stronger the contradictions people experienced, the more contrary facts had to be resisted and the more these collective patterns had to be reinforced. From a social perspective, these collective patterns were necessary as an external equivalent of, and complement to, what little remained of the culture-based connectedness of American life. Given their existential and cultural necessity in the emerging American mass society with its other-directed members, such collective patterns had to be broadcast by the mass media so as to reach everyone. Moreover, what "worked" on these media became directly related to the meeting of such existential and cultural needs. All this was accomplished with a much greater measure of spontaneity than we can readily admit. The existential and cultural needs of people converged with the need to adapt their lives to the technique-based connectedness of society, and this need in turn converged with those of the corporate "citizens." No doubt the growing concentration of ownership of these media helped this along, but underneath were needs deeply felt by all parties, and these converged extraordinarily well.

Without any overall coordination and little direct coercion, the collective patterns required by this mass society were thus put together by a variety of sources, all spontaneously contributing to a kind of social integration propaganda accomplishing what no totalitarian society had been able to achieve.[66] This social integration was facilitated by the rise of new heroes: movie stars, who amplified and celebrated physical beauty and played down the beauty of character and culture; champion athletes, who exemplified performance and efficiency and helped to play down their disordering effects on human lives (including those of the athletes themselves); and the political father or mother of the nation, who exemplified and celebrated the heroic leader, playing down people's being immersed in the crowd and the loss of the cultural support of democracy. The mass media also helped to deal with the disordering effects technique had on human life by dissociating public opinion from private opinion, statistical morality from personal moral convictions, the brain from the hand, and thus thought from action.

The cult of the fact also helped human techniques to create and sustain the kinds of social groups necessary for the maintenance and evolution of technique-based connectedness and everything that came with it, which included the suppression of what remained of culture-based connectedness. Human techniques such as group dynamics, human relations, public relations, pedagogy, and integration propaganda were developed, and their decisive roles continue today. People have to face the facts: individuals do not exist apart from the group, by means of which they participate in society. They thrive in well-functioning groups, and, at the same time, the collective functioning of the group owes everything to well-adapted members. When one member of a group is selfish, aggressive, or anti-social in other ways, the group suffers. In turn, this individual will be unhappy because of his or her maladjustment, while the other members may feel frustrated when they cannot reach their full potential. Moreover, it is the entire person that must be considered, and thus the emphasis has to be put on openness and trust. The conditions under which these groups operate in relation to a technique-based connectedness are never acknowledged. The context is simply ignored.

The human techniques designed to adapt the individual to the group, and the group to the needs of its members, had and continue to have the kinds of effects no technique can avoid. In the previous discussions of the rational approach as well as the technical approach, we noted that goal-directed behaviour, such as the improvement of efficiency, necessarily distorts the context taken into account. This problem also occurs in relation to the group: it is impossible to respect the integrity of both the members and the group. Whatever aspects are important for the group accomplishing its goals will receive all the attention and everything else will be ignored. For example, the qualities of the extroverted personality became regarded as essential for well-functioning groups, while the group functions that would have benefited from careful individual thought and reflection by introverted personalities were all but overlooked. Contemporary business schools are bogged down in this kind of optics, which helps to explain why one management fad follows the other while the underlying structural issues are ignored.[67] CEOs who are introverts and quietly deal with these structural issues tend to be much more successful, but little attention is paid to these findings.

The same is now happening in our schools, with a vengeance. With the emphasis on group learning, individual learners have a "problem," and introverted children have a kind of "social disorder" related to adapting and fitting in. Original thinking, broader reading, and critical reflection, which make people resistant to the mould created by the latest pedagogical fashions, are considered to be interfering with the group processes of the classroom. The greatest possible psychological and social plasticity must be inculcated by the schools, with devastating consequences for many children. It does, however, prepare them for dealing with the difficulties they will face as members of a mass society in uncritically adapting to a technique-based connectedness with a cultural support that has become highly desymbolized. It is this latter problem that is entirely overlooked in our schools and our mass societies. This is no accident, because it makes it possible to assume that the required adaptations are peripheral to the human character and thus can be readily made, while in effect they are so large that the entire person may be called into question, resulting in profound psychic difficulties. These human techniques all assume that the only difficulty people encounter is an inadequate adaptation to the life-milieu in which human life is lived. In this context, happiness results when the reciprocal interdependence between a person and that life-milieu is constantly being adjusted by means of appropriate techniques. Moreover, there can be no individual happiness apart from that of the members of the group through which individuals participate in a life-milieu. The individual must therefore reproduce what that life-milieu can offer, and doing so becomes a matter of evolving "correct" relationships.

Human techniques thus rest on a particular conception of human life: the individual helps to form the group and the group in turn helps to form the individual. We proceed as if there is no meaning for human life other than what is associated with participation in the life-milieu in relation to which human life is lived.[68] Implicit in these kinds of approaches is a fundamental ethical and spiritual choice: human life is now restricted to what can be observed by means of the cult of the fact; the person is entirely shaped by needs; and anything outside of the maintenance and evolution of technique-based connectedness is an aspect of human life that can safely be ignored. In this way, everything is reduced to the production and consumption of what is directly or indirectly necessary for this technique-based connectedness. In this context, people produce to consume and consume to produce.

Making sense of and dealing with life in this manner implies a variety of assumptions, according to Jacques Ellul.[69] Psychic phenomena always have their roots in a life-milieu, and this life-milieu is taken as a given without any critical examination. Contemporary mass societies live in the life-milieu of technique, which we thus take for granted, as opposed to taking responsibility for what we have created. Whatever of this psychic condition does not translate into human behaviour is of no concern. As we have noted for the scientific approach to knowing and the technical approach to doing, whether a phenomenon is of a psychic (living) or a physical (non-living) character makes no difference in the methods to be used. Finally, the way the relationship between the individual and the group is conceptualized is not critically examined in order to determine whether, from a social and historical perspective, it is relatively normal. In other words, the present situation created by technique is implicitly taken as normal, or as the best humanity has been able to make it thus far. In sum, these assumptions take the developments of the last half-century for granted, which means that everything related to the technique-based connectedness is important while everything related to what little remains of the culture-based connectedness of human life is of little consequence. The bias in favour of the extroverted personality type at the expense of the introverted personality type is symptomatic. Similarly, if the personal convictions of a member of a group interfere with its functioning, it is in the best interest of all parties to have these convictions "adjusted." The bottom line is that the historical journeys of all civilizations have been found wanting, leaving humanity with a single universally valid journey; since this is the case, nothing should be allowed to interfere with it because nothing good can come of it. This puts our creations beyond critical scrutiny, leaving no room for preventive approaches or other equivalents of practical wisdom.

Schools provide us with another good example. To varying degrees, they emphasize the need for children to fit in and belong, which implies an uncritical adaptation to technique and everything that comes with it. One of the most powerful ways of socializing children to live as other-directed persons is by means of the cult of the fact and a statistical morality. In the final analysis, it is more essential to fit a child into the group and adapt the needs of the members to those of the group than to ask questions about the direction in which the group is moving and to evaluate this orientation by means of symbolizing it. Proceeding in this way means that no external reference points will be required because children will learn to find their way while being minimally sustained by symbolization, experience, and culture. In the extreme, as long as people are adapted to their groups as good team players, the possibility that these groups may be moving in directions that are harmful to themselves, their members, and the future of a symbolic species, its communities, and the biosphere may form the subject of fascinating conversations in ivory towers but will not interfere with the all-important business of evolving and adapting our technique-based connectedness.[70] This connectedness demands that all other-directed persons learn to participate in the cult of the fact by withdrawing their earlier reliance on symbolization and culture. It implies an uncritical commitment and trust in discipline-based approaches to knowing and doing and a complete conformity to *what is* as opposed to what ought to be. It is the human technique of group dynamics which helps to engineer these commitments and ensure that everyone freely enters into them because everyone needs to belong and to fit in.

The human techniques of group dynamics are complemented by those of public relations. Once again, the well-intended reason for developing these techniques can readily be understood. As noted, the mass media insert people into a world that is radically different from the symbolic universe of a culture. Events come out of nowhere and disappear into nowhere several days later, and yet they can potentially have a significant influence on our lives. Our access to most of these events by means of symbolization, experience, and culture is virtually non-existent. This makes the world of the mass media threatening and, at the same time, necessary when our cultural resources have been highly desymbolized. Without further intervention, many of us would likely feel unhappy and uprooted by not really knowing what is happening to our lives and our world. It is impossible to relate to "reality" the way we relate to what little remains of the symbolic universe jointly created through our personal lives. No dialectical enfolding takes place, and this must be technically compensated for by means of public relations techniques. They "engineer" a great many events on the news to have a particular effect. What matters is not what our organizations and institutions are really doing, but the perceptions people have of what they are doing. The former would require a great deal of experience derived from conversations with people who have had direct experience, the reading of reports that summarize detailed studies that have been done on these events, checking some original sources directly, critically reflecting on all this information, and finally arriving at a private opinion of them. Following this kind of procedure is the closest the members of a mass society can come to symbolizing something in the world created by the mass media and social media. The problem is that no one has the time or the resources to undertake this process, even for a tiny fraction of all the "event-dots" they "experience." It is the perceptions that matter, as opposed to the complex web of relationships the "event-dots" obscure.

The gap between the "event-dots" and what is really going on is the locus in which public relations techniques operate, as follows: if only people had more information about what is really happening, they would understand that it is all reasonable and just, that they would have done things in much the same way, and that in this way their lives in the world would be justified. The provision of this assurance is the task taken on by public relations techniques. They "re-engineer" the kinds of "event-dots" affecting their employer or client. For example, many organizations and institutions of a mass society are unable to gain the trust necessary for their legitimacy. Corporations, governments, non-governmental agencies, churches, and charities all require the trust of the general public, but this cannot grow out of people's daily-life experiences with these organizations and institutions. For example, by means of public relations techniques, corporations inform the public that what they are doing is a consequence of the genuine influence and real power the public exercises over them. It is explained that, as consumers, members of the public make the decisions that create the demands for the products in the corresponding markets and provide the necessary feedback by responding to surveys; now, thanks to the social media, they can provide even more feedback and help to design the new products themselves. By clarifying what the company is doing and relating it to people's concerns and aspirations, public relations techniques create images that imply that these companies are there to serve their customers. Corporations are no longer the all-powerful legal persons who exercise the decisive influence on governments and societies at large. It is all about creating impressions as opposed to dealing with what is really happening, and ensuring that these impressions meet the needs of the corporations as well as the needs of the general public in a mass society with a highly desymbolized culture. Public relations firms and their spokespersons ensure that not only do the impressions of the general public become reality in the sense we have referred to it, but also that these impressions serve the needs of both parties.

It goes without saying that these public relations efforts never reveal anything regarding the business strategy of a corporation, how it makes critical decisions, where and how it manufactures its products, or anything else essential to its operations. The general public will be consulted only about relatively trivial details related to packaging or something else non-essential. Nevertheless, the public has the impression of having been consulted. Every organization involved in controversial activities will have mastered the art of conducting public consultations without having any intentions of changing anything essential. If the specialists in public relations do their homework, the general public will either walk away with a positive impression of a highly technical and sophisticated operation which they cannot help but admire, or they will accept it out of necessity: we have no choice other than being realistic.

The general public cannot expect that corporations reveal anything that is likely to meet with disapproval, such as the working conditions of the people manufacturing the product overseas, the effects this has on the families of these people and on their communities, the deterioration of local ecosystems, or anything else that may disturb the technical planning of the corporation. Public relations techniques manipulate the context in such a manner that everything revealed will be positive and helpful in gaining the confidence and trust of the general public. We are dealing with the diffusion of illusions, but ones that are absolutely necessary because they meet the needs of both parties.

Take the example of the movie entitled *Super Size Me*. It essentially told the story of a human being taking the place of a laboratory animal to test high dosages of "McFoods" instead of drugs. A very short period of exposure revealed alarming effects on his health, to the point that his medical team urged him to stop the experiment. If medical drugs had these kinds of effects during trials, there would not be the slightest chance of their being approved by the regulatory agencies. Very different impressions about the movie were "engineered" by the public relations firms representing the fast food industry, and these carried the day. Similar results can be observed when garment factories collapse with significant loss of life, when people have to move out of their homes because of the proximity of wind farms, when lotteries essentially tax the poor and create gambling addictions, and when real estate monopolies siphon enormous resources from the public by compelling the seller to pay the commissions of the agent representing the buyer, effectively blocking any real competition, to mention only a few. The reconciliation of the needs of powerful organizations and those of the general public in a mass society is accomplished metaconsciously, thus making it unnoticeable. That this is the case is evidenced by the widespread acknowledgment that a great many events on the news shows are directly or indirectly "engineered" by public relations techniques to create the desired effects for everyone concerned. By making certain associations and by ignoring others, public relations techniques all contribute to the creation of the collective patterns described above because of the deeper underlying need to converge life in a mass society with the actions of corporations, both related to an underlying technique-based connectedness.

There are no adequate concepts to describe the way the context of everything is being manipulated by including some associations and ignoring others in order to create certain feelings, images, and impressions. It is difficult to call this outright lying. It is a selective way of exercising the cult of the fact in order to externally create the equivalent of meanings and values that are no longer established by means of symbolization, experience, and culture. For the same reasons, it cannot be regarded as a deliberate and conscious manipulation of people in order to induce certain behaviour patterns, as was observed in the totalitarian societies of the twentieth century making use of propaganda.[71] This manipulation is a question of creating false associations that lead to impressions that are not true. It creates the kinds of spectacles we all need as members of a mass society with highly desymbolized cultures, who can no longer fully participate in our lives as a consequence of desymbolization. A lack of embodiment, participation, commitment, and freedom has diminished our ability to participate in the moments of our lives, to the point that we have become spectators of those lives to a historically unprecedented level.

Human techniques have entered, or are increasingly entering, a stage in their development that may be described as the battle for the human spirit. We have thus far examined the need for an external replacement for the symbolic universe of a culture into which people were dialectically enfolded.

Compensations for a highly desymbolized culture continue to be discovered in a relatively spontaneous fashion. For example, progressive high technology firms in California found that it was no longer sufficient to exercise the usual discipline over their employees and trust that society would do the rest by means of its culture. They turned themselves into a community of work to replace the community of life (society). A level of social control had to be established, which mass societies could no longer furnish. Progressive corporations discovered that providing their employees with the stereotypes, value judgments, and lifestyles to create a community of work was advantageous. Employees had to be formed for these new kinds of work communities. Employers unintentionally discovered that they had to perform a sociocultural role greater than the one traditional cultures had assigned to the extended family or to the guilds. A total integration of the person was required, not seen in human history since societies displaced clans, except that techniques now had to accomplish what cultures had done for these societies. It was also discovered that this battle for the allegiance of the employees could not succeed unless it was begun with their children as future employees and customers. Other institutions and organizations discovered much the same thing. Progressive schools were formed as well as other progressive organizations. It became even more important to combat introversion as taking distance from the group, genuine education as taking distance from social currents, and so on. Organizations dedicated to a single issue, however, could not follow this path because most of them put everything other than the issue into the background, to be largely ignored.

The rapid proliferation of human techniques is evidence of an implicit choice we have made in our civilization. We have placed all our trust in the discipline-based approaches of knowing and doing, and we expect the state to perform all the other roles that can no longer be taken care of by means of symbolization, experience, and culture, as a consequence of desymbolization. Furthermore, as noted earlier, another implicit decision was made to adapt individual and collective human life to the technical order that resulted as a consequence of this commitment. Human techniques of all kinds had to press human life into the mould made up from the necessities imposed by living as if discipline-based approaches to knowing and doing and the ordering of human life by the state had no limits. These developments led to a further implicit decision that we must attempt anything and everything, almost regardless of the risks to human life, in order to make the technical order work. Layer upon layer of compensating techniques created a top-heavy system that turned technologies from means into ends (to improve efficiency), economies into anti-economies, societies into anti-societies, and a self-regulating biosphere into something we must manage externally. We have also transformed our belief that it is society that corrupts people into the conviction that it is individuals who corrupt the group, with the result that being an introvert or being anti-social is a very serious matter. These kinds of people must be removed from the body social in order to be healed. The first step in this adaptation is to accept the cult of the fact, that is, to accept what is and to resist the temptation to interpret it or judge it by means of some external reference point. When we thus uncritically accept our condition, the fact becomes a value. The technical order is; and we must learn to justify our work, company, society, and the state. In addition, we must learn to accept the cult of efficiency and the cult of politics. Together they have contributed to the creation of all the external patterns to sustain human life in obedience to the necessities of technique. It amounts to a comprehensive and metaconscious choice of non-life over life.

As argued in the previous chapters, the opposite choice, for life, would require a combination of the resymbolization of science, the imposition of design exemplars and preventive approaches on techniques, a preventively oriented economic strategy, and additional alternatives yet to be discussed. The choice for life accepts that humanity has a history which has always been driven by the ideal that what ought to be should be imposed on what is. In our situation, what ought to be must challenge what is in discipline-based domains in order to allow back into our human life what today is kept at bay by discipline-based science and technique. By resymbolization, we can distance ourselves from the life-milieu of technique and create a modest sphere of opportunity in which we can once again live out a small measure of freedom in relation to what alienates us.

As we await this human reawakening to our condition, public relations will continue to tell us that our frustrations, such as not having our applications promptly approved by the government, are simply due to our failing to understand how many different offices have to approve it and thus our lack of appreciation of the complexity and the hard work that is being done by our government on our behalf. When an oil company encounters public outrage over some of its practices, it will have an engineer explain how complex the problem of extracting oil is; once we understand this reality, we will be reasonable and appreciate that we would have done much the same thing. In other words, human techniques of all kinds constantly turn our sense of what should be into an appreciation of what is. In this way, tensions between human life and the order of non-life are systematically eliminated and our alienation and reification are deepened as a result. We must live in and for the new technical order because it alone can make us smile and be happy. Thanks to human techniques, technique and the nation-state will be "all in all." They will accomplish this by taking human life out of the context of what little is left of our culture-based connectedness in order to insert us into the technical order with its technique-based connectedness. Nevertheless, technique ultimately depends on our being a symbolic species, and this means that it will never be able to absorb humanity into the equivalent of a mega-machine. In no way do anti-societies resemble a mega-machine.

Living in Anti-Societies

We have noted that in the period commonly referred to as prehistory, food-gathering and hunting clans lived in the life-milieu of nature. During the first phase of human history, societies emerged that became the primary life-milieu within the secondary one of nature. During the second phase of human history, the life-milieu of technique emerged, which mediated virtually all relationships with the secondary life-milieu of society and the tertiary life-milieu of nature. This development soon turned societies into anti-societies with the opposite characteristics.

Anti-societies minimally rely on symbolization, experience, and culture. As a symbolic species, humanity had always symbolized everything in its surroundings, turning "what is" into "what ought to be" according to a cultural ethos anchored and expressed through metaconscious myths. Everything was thus drawn into the symbolic universe of the society in which its members lived. Roughly until the Industrial Revolution, the cultural ethos of societies tended to have a traditional moral and religious expression related to the primary life-milieu of society and the secondary life-milieu of nature. In the same way, human consciousness reflected these life-milieus.

Industrialization, beginning with the introduction of the technical division of labour, gradually turned all this around because it implied a very different approach to the material and energy constraints of human life expressed by the first and second laws of thermodynamics. A much greater conformity to these constraints brought forth one necessity after another: the mechanization of the spinning of yarn necessitated the mechanization of all the other production steps in textile making. Once this had been accomplished, there was no choice but to mechanize all other industries. In order to make this possible, the population had to be concentrated in the industrial centres, and so on. The larger these patterns of necessities became, the more rigorously they had to be imposed because any non-conforming elements could disturb these patterns.[72] Cultures became more and more desymbolized in this process as a wedge was driven between the technology-based connectedness and the culture-based connectedness of human life. The greater the disordering effects of driving this wedge became, the more the cultural functions had to be replaced by a variety of human techniques; integration propaganda became the necessary complement to highly desymbolized cultures.

Together all these necessities ensured that "what is" all but eliminated "what ought to be." Had this happened during the human journey, a fallen branch on the forest floor would have remained exactly that, and nothing human and cultural would ever have emerged. Humanity would not have become a symbolic species. The reversal of "what is" and "what ought to be" embodied within a cultural ethos made it necessary for science to be autonomous in the sphere of knowing, thereby excluding all other forms of knowing from public life. We serve this science by means of the cult of the fact, which dissociates facts from the human context and acts as an intellectual framework that validates them. Facts thus become facets of an absolute reality that *is*. It was also necessary to make technique autonomous in the sphere of human doing by enclosing it in disciplines separated from all culture-based approaches. We serve it by means of the cult of efficiency, which dissociates domains from a symbolic universe in which something must be done according to its meaning and value in order to participate in the evolution and adaptation of a way of life as guided by ultimate meanings and values embodied in metaconscious myths. Technical doing had to be guided by output-input ratios following the separation and division from everything else, thereby making it appear objective and absolute.

When everything became political, the state became absolute as well and was served by secular political religions. Politics has been separated from relative political action, made ephemeral, and given over to the need to maintain power.[73] In all these spheres, the way everything is related to everything else and evolves in relation to everything else is ignored in favour of what we have referred to as reality, in which everything can be divided, separated, measured, quantified, and improved on its own terms. History was also dissociated from symbolization, experience, and culture to be entrusted to universally valid alternatives with the result that anything radically other became unthinkable and unliveable. We are now assumed to be travelling in the direction of history, and this history will be revealed by economic growth founded on discipline-based approaches to knowing and doing with limitless potential, thanks to global competition and free trade presided over by democracy. All this implies a technical ethos that is psychopathic in character because everything is known and improved on its own terms without any regard to the consequences for everything else. Technique and the nation-state incarnate this psychopathic ethos, which has become the crowning glory of our anti-societies.

These developments have made the living of human life in these societies extremely difficult. We have been completely marginalized as a symbolic species, and can participate only as human resources for science, technique, and the nation-state. The previously noted three contradictions lived by the American people just prior to the Second World War have been intensified, beginning with the rise of the phenomenon of technique and then its linking into a system by means of information technology. A fourth contradiction has been added, related to the growing unity of this system of non-sense – a unity achieved by disordering human lives, communities, and ecosystems. This simultaneous technical ordering and the disordering of all life are indissociably linked. Human techniques engineer our commitment to this ordering, while at the same time compensating for the disordering of our lives.

Nevertheless, there remain many signs that we find it extremely difficult to live in a reality of non-sense.

If, for reasons of our career and work, we must make the best of the non-sense of discipline-based approaches, we will experience a profound inner meaninglessness and loneliness. It will divide our multiple metaconscious social selves from our metaconscious psychic selves and create additional tensions between them and our metaconscious spiritual selves. If we have the opportunity to choose our work and live our lives so as to have as much meaning as possible, we will still face the difficulties of integrating the many screen-based devices into our daily-life activities. If we can do this along with other kindred spirits, we may well be less lonely, but we will internalize other conflicts as different tensions between our metaconscious psychic and social selves on the one hand and our metaconscious spiritual selves on the other hand. Technique has also heightened the tensions between ourselves and others, who may be our competitors while we also depend on them as neighbours. It has vastly increased the efficiency with which our needs are endlessly stimulated, while at the same time it undermines our ability to satisfy them. Moreover, it has vastly increased the powers of organizations over our lives, which cannot be offset by endless talk about human freedom. It all adds up to a great deal of frustration and helplessness, which is partly vented on those whose job it is to link us with one or another form of technical power. The resulting profound anxiety about others must be repressed, since we are expected to be "nice."

In order to escape the growing invasion of our lives by the technical order of non-sense, we are all escaping into the irrational. The more cultures are being desymbolized by technique, the more difficult it becomes to use any cultural resources (including morality and religion) to master and channel our physical and biological drives. It is as if we metaconsciously recognize that, as technique colonizes more and more of our lives, we have no choice but to take a stand in our physical and biological selves and express ourselves through them. It boosts the use of our emotions and sexuality as expressions of our subjective selves.

Following Richard Stivers, we may conclude that human life under technique is characterized by the following four contradictions: a compulsive need for power as well as a compulsive need for love; a dependence on technical rationality as well as a need to escape into irrationality; a trust in the powers of non-sense, which produces false meanings, but an inability to give up sense along with symbolization; and the empowering unity of the technical order along with the debilitating fragmentation of all life.[74] He shows how the influence of technique on human life is the context in which all psychopathologies in contemporary mass societies can be understood. His illuminating synthesis is based on showing a convergence between the previously cited works of Karen Horney, J.H. van den Berg, and Jacques Ellul. A very brief and inadequate overview follows below.

Technique has produced an erotic need for power, and this is offset by a neurotic need for affection. It has transformed all intimate relationships. There is a disturbing tendency for men to use sexual relationships to hurt women and for both sexes to avoid passionate feelings. It amounts to a kind of technical pursuit of pleasurable experiences that satisfy an erotic need for power as well as an erotic need for love. Within families, parents have an erotic need for the love of their children as the children need that of their parents, but this love is frequently used by both sides to control and manipulate in order to satisfy a need for power. To make life somewhat liveable in this social milieu, many people cultivate a narcissistic social self that directs aggression and portrays self-importance towards others. As this confrontational behaviour is internalized, multiple metaconscious social selves are in conflict with metaconscious psychic selves, yielding hopelessness and despair. Others are simply there to be consumed, like everything else in a consumer society. For those who, for one reason or another, cannot cultivate this kind of pathological narcissistic self, a state of depression frequently results as a kind of mourning over a meaningless life lived as non-life by an empty self.

In the course of industrialization, Western civilization substituted happiness for love. It also substituted power for hope, as technique imprisoned what it could not desymbolize. Eventually, human techniques engineered a commitment by default: robbing people of the will to resist and to fight for a more liveable and sustainable future. Other civilizations had regarded human suffering, disease, and death as being integral to human life, and culturally gave them the kind of meaning and value that made human life liveable. Happiness was an occasional relief from adversity. Under the influence of science and technique, we have reversed all this and convinced ourselves that happiness is the norm. Medical techniques will rescue us from disease and suffering and will postpone death, while medical enhancement techniques will go even further and reveal our true selves (as we will see in the next chapter). In other words, happiness is something that science and technique can deliver. It makes love, freedom, and compassion almost superfluous, because people can buy happiness on their own without any outside help.

Technique has replaced the "cultural niche" of a symbolic species with "human resource niches" of a technical order, which suppress this symbolic character as much as possible. The baths of images that surround us in everything we do and everywhere we go provide us with a kind of technically structured second-hand experience that efficiently paralyses our symbolic functions to the greatest extent possible. Our reliance on discipline-based approaches to knowing and doing dispensed by experts disempowers us in relation to the technical order. The "system" that has been built up with this knowing and doing cannot easily be compared with the "systems" of the past, which were controlled and regulated by a religious authority, a political elite, or an oligarchy of wealthy land owners and merchants, captains of industry, or capitalists. We still do not have a popular mythology for understanding how rationality could possibly imprison humanity, or how technique can function as a system that has a much greater influence on us than we can have on it.

A growing number of people in North America are beginning to realize that their lives are no longer improving; that their jobs are increasingly less secure; that they may not have any pensions for their old age; that the education of their children is increasingly unaffordable and no longer a key to a steady and well-paid job; that everyone must increasingly compete for what is shrinking or disappearing under free trade; that to have any chance at all, every possible technique must be relied on to make the best impression; and that, nevertheless, we continue to be defined by what we consume. Somehow, the American way of life has been reversed into a race towards a global bottom line, but no one appears to have a good explanation. Because human techniques continue to engineer commitments to the "system," it is pointless to vote for political parties that will be unable to deliver because of the workings of this system. The growing portion of the population that has been robbed of any hope by the system have resigned themselves to the fact that their votes make no difference whatsoever: since none of the political parties can make any difference to their lives, voting has become a senseless activity. Somewhere, someone must be manipulating the "system" for their ends; conspiracy theories of all kinds come and go in an endless procession. In the massive disordering of human lives and communities, many people find themselves alone with little or no effective social support. Some people may have the inner resources to go into overdrive and fight everyone else in order to have a good life or at least maintain the status quo, but most people feel so overwhelmed that they give up. The result is a widespread and profound anxiety about our lives and the future. The disordering forces within the technical niche are so powerful as to push us towards neurotic behaviour of one kind or another; many people require antidepressants, stimulants, enhancement technologies, and a great deal else in order to make their lives liveable.

The multiple niches of the technical order are thus characterized by high levels of desymbolization that produce meaninglessness and normlessness, while a will to power, human techniques, and integration propaganda attempt to hold the door shut to relativism, nihilism, and anomie. It would appear that people either adopt a will to power or withdraw into depression. In the extreme, life becomes unliveable for some people, who succumb to psychotic illnesses. It is remarkable how the symptoms of these illnesses reflect what we have referred to as "technique changing people." Potentially these people ought to have been performing the functions canaries used to provide to coal miners, but since we lack an adequate integration of psychology, psychiatry, the social sciences, and history, the signs of the times are readable to very few people. I will rely on one of these voices[75] to briefly highlight some of the symptoms and conditions associated with paranoia and schizophrenia.

Both these illnesses may be associated with the way others represent to us the unity and power of the technical order and its simultaneous power to disorder and fragment human lives. The resulting metaconscious knowledge may lead to intuitions that, by creating a defensive shield against others, people can make their lives somewhat liveable. Externalizing such intuitions can lead to mistrusting and being guarded against others, doubting their loyalty and friendship, being on the alert for whatsoever may be threatening in their behaviour, holding grudges, and being easily aroused to jealousy. What may have begun as neurotic behaviour can thus mushroom into suspecting everyone's motives and actions. At this point, people may either become aggressive towards others and make the general mistrust explicit, or become passive and withdraw. In either case, they rigidly stick to their convictions that others must not be trusted, thus requiring them to be constantly on the lookout for any sign of indifference, rejection, or hostility. They must withdraw from the world they share with others and take refuge in a subjective "world" of their own. Since this "world" no longer evolves and adapts within the social world of others, it becomes rigid, like a machine that is invulnerable. This machine is equipped with a whole repertoire of defence mechanisms. In a sense, paranoid people have internalized the profound anxiety associated with the confrontation of technique and the way it changes people. They have taken it to a neurotic extreme to create metaconscious paranoid social selves that aggressively or passively deflect these developments back to others in their behaviour. Paranoia constitutes a kind of defensive unity to protect the metaconscious paranoid social selves living in their own subjective "worlds."

In contrast, schizophrenic people appear to take the anxiety associated with "technique changing people" to the opposite neurotic extreme, to create a fragmentation and even loss of the self, which also involves the loss of the social world shared with others. Schizophrenia amounts to a kind of collapse of their being, manifested by a profound fear of becoming a dead thing under the control of others, not unlike the reification by technique. Schizophrenic people appear to live a privatized existence isolated from others, thereby incarnating the disordering effects of technique. Any cultural mediation with their surroundings is almost nonexistent, again reflecting the desymbolizing influences of technique. Since time and space are included in this desymbolization, these people lose any conception of living in time and space, and this makes the living of their lives by a unified self impossible. They are one fragment and then another, without any continuity. The body may become one object among many others, but the inner self may be hyper-conscious of the body, which is regarded as the false self that must constantly be watched by the disembodied inner self. All this is further complicated by the relationships with others that occur via the false self.

These are but very brief, and thus inadequate, interpretations of what can happen when the influence of technique makes the living of human lives impossible for some people. We will return to the subject of living under the regime of technique in the next chapter.

Resymbolization, Societal Renewal, and Equality

As we begin to resymbolize human lives and societies, doors will be opened to policies now deemed "unrealistic" and to others that have been made unthinkable and unliveable by our deep metaconscious commitments. We will begin challenging "what is" by resymbolizing the relationship between economic growth and our lives. As noted earlier, during the last two centuries we gradually began to live as if there was a direct correlation between economic growth and the quality of our lives and communities. When we turn to the findings of social epidemiology, it becomes apparent that no such correlation exists. For the poor, raising the standard of living initially translates into a beneficial influence on a great many things in their lives. Families can increasingly feed themselves and take care of a growing number of their needs. The pressure to do whatever it takes to feed your children and survive subsides. Gradually, more and more people have presentable clothing, and a measure of self-respect accompanies all this. Better nutrition improves health, and the community may slowly gain access to clean water, basic sanitation, and adequate shelter. Social relations with more affluent people may become a less painful reminder of who they are, and lower the shame and humiliation that comes with poverty. Within this limited context, the secular myths of the industrializing societies of the nineteenth century were correct, and the evidence from social epidemiology tends to confirm this.

The correlation between economic growth and the quality of individual and collective human life becomes a great deal less clear once a minimal standard of living has been achieved as evaluated by the culture of the people involved. For example, life expectancy and self-reported levels of happiness begin to level off, as do measures of economic welfare such as the genuine progress indicator (GPI), which seek to calculate the net benefits of growth by subtracting the costs.[76]

When standards of living grow still further to reach the levels of affluence found in many contemporary mass societies, the relationship between economic growth and the quality of individual and collective human life appears to undergo a fundamental transformation. From a historical perspective, these societies represent the pinnacle of technical, material, and economic achievement. There is considerable evidence to suggest that our efforts to meet our essential needs for food, clothing, shelter, and physical care are gradually replaced by a growing struggle to survive psychically, emotionally, socially, and spiritually. It is likely that the metaconscious knowledge of separation, disconnection, non-life, and death has grown to the point that the metaconscious knowledge of vitality and life is threatened. For example, working in the image of non-life requires the postponement of life to the weekend; but it is frequently turned into an escape by means of alcohol and drugs, either in the accepted context of parties or a lonely solitude at home. Many people appear to be driven by a deep metaconscious sense that what really matters in their lives is how others perceive them, as opposed to who they really are. This results in grasping every opportunity to mark their status by means of a variety of consumer objects, by creating their own "brand" on the social media, and by doing everything possible to heighten their importance by listing all their "friends" and social engagements. Other people seek comfort in eating, which has been turned from a nutritional or social occasion into a compensation, consolation, and reassurance for what is left of a life. The increase in the percentage of the population that are overweight or obese continues to attract a great deal of concern. To exploit this situation, food has been made a great deal more addictive through the addition of sugar, salt, fat, and caffeine.[77]

These kinds of developments are almost certainly an indicator that the falling away of the traditional social support based on a culture as a consequence of desymbolization is only partly compensated for by human techniques and integration propaganda. It would appear that many of us no longer are adequately sustained by a network of social relationships, including friendships and family ties, that is stable and viable enough to provide us with a social mirror in which we can see ourselves as persons who are cherished, loved, and accepted for who we really are despite our indissociably linked strengths, weaknesses, and vulnerabilities. Consequently, the required support must be of a dialectically enfolded kind found in the more conventional social relationships and groups, but these are under the heavy pressure of desymbolization. In compensation, we have created a culture of narcissism.[78] It is hardly surprising that with the mounting metaconscious knowledge of separation and non-life, there is an attempt to affirm our psychic metaconscious knowledge of vitality by means of every technical compensation available. Again, the needs of people in a mass society have been made to converge with the requirements of technique-based connectedness. Our insecurity driving on multi-lane highways and in urban settings has become transformed into a need for pickup trucks and sport utility vehicles. The more such vehicles project an image of power and aggression, the more they compensate for a heightened sense of insecurity. The more the social fabric became unable to sustain the individual, the more cell phones had to be transformed into mobile computers to handle every possible technical extension of the other-directed personality. The more many of us needed alcohol and drugs, the more a war on drugs was required to draw attention away from the real issue by dissociating the supply of drugs from the demand. By ignoring the latter, we could quietly let it grow without any embarrassing questions being asked. The growing needs for comfort food are met by the increasingly addictive products of the McFoods of this world with industrial precision and efficiency, entirely dissociated from concerns about nutrition and what drives these demands.

The lives of many people are now acknowledged to be in the grip of anxiety, if not depression.[79] The desymbolization of human life and society has resulted in a lack of meaning and direction, which makes it more difficult to cope with all the demands of our lives. This has translated into an epidemic use of antidepressants as well as stimulants to boost our performance at interviews, important meetings, or exams. Drawing these kinds of correlations can be supported with a great deal of evidence from social epidemiology. In the following pages, I will interpret the findings as reported by Richard Wilkinson and Kate Pickett in their important book, *The Spirit Level: Why More Equal Societies Almost Always Do Better*.[80] Their interpretation will be complemented by the intellectual framework developed here.

The subtitle represents an excellent summary of the findings of social epidemiology. This can be understood in terms of the sustaining role societies have always played in human life. The withdrawal of this support as a consequence of desymbolization and our being undermined as a symbolic species by technique has created fabrics of social relationships of a completely different kind from those found prior to industrialization. We must not forget that technique achieves its mesmerizing performance by disordering human lives, societies, and the biosphere.

The subtitle also sums up some good news for humanity and the planet. Although we are pushing every possible limit in our quest for performance and efficiency, we have had to give up a great deal for the peculiar kind of affluence produced by technique. It may therefore be possible to divert productive capacity away from the creation of compensating technical affluence to meeting the material needs of the poorest third of humanity. Such a diversion could increase the quality of human life of the relatively affluent two-thirds by showing the limits of what material things can do for us – a violation of our secular economic mythology. At the same time, such a diversion would make contemporary ways of life more sustainable by the biosphere. The mindless growth proposed by many in order to gain a future prosperity could be replaced by a genuine future for our children and grandchildren. All we need to do is stop serving our secular masters and try, as best we can, to see ourselves and our world for who and what we really are.

To begin with, a distinction must be made between *lived* wealth and poverty and *economic* wealth and poverty. The former are the result of symbolizing the experiences of our lives, which includes relating wealth and poverty to the social inequalities experienced in our communities and societies. Doing so remains somewhat possible in contemporary mass societies with highly desymbolized cultures because the dominant other-directed personality has a heightened awareness of the relative importance of a lived social equality, or lack thereof. It is also possible to understand wealth and poverty in relation to the demands these ways of life impose on people's lives. For example, a mobile computer may be a relative necessity for other-directed teenagers in desperate need to stay in touch with their peer group and the world, from which they would be excluded if they could not promptly respond to a constant stream of text messages, postings on social media, and so on. It is a question of social life or death for them. It is essential, therefore, to interpret the findings of social epidemiology in the context of the limited ability of contemporary mass societies to sustain their members and what the consequences of desymbolization are in such a situation. All this must be distinguished from the statistics gathered within the discipline of social epidemiology.

The lonely and other-directed persons living in contemporary mass societies have a relative sense of their wealth and poverty despite extensive desymbolization. That this is the case becomes evident when we compare the differences that people experience in their own society with those between societies, which people can experience only if they spend a great deal of time in both of them. The higher material standards of living of the wealthier part of a society translate into these members being happier and healthier than those with lower standards. In contrast, when comparisons are made between different countries, this correlation does not hold. The wealthy in one society may be several times richer than those in another society, but their health and happiness are usually little different. In other words, differences experienced within one's own society matter a great deal because they are symbolized and lived, while statistical differences between different countries matter very little. It is the experienced and lived differences that matter a great deal more than economic and statistical inequalities. The exception occurs where the latter become an integral part of a variety of human techniques such as integration propaganda, thereby becoming "experiences" on the screen. The following lived measures of inequality have been reported in the literature of social epidemiology:

the level of trust as a quality of the social fabric of relationships; mental illness and the widespread use of antidepressants, stimulants, alcohol, and drugs; the incidence of obesity and people being overweight; the academic performance of children in schools; teenage births; homicide; and social mobility. For many people, the statistical indicators also include life expectancy, infant mortality, and rates of imprisonment. From these indicators of inequality, a composite index of health and social problems has been computed, which increases as levels of inequality rise. An alternative indicator of inequality in a society is the ratio of the income of the top 20 per cent and that of the bottom 20 per cent. The index for health and social problems relates closely to the levels of inequality in each US state, but there is no correlation between the index and average income levels. The same picture emerges from international comparisons. It may be concluded, therefore, that beyond a certain initial phase, economic growth contributes less and less to well-being, since the richer states or nations do not do better than the poorer ones.

The findings from social epidemiology also rule out the commonly accepted explanations of why the poor find themselves in the situations they are in. Their poverty is commonly attributed to the circumstances themselves or to their person as a result of attitudes, vulnerabilities, and "inner resources." If it were simply a question of inadequate diet, housing, educational opportunities, and availability of decent jobs, the wealthier and more technically advanced societies should do better, but this is not the case. If personal differences are to blame, it would be difficult to explain that highly unequal societies have more of the associated problems than more equal societies. Moreover, these kinds of explanations overlook the reciprocal character of our relationships with our surroundings: we affect these surroundings as they affect us. If the latter interaction is more decisive than the former, many people may feel trapped in this lived reciprocity, and only those who have access to influence and power may be able to escape, as is common in highly unequal societies. In the extreme, a social group may develop a sense that society is not there for them. They have effectively been excluded, which is necessarily accompanied by the remainder of society legitimating the situation in a variety of ways. The members of the group may be seen as lacking the kinds of talents required for their participation in its way of life, and this lack may in turn be linked to race, creed, or sex. The group may also become a scapegoat for their society. Whatever the reason, human beings are then essentially being treated as refuse or worse.

Every social relationship between such people and the members of the society from which they have been excluded will do a great deal of damage to their metaconscious social selves, and over time will turn them into the very people society, through its stereotypes, expects them to be. Before this happens, many people (but especially young men) may assert their vitality by attacking the others rather than themselves. An aggressive posture to others may be adopted because anyone can be the enemy. It is only a small step to turn this anger about the injustice of it all into violence to try to demonstrate and re-establish autonomy, pride, and self-respect. It greatly facilitates the formation of street gangs, who feel entirely justified to "get back" at society. These are all common responses to social relationships which lead to a kind of inner death and the disintegration of the group. For Robert Lifton,[81] anger, rage, and violence are progressive stages for coping with psychic images of disconnection and death resulting from these kinds of social conditions.

For a society, however, such forms of extreme rage and volatility are seen as confirming its impressions and legitimating its treatment of these people. The resulting vicious circle turns life into an endless repetition, from which the society cannot evolve and adapt. Examples include the treatment of indigenous people and the African slaves in North America. With notable exceptions, many West European settlers could simply not accept these people as fellow members of a symbolic species, although with a different culture. Lifton[82] interprets the extreme violence perpetrated by the Nazis as being rooted in Hitler's attempts at revitalizing the German nation. This revitalization required a scapegoat, found in everyone who was different, but especially the Jews, who were cast in a mould of moral decay and degeneration by means of totalitarian propaganda. As such, they constituted the ultimate threat to the nation, which was simultaneously cast in the mould of vitality and immortal destiny. The violence was thus connected to psychic images of ultimate vitality. A great many ordinary men and women experienced new energy and a sense of purpose and meaning as a form of personal revitalization through the Hitler movement.

Under the influence of "technique changing people," metaconscious knowledge of non-life abounds in contemporary anti-societies. At the same time, new secular myths have created metaconscious knowledge of a new kind of immortality rooted in the limitless powers of science, technique, and the nation-state. Having been weakened by desymbolization, people can now compensate for this by the cultivation of self-importance and a will to power, in part marked by technical status symbols. Even among many people living below the poverty line, life appears to be unbearable without televisions, computers, and cars, even when this necessitates visits to food banks. The ability to compensate for the effects of technique on our social selves by means of a culture of narcissism varies enormously from the top to the bottom of contemporary mass societies. Moreover, there no longer is any pushback from their highly desymbolized cultures against those who feel they are worth everything they can get, thus living life as a war against everyone else. Hence the enormous popularity of aggressive-looking, powerful, and oversized vehicles, power suits, and other such "symbols." In turn, the more the gaps of every kind widen, the more the other-directed personality will become aware of this, thereby confirming the sense that they are engaged in a war in which they are either getting ahead or losing. These developments are becoming self-reinforcing as human life gets stuck in repeating the same cycle instead of evolving by symbolizing the situation for what it is. Others have been transformed from neighbours into competitors, and this has completely and qualitatively changed the fabric of social relationships. All of us still need to be sustained through our relationships with others, even though we do everything we can to hide behind every possible action of self-importance. There appears to be no other way of reconciling a great deal of narcissistic behaviour with an epidemic of anxiety and depression.

The growing scale of income inequality drives the establishment of social status differences. These differences are gradually translated into others related to nutrition, clothing, shelter, and education; these in turn are expressed in many other markers of identity. Status differences may also be linked to the ability to compensate for the desymbolizing influences of technique on our lives, which profoundly affect the psychic resources we need to stay in the race.

The above developments vary from country to country, with the result that social epidemiology provides us with a key to understanding the implications for individual and collective human life. The more egalitarian societies have a social fabric capable of sustaining their members, to the point that people are healthier, children do better in school, violence is more effectively delimited, people are happier, levels of incarceration are relatively low, and people are more likely to trust one another.[83] It may well be that this greater trust creates a social milieu where people can more easily turn to others for support; where that support is likely to be helpful because the other is more like yourself; where people can spend more time sustaining and being sustained by others, since they are less occupied by all the duties directly and indirectly related to marking their status; and where friendships, marriages, and family ties are more likely to survive a stormy period. People's identities are better anchored in social relationships, and they can be more confident about how others feel about them. They are no longer constantly on trial by others and can pay less attention to status. It is hardly surprising, therefore, that the data from social epidemiology shows that the advantages of some societies over others can be predicted by their greater levels of equality.[84]

Unfortunately, we have gradually grown accustomed to a kind of social Darwinian explanation of human history. It holds that societies inevitably establish pecking orders based on power and coercion. The males at the top have preferred access to resources and females, regardless of the needs of those at the bottom. For these reasons, those at the bottom strive to get to the top. Might is right, and the weak eat last or not at all. As a symbolic species, we did not accept this situation and transformed it by means of symbolization, experience, and culture. This transformation created a margin of freedom in relation to these necessities of survival and occasionally gave birth to much more egalitarian societies, in which sustaining others and in turn being sustained by them was regarded as being in everyone's best interest and in conformity to lived values. Social epidemiology has confirmed the merit of these cultural achievements based on cooperation and the recognition of everyone's material and psychic needs. By means of their culture, such societies attempted to put everything in its rightful place relative to everything else, according to their values. Such symbolic and cultural achievements had to be constantly defended against material, social, and spiritual necessities in order to have "what ought to be" triumph over "what is." Social Darwinism may therefore be regarded as a triumph of desymbolization and an acceptance of a partial relinquishing of our being a symbolic species in the face of the thermodynamic necessities imposed by industrialization. We are sinking back to a kind of animal level. All this is justified and even celebrated by a secular economic mythology. It compels the acceptance of what is, and in the name of being realistic, we give up our expectations of conviviality. Like baboons, we spend more time avoiding being hurt by dominant ones than on avoiding what can really kill us.

In highly unequal human societies, there is a constant competition for status. People are endlessly confronted with situations in which they must decide whether to back off or challenge the status of the other. The result is a constant stream of experiences of downward discrimination, which is perpetuated on those who back down and acknowledge their lower status. Everyone constantly has to watch their back, treat others with suspicion, and fight for everything they get. Our educational systems may well talk endlessly about backing away from all this, but, by embracing one generation of educational technologies and human techniques after another, our schools unintentionally conform their students to technique based more and more on power and efficiency, and thus prepare them for guidance by integration propaganda. The difficulties schools face when they attempt to deal with bullying is but one example of the near impossibility of limiting the social battle for inclusion or exclusion in a group as it competes for status with others. There is tremendous pressure to acquire status symbols because, once affluence is reached, it is the relativity of these symbols to each other that matters, as opposed to their meaning and value for human life; this is largely determined by technique and our implicit choice of non-life over life.

Are the differences in social inequality largely the consequence of how deeply a mass society has been transformed by "technique changing people"? The ongoing increase in the incidences of anxiety and depression predates the sharply rising levels of inequality in contemporary mass socities.[85] This sequence of events suggests that we need to pay a great deal more attention to the desymbolizing effects of technique on our lives and cultures, and interpret growing social inequalities as a consequence of technique. Nevertheless, growing levels of inequality undoubtedly contribute to higher levels of anxiety and depression and thus stimulate end-of-pipe technical solutions such as more effective drugs and therapies. They have also led to a variety of self-help seminars, programs, and books to boost self-esteem and reap the rewards of getting ahead of others. It is remarkable how well Christianity has been adapted to the American way of life, and we can marvel at the intellectual acrobatics of many theologians in justifying all of this.[86] They fail to acknowledge that, as human creatures having no life in ourselves, we are tempted by our endeavours to open ourselves up to everything promising life and immortality. This has made us particularly vulnerable to adapting thought and action to limitless knowing, doing, and political organizing.

Contemporary mass societies have created "technical human resource niches" which generate a great deal of anxiety and depression. At the same time, people have had to boost their confidence in themselves in the face of being constantly tested by others. Underneath these and other phenomena lies a deeply ambivalent metaconscious knowledge with an imbalance between life and non-life. This makes it next to impossible for many people to live their lives on the basis of who they really are and what, according to their desymbolized culture, "ought to be" as opposed to what others think of them and "what is." From the very outset, this was probably a losing battle for many because they were other-directed persons. For a few, the battle may continue; but they must often struggle alone without social support, and this is difficult to sustain. In any case, all of us are deeply affected in our being and our lives. Our experiences provide us with increasingly less reliable access to our own lives and our world as the perceptions of others and ongoing desymbolization increase our reliance on human techniques and integration propaganda. The same "technical human resource niches" have also prepared us for the explosive growth of social media. These media are increasingly necessary to extend our other-directed personalities by using a highly desymbolized organization of the brain-mind as a radar to help us make friends more easily, better cope with the criticism of others, and wear a mask of self-confidence. For many members at the bottom of society, this mask of self-confidence is maintained by aggression and violence towards others. All measures of inequalities within our societies are signs of a breakdown of social cohesion, creating what we have referred to as anti-societies.

There is a pressing need for a resymbolization of science, technique, the economy, and the state, which will contribute to a possible recreation of a fabric of social relationships capable of sustaining human lives. A variety of possible initiatives may be considered. Should an annual guaranteed income be paid to those people to whom society can no longer offer work because of structural unemployment? Should the use of alternative currencies be used to build a culture of genuine local development? Could a redesign of the tax system, away from human work and towards the throughput of matter and energy, redistribute wealth to create more equal societies? Should this tax system be used in a kind of end-of-pipe approach to redistributing wealth to create such greater equality (as has been successfully achieved by Sweden), or should it be used in a preventive manner by restricting the disparity between the top and bottom salaries in organizations (as it is very successfully practised in Japan)? All of this will appear to be completely unrealistic to a great many people, but this is exactly the point. The resymbolization of human life is unrealistic in the face of science and technique because it seeks to once again transcend "what is" by means of symbolization. It involves a struggle against all forms of desymbolization and the human attitudes it engenders (such as, if nothing is in its right place, we might as well live our lives accordingly and get ahead). From a historical perspective, it has been symbolization that has made us human; undoing this will simply make us into another animal species. Resymbolization is the only way of keeping open the door to a liveable and sustainable future.

A few predictions based on social epidemiological data may be sobering. Wilkinson and Pickett[87] estimate that if the United States reduced its huge income inequality to something close to that of the four most equal countries in the world (Japan, Norway, Sweden, and Finland), then the proportion of the population who feel that one can trust others might rise by 75 per cent, rates in mental illness and obesity might be cut by two-thirds, teenage births could be halved, prisons could be emptied out by 75 per cent, and people would have to work less each year. It must be emphasized that these are but some indicators of what would constitute a vast improvement in US civil society. Wilkinson and Pickett clearly demonstrate the enormous corrosive effects that the wealthy have on society.[88] Moreover, it would appear that it is technique that has transformed the United States into one of the most unequal societies during the last half-century, leaving its civil society almost completely defenceless against these deep structural changes. Much of the influence of technique on civil society occurred through the financial systems in general and the role of money in particular, and it is these factors we will examine next.

Alternative Currencies and Social Cooperation

As with other issues, the discipline of economics is incapable of shedding light on the role of money in human life and society. It is limited to understanding money as one economic phenomenon among many others, and thus to what money does in this domain. It makes the role money plays in human history invisible, with the implication that money is sufficiently neutral that any broader questions can safely be left to the philosophers. Once more, this is rather surprising, given that both the Jewish and Christian traditions, which played such an influential role in Western civilization, had a great deal to say about the non-neutral character of money. It is depicted as a spiritual power in human history capable of possessing people to the very depth of their being.[89]

Even the most systematic and comprehensive analyses of the phenomenon of money have left people with a sense that something remains unaccounted for, and some economists have openly acknowledged its mystery and its secular magical character. Our current financial systems were created just prior to the Industrial Revolution in order to deal with growing trade. Commerce and industrialization then extended their influence into the daily lives of everyone who became a wage earner. These people's lives became almost entirely dependent on the technology-based connectedness and the economic order which regulated it, and the role of the culture-based connectedness became secondary in their lives. Social exchanges became mediated by money, and this mediation was so lacking in neutrality that money's ability to alienate human lives and communities was widely recognized. This power over human life is hardly surprising when we examine how money is created and the functions it performs. What is surprising is that relatively little attention has been paid to the way it has transformed everything because it is far more than a neutral medium of exchange.

The "creation" of money may involve an injection of funds by the central bank of a nation-state into the reserve accounts of the commercial banks it serves.[90] With these reserves and the deposits it has received, a commercial bank can make loans, except for a small fraction of the funds which has to be kept in reserve. If this rate is 10 per cent, for example, one hundred million dollars of deposits can be turned into 90 million dollars of loans. Borrowers use these loans to pay for goods and services, and these payments end up as deposits in banks. All but 10 per cent of these deposits can then be turned into additional loans that result in deposits, from which more loans can be made, and so on. If all the borrowers were to default on their loans, the banks would still owe their depositors. Consequently, these banks have (collectively) essentially created 90 million dollars of fiat money on the first round, 81 million dollars on the second round, 72.9 million dollars on the third round, and so on. All this fiat money disappears when the process is reversed and all loans repaid. Such a fractional reserve system can trigger bank crises and failures when, for whatever reason, many of the clients of these banks decide to withdraw their money.

Because interest is charged on all these loans, borrowers must compete with each other to procure this interest, because the financial system deliberately does not create the matching funds *ex nihilo*, in order to make the fiat money scarce and thus valuable. For example, if all the loans were repaid after one year and all deposits withdrawn a few days later, in addition to likely bank failures, some borrowers would have to go bankrupt unless the central bank added to the money supply *ex nihilo*, or unless the economy grew the stock of goods and services. However, the thermodynamic constraints imposed by a finite planet make this impossible in the long term.[91] In other words, the financial systems necessary to sustain ongoing industrialization have turned societies from fabrics of reciprocal obligations between people into a competition of everyone against everyone for currency that is made scarce by banks charging interest and by governments levying taxes. Everyone's credit rating is essentially determined by their ability to compete against everyone else for scarce currency in order to avoid bankruptcy. All this is further complicated by an entirely new kind of economic activity that has been growing for a half-century, which Herman Daly has described as the making of money from money without any intervening economic activities.[92] As noted, Bernard Lietaer reports that this activity accounts for about 97 per cent of the global economy.[93] It is one of the ways in which technique has turned modern economies from wealth creators into wealth extractors and thus anti-economies. The destruction of the social fabrics of the industrializing societies and their replacement by competition has led to the widespread acceptance of a social Darwinian explanation of human life, making the much more cooperative character of a life sustained by symbolization, experience, and culture all but inconceivable.

The compounding of interest strongly skews all human decisions because it devalues the future relative to the present. In other words, the worth of society and the planet for our children and grandchildren is greatly devalued; according to the logic of discounting, our present wealth takes priority. Unfortunately, even if all that wealth were passed on to future generations, they could not breathe, eat, and drink this money.[94]

Moreover, the current financial system is highly unstable. According to the International Monetary Fund (IMF), between 1970 and 2010 there have been 145 banking crises, 208 monetary crashes, and 72 sovereign debt crises, which amounts to 425 crises in forty years.[95] The human misery this represents is almost unimaginable and rivals that of a great many wars.

Compound interest and currency speculation make the ultra-rich even wealthier, causing them to appropriate an ever greater share of the resources of their nations. For example, consider the influence corporations like Walmart are having on local American communities.

Not counting the enormous salaries of their top executives, the family that owns Walmart earns the same as 150 million Americans, thus withdrawing the equivalent of the incomes of 150 million Americans from local communities. Stores have frequently opened with subsidies from local governments; wages are sometimes subsidized by grants to "train" employees; local businesses have been ruined and downtowns destroyed.[96] The pressure on suppliers is often so great that they must imitate this behaviour with additional social and environmental dislocations. Moreover, some employees and those who lose their jobs often become burdens on the social services of local communities. In the case of the former, they are part-time employees, have few or no benefits, and do not earn a living wage, thus requiring compensation of one kind or another. Many Americans may recall that in the Jewish and Christian traditions, this kind of behaviour was regarded as an offence against God and a violation of the second great commandment.[97] It is no wonder that the ultra-rich must segregate themselves from society in exclusive hideaways surrounded by bodyguards. Otherwise they might not be able to sleep at night if confronted with the results of their decisions. In addition, the exponential growth of the interest-bearing portion of their wealth and their returns on "investment banking" all increase the claims against the stock of goods and services of the community, thus imposing a further burden on local communities.

Why do we have so much difficulty connecting the growing number of children who go to school without breakfast and other symptoms of a growing poverty to the characteristics of our financial system and the inequalities it helps produce? It is astounding that the Jewish and Christian traditions have been so subverted by the American way of life that many now think of the above kinds of arguments as politically left-wing. They are nothing of the kind. They are simply the kinds of arguments that seek to protect the public good and our common future. There can be no common good or public future when 441 billionaires have assets equivalent to the incomes of half of humanity, while the top three own more wealth than the forty-eight poorest countries on our planet.[98] Lietaer cites a German study showing that this transfer of wealth from the bottom to the top is exclusively the result of our monetary system.[99] In the United States, the top 1 per cent owns more wealth than the bottom 92 per cent of its people.[100] Our current financial systems are distorting all the relations in society and the biosphere as a consequence of the charging of interest, necessitating perpetual growth on a finite planet. They have made the notion of a public good and a common future meaningless in the face of historically unprecedented concentrations of wealth. It is undeniable that these systems have also generated a great deal of genuine development, but we are now at a point where all the gains are steadily being eroded.

It is evident, therefore, that the economic policies proposed in the previous chapter will have to be complemented by two additional ones. The first is to impose a modest tax on all short-term speculation, which would reduce it to the levels that were common prior to the creation of the anti-economies, or what was current during the decades following the Second World War. The second is to reduce income inequalities between rich and poor to the status quo at that time. The wealthy and the speculators of this world must recognize that they are helping to destroy the very system that has made them who and what they are, and that these modest measures will still leave them extremely wealthy without any negative effects on their happiness. It is a question of containing the extreme consequences of variations in the performance of technique.

As a medium of exchange, the current monetary systems exclude the needs of the poor, the working poor, the unemployed, and the underemployed, who cannot translate some or most of their needs into market demands for lack of sufficient currency. Fortunately, alternative currencies are able to perform this task, as we will see. Many badly needed public services can no longer be financed by local, regional, or national governments; once again, alternative currencies could have an important role to play. In terms of the present financial system's ability to measure value, we have noted the gap that exists between the meaning and value of something for our lives as established through symbolization and culture and its value established in a market. Consequently, we are no longer able to distinguish between two techniques with roughly equal efficiency, productivity, and profitability but significantly different implications for human life, society, and the biosphere. Similarly, we no longer distinguish between economies producing roughly the same gross domestic product but with vastly different effects on society and the planet. Our current financial system is therefore unable to make the kinds of value distinctions that we badly need.

On a global level, the national central banks compete with each other to maintain the scarcity of their national currencies relative to one another. They must carefully monitor the supply and demand for their national currencies in the economy and financial markets in order to accomplish this task. As noted, every unit of currency in circulation is a debt incurred by individual persons or entities such as a state, a corporation, a municipality, or a non-governmental organization. In deregulated financial markets, the abilities of central banks have been significantly restricted, but there remains an aspect essential to any fiat currency. Following the Great Depression, measures were taken to prevent a reoccurrence. These included a rigid separation between investment banking and commercial banking, which was later relaxed, with near disastrous results in 2008. The alternative so-called Chicago plan, favoured by many academic economists, would have made bank debt money illegal, and thus bank loans could not exceed the deposits on hand. It would have eliminated bank crises but not monetary crises. It would also have returned the power to create money from the central banks to the national governments.

A key role of any national currency is to permit the citizens of a nation-state to pay the taxes legislated by their governments. It obliges citizens to work, trade, and invest to obtain the necessary currency, and this may take two-thirds of their earnings for each and every year of their lives. According to Lietaer, the primary purpose of taxes thus appears to be to help make the national currency scarce and thus valuable.[101] After all, when it comes to fiat money, the ultimate beginning and end of any national currency is the state. Its central bank must ensure that there is enough currency in circulation for economic exchanges to take place, in a manner that avoids inflation or a recession. In addition to acting as a medium of exchange, this currency also functions as a store of value, a unit of account, and an instrument of massive global speculation. These functions can easily be at cross purposes from one another. Since governments and central banks generally take their advice from professional economists, we face an endless stream of erroneous economic diagnoses and seem to stumble from crisis to crisis. After all, economies are not closed systems, they are not static, and they are not linear. Their resilience is ultimately much more important than their efficiency, but this issue cannot be tackled by a discipline-based approach. It is for this reason that the need for complementary currencies is necessary in order to transform a financial monoculture into a more resilient diversity. Nevertheless, such complementary currencies are at best an end-of-pipe solution to the problems created by technique in some cases, or a temporary means for dealing with the crises created by technique in conjunction with demographic changes. For this reason, the four scenarios for the future of money set out by Lietaer in his earlier book[102] are much more persuasive than the claim that alternative currencies can turn scarcities (largely created by technique) into prosperity, as he claimed in a later co-authored work.[103] Nevertheless, there is no question that alternative currencies can bring enormous relief to the unemployed, underemployed, working poor, and elderly, as well as making a variety of public services affordable under the current constraints. Moreover, such currencies will be a great deal more effective when they are understood and designed to respond to the crises aggravated largely by technique. In other words, the creation of alternative currencies is but one strategy in the portfolio we are developing in this work.

It is now possible to appreciate the non-neutral mediation exercised by a national currency. Suppose that a village in a traditional society relies exclusively on barter exchanges, frequently coinciding with bonds of kinship, friendship, and good neighbourliness. Such networks of reciprocal and cooperative interdependence would be significantly altered by a feudal regime, or any other for that matter. Imposing a regime involving a national currency and forbidding all barter exchanges would unleash the kinds of economic and social forces on this community that were described above. It would completely transform the social fabric and almost everything else along with it.

Next, suppose that an alternative currency is made available for the exchange of unused human talents and resources. Anyone who is underemployed or unemployed could then use this currency to facilitate exchanges: a haircut for some vegetables; a repair to a vehicle, with parts paid for in the national currency and labour in the alternative currency; a summer's cutting of the lawn for a watercolour painting; daycare for groceries, and so on. Businesses would be encouraged to join, to become more productive by filling unused capacities. For example, local restaurants could accept the alternative currency except for the two nights during the week when they are generally extremely busy. This model can be further extended by municipal governments as a partial payment of local taxes, with which they could procure the services of cooperating businesses or of individuals in cases of unusual demands. What is required is a local network of reciprocal and cooperative interdependencies within which goods and services are exchanged by means of the alternative currency. It is estimated that there are already some four thousand such currencies in existence today.[104]

These alternative currencies work alongside the official currency, which is why they are also called complementary currencies. They are voluntary and thus also referred to as cooperative currencies. They are designed to function only as a medium of exchange; this is frequently encouraged by a kind of negative interest or demurrage fee, which results in a modest charge when the currency is not spent within a certain time, thus making it unavailable to the community to facilitate further exchanges. The fees charged can be applied to cover the overhead of running the system. It is impossible to discount the future, so that the concentration of wealth is avoided. Alternative currencies can greatly strengthen local economies and make them as self-reliant as possible, thereby making the national economy more resilient.[105]

From a historical perspective, these complementary currencies have frequently sprung up during local or national crises. For example, the fact that the principal employer in a town closes its doors does not change anything other than that the former employees are no longer able to back their needs with their wages. In such cases, these needs can be backed by a complementary currency acting as a mutual credit system. The Irish created such a system spontaneously when their banks regularly went on strike for a total of twelve months during the period from 1966 to 1976.[106] Following the Great Depression, the Swiss created a dual currency through which one-quarter of the participating businesses became one another's mutual creditors, and it has continued to stabilize the Swiss economy to the present day by offsetting business cycles and reducing unemployment, especially during times of tight credit.[107] Liberation theology in Latin America and the Mondragon network of cooperatives in the Basque country in Spain each inspired a system of mutual credit, extending micro-loans to help households through difficult times and to start up small businesses.[108] In Sweden and Denmark, a mutual credit system allows people to purchase homes and help finance others to do the same.[109] The city of Curitiba in Brazil used cooperative currencies to address many different problems faced by the city and its poor. It transformed the lives of many of its people without any increase in taxes.[110] This example alone ought to silence every city politician who abandons his or her responsibility with the excuse that no further tax increases are possible. In sum, there is a host of examples showing how a kind of Keynesian stimulation is possible through complementary currencies without any debts being encouraged. These examples show the complete ineffectiveness of our political parties and why the right has been able to capitalize most effectively on the cult of the fact. No imagination, responsibility, or care is required; in an anti-society, they know best how to exercise this cult. Resymbolization is the only genuine political path to regaining a political centre based on moderation and a concern for the public good.

It should be noted that Germany, Austria, and the United States all had complementary currency systems in operation following the Great Depression. In the first two countries they were ruled illegal, which left little alternative to Hitler's Nazis, and in the United States they were displaced by the New Deal.[111]

There are very few local needs that have not been addressed by means of complementary currencies: education, employment, poverty, health care, elder care, ecological buildings, environmental cleanup programs, making cities more liveable and beautiful, extending municipal services, and more. What has been accomplished is profoundly moving, but it remains a drop in the bucket against the devastating effects of technique on our humanity and consciousness. It all comes down to the choices we make. Why do we flock to the Wal-boxes of this world at the expense of our local businesses and communities, thus depriving our neighbours of jobs and our children of a future, when we could mobilize to create better alternatives? The answers have always been the same throughout human history: our enslavement to traditional myths or contemporary secular myths complemented by integration propaganda, making any alternatives almost unthinkable and unliveable without strong personal convictions based on what is radically other.

Every day the media draw on the cult of the fact. Of course, unemployment and underemployment are very serious. Our inability to balance budgets will get much more serious as the population ages and the costs of pensions, health care, and social services become increasingly unaffordable. However, the situation is considered to be temporary because our governments are creating jobs, bringing their budgets under control, negotiating trade deals, and finding new efficiencies everywhere. Hence, we must steer the course, and by growing the economy we will create the wealth required to deal with all these issues. These are the "facts" as established through discipline-based approaches to knowing and doing. If instead we rely on our experiences and listen to our neighbours tell us what is happening in their lives, and if we check all this against the intellectual efforts of resymbolizing our knowing and doing, the situation is the exact opposite. People are experiencing the very beginning of the consequences of the unfunded liabilities that come from pension funds, health care obligations, the care of the elderly, educational opportunities, social breakdown, and environmental deficits. Depending on one's socio-economic stratum, net wealth production has not grown for half a century or has declined steadily. All these issues and more are directly attributable to the deep structural changes caused by our discipline-based approaches to knowing and doing and everything that comes with it. Improving the efficiency of everything on its own terms increases efficiency at the expense of everything that matters to our lives. There will be no recovery of jobs and there will be no balancing of budgets, unless this is done at the expense of the weakest and most vulnerable members of society. The human suffering can be expected to steadily grow for decades to come. Many of our gains will be reversed.

I do not know of any analysis attempting, in one way or another, to transcend some of the limitations of the desymbolization of our individual and collective lives that does not confirm this in some way. It is not a question of pessimism but of realizing that we have been alienated by the cults of the fact, of efficiency, of economic growth, and the mythology of our being secular individuals liberated from all social taboos. It is a question of choosing life over non-life as embodied in science, technique, and the nation-state. Perhaps complementary currencies can be a vehicle to help us re-experience cooperation and a sustaining social fabric woven from meaningful interdependence and support. However, such attempts must be an integral part of a much broader strategy. We are facing one of the greatest aftershocks of the Second World War, which opened the doors to technique on an unprecedented scale, resulting in a huge demographic transition and the undermining of the biosphere by global warming, among other catastrophic effects. When we need one another more than ever, the desymbolization of our lives and our societies has been added to our difficulties. Nevertheless, we must set our feet on the path of resymbolization and life.

5 The Cult of Disembodied Personal Life: Epidemics of Anxiety and Depression

Reality and Daily Life

Our being a symbolic species has been turned inside out by the way Western civilization has transformed the relationship between the visual and oral dimensions of experience during the last five centuries. This transformation significantly affects the way our lives work in the background as a result of symbolization. It is responsible for our experiencing more than our senses detect and for communicating more than we are aware of. For example, what we see of the world is not surrounded by a border corresponding to the limits of our retinas, because whatever lies beyond it has been interpreted out as having no meaning when metaconsciously fitted into all the other experiences of our lives. Similarly, every moment is lived with the contributions of the metaconscious knowledge of the way of life of our community and its culture, which allow us to anticipate the kinds of situations we face in our daily lives and to deal with them appropriately and effortlessly. We may also have feelings or intuitions about the deeper significance of these situations, which would be impossible if we did not live them as moments of our lives. Consequently, desymbolization can substantially diminish our ability to live our lives.

One important way in which our lives work in the background is that we habitually hierarchize what we derive from our five senses. In most situations, it appears that we rely on one of the senses more than the others in order to get a good grip on things. Doing so implies a great deal of metaconscious knowledge anticipating what other people and the world are like. For example, every culture implies a way of integrating what we hear and what we see to best put us in touch with others and the world. We have already suggested that this is not simple, because our vision presents us with what is real in our world and our listening with what is true about our lives in the world. Getting a grip on the structure of the former begins with separating its elements, and dealing with each one by defining, measuring, quantifying, and mathematically representing it. Getting a grip on the structure of a symbolic universe involves symbolically interrelating everything to everything else in individual and collective human life in the world. Each element is now understood not by separating it, but in the opposite manner, namely, in its relatedness to all the others. Doing so involves a language and a culture in order to participate in the adaptation and evolution of what is true for human life in the world. In the one case, understanding others and the world involves the making of observations by a person who is detached from them as another distinct and separate element of that world; in the other, understanding comes from integrally participating in its dialectically enfolded character, where everything is what everything else is not.

Living as if the one or the other is the real world permeates all of daily life. Take the example of a teenage boy agreeing to go on a blind date arranged by friends, with everyone belonging to a contemporary mass society. When he first meets his date, there are two ways in which she presents herself to him. There is the person he sees and there is the person with a life that he is about to enter as they begin talking together. Since in these cultures an image is worth a thousand words, the physical person may be much more important than the social person. Without this cultural bias, the teenage boy reflecting on the date the next day may wonder who the real person is: the one he saw, danced with, and kissed or the one who lives the life of which she spoke. Reciprocally, his date may be wondering the same kinds of things. Visual appearances may well count a great deal more than what they revealed to one another about their lives by talking and listening. In the extreme, a "trophy date" may boost the boy's ego by making the two of them the centre of attention, but she may later turn out to be much less interesting as a person. In contrast, someone who does not physically conform to whatever the current fashions of beauty happen to be may well turn out to be the kind of person he would love to spend his life with, and she with him.

The same kind of dilemma is involved in the traditional painting of a portrait.[1] In pre-industrial Western cultures, this was not first and foremost an attempt at making an exact reproduction of someone's face, as was later accomplished by means of photography. A good portrait was intended to reveal the person during that particular time of his or her life. In other words, in those cultures the image was supposed to symbolize something of that life because the word dominated the image. The painter had to get to know the subject, and the better the acquaintance, the more likely the portrait would symbolize something of that person's life by the way the eyes, facial expression, and posture were selected and painted. A good painter was thus a good interpreter of someone's person and life as well as one who had the artistic and technical abilities to get this down on a canvas. It was as if, with a good knowledge of the subject, the painter had selected from among millions of photographs of the person engaged in a variety of daily-life activities the one that stood out as truly representing him or her, and then used it as a point of departure for posing the subject for the painting. It may be argued that such a painting would have a kind of objective basis since the photographer played no role after the shutter was pressed, but that it was subjectively true in the context of that period of the subject's life. It would be a portrait not simply connected to a single moment but to all the moments, of which this one was deemed to be the truest representation of them all. Of course, what was true of the traditional portrait painter also holds for the greatest portrait photographers of our days.

At the same time, such a portrait also expresses something of how the person is typical of that time, place, and culture. For example, typical portraits of a tradition-directed person, an inner-directed person, and an other-directed person are likely to be very different. The eyes may reveal a different inner world and a different relationship with the other in the form of the master painter or photographer.

Initially, what is true and what is real manifest themselves very differently to blind persons because their experiences begin with sounds that lack the stability of visual images. Sounds are often mobile, they come and go, and they are often threatened by a great deal of noise in urban settings. Hence, the world initially presents itself to blind people as more fluid, dynamic, mobile, and non-repetitive in comparison to the seen world. Its sounds are indicative of ever-changing relationships and events: a dog that suddenly barks and makes its presence known; people whose footsteps cannot be heard over the noise of traffic and whose presence is revealed when they stop to talk; bells that chime the passing hour and mark a nearby church building; the traffic at an intersection indicating the presence of stoplights; and so on. This world of sounds somewhat resembles the world of people's relationships and lives. To it must be added the sensations from touch, smell, and taste, supplemented by a person's imagination.

It is difficult for us as members of secular mass societies to imagine what it must have been like to live in a symbolic universe whose visual manifestation was dominated by the symbolic significance of everything. To put it simply: what you saw was not what you got. To explain this, let us once again return to the development in Western civilization that led up to our kind of world, one closely identified with discipline-based approaches to knowing and doing, anti-economies, and anti-societies. When Christianity began to spread to the point that it became the official religion of the Roman Empire, its teaching that the universe was a creation had profound implications. First of all, it emptied out the world of all the spirits, leaving nothing but created entities and life. Whatever gods or eternal souls remained were thus cultural fabrications corresponding to what we now understand to be the explicit cultural elaborations of intuitions regarding the deepest metaconscious knowledge or myths. Equally important was the fact that the meanings and values of everything could not be contained within a symbolic universe, not even by its myths, but ultimately had to be referred to the creator of everything. For example, the meaning and the value of an animal could not be established by differentiating it from all other animals and these animals from everything else in a symbolic universe. An animal was a creature, and this had fundamental implications for what people could and could not do with it. All this gets very complex because, under the influence of Greek philosophy and the Roman political interpretation of human life and society, Christianity moved away from its Jewish roots, thus causing an explosion of misunderstandings and contradictions that eventually contributed to the medieval ways of life becoming unliveable.

A variety of later developments would have simply been unthinkable and unliveable in the symbolic universes of medieval societies. The structure of the medieval university explains some of this. Its central faculty was that of theology, followed in order of importance by that of philosophy. In other words, people could make all the scientific observations they liked but they would never get to the bottom of anything without an understanding of its being part of God's creation. Similarly, technical doing would have been unthinkable because the concept of an absolute efficiency could have no place in a dialectically enfolded symbolic universe. It was also impossible to conceptualize the dependence of humanity on matter and energy apart from a cultural mediation based on values.

Until the eighteenth century, West Europeans lived in the symbolic universe of the culture of their societies. Arguably the most significant event of that century was an exodus from these symbolic universes towards a reality-like world of human life.[2] The emergence of discipline-based physical sciences and the introduction of the technical division of labour, followed by mechanization, may have received too much credit for causing these changes. It is more likely that these developments were expressions of these changes as what was becoming knowable and doable in terms of this cultural shift. Much of it appears to be rooted in humanity's relationship with matter and energy. As noted in the Introduction, the Middle Ages had failed to locate the boundaries between matter and spirit, body and soul, and life and machine. If no such boundaries existed, and if life was not entirely spiritual, it had to be a machine (depicted by the most perfect specimen of that time, namely, the clock). This perception marked the beginning of one of the greatest cultural transformations of human history. Symbolic universes could no longer be what they had been, and they became gradually conceptualized in terms of a reality. It had already become thinkable and doable to have physical scientists make observations in an objective manner and to discover facts. Now it also became thinkable to introduce a technical division of labour as a superior way of conceptualizing human work in the image of the machine. God could now be portrayed as the great clockmaker, having put the universe-clock together and allowing it to function according to mechanical laws. At the same time, God became the planner of everyone's destiny, thus making everyone a cog in the grand scheme of predestination. No longer did this God enter into human history to accompany humanity, as portrayed in the Jewish and Christian Bibles. The soul was the only "spiritual" remnant, and it was conceptualized more or less according to Greek philosophy. Finally, the relationship between the individual and society had to be reconceptualized, since individuals as "parts" of society could be assembled more rationally by means of a social contract. All such developments and many more point to a complete transformation of the deepest metaconscious knowledge of the people of that time, a desymbolization of their symbolic universes and a mutation of the human spirit.

I am not in the least suggesting that this metaconscious knowledge constituted a kind of world view or a philosophy. It can best be imagined as the spirit of an age inhabiting the people of that time, as their absolutized commitment and point of orientation for living in an ultimately unknowable universe. In other words, the spirit of an age may be likened to a kind of collective personality of a people that is anchored in and oriented by myths. As such, this spirit determines the form alienation takes in the communities it governs. People obey and serve the spirit of their time because of their need for meaning, order, and direction.

As a consequence of these kinds of developments, the Christian creation fell apart into nature and humanity. Humanity could now stare at nature, thus eliminating all reciprocity. Separate landscapes could be painted and the separation between the "inner" world of human beings and a distinct landscape in the background could be portrayed, as in the painting of the Mona Lisa.[3] Tourism also became thinkable and doable. These thought patterns endure to the present. Environmentalism misses the obvious point, believing that there can be environment "out there" that we can study in its own right. On the contrary, I am a part of your environment as you are of mine, and the same is true of all other life forms.

The exodus from the symbolic universes of the cultures of Western European societies in the eighteenth century raised a multitude of explicit and implicit questions. If the universe was no longer a creation, then what was its "nature"? How can anything be known in that universe? What would be the best way of doing anything? How could an economy best be organized to ensure a better future? Why stick with existing legal systems when reason (and later, rationality) could improve them? To their very depths, symbolization, experience, and culture were affected. For example, in order to understand something, a person had to observe it. Doing anything could no longer be guided by symbolization either, and was eventually handed over to the guidance of absolute efficiency based on improving a divided reality one element at a time and assessing the results on their own terms independently of all other such attempts. It is clear that the concept of efficiency is neither thinkable nor doable in a dialectically enfolded symbolic universe. Hence, a great many developments of the eighteenth century appear to mark an exodus from the symbolic universes humanity had been creating as a symbolic species, as well as marking a parallel genesis of the desymbolization of human life in the world. These symbolic universes were being remade in the image of what we have referred to as reality.

The interpretation of a growing number of aspects of human life in the world in terms of machines encouraged a new relationship between the visual dimension of experience and the aural dimension.[4] There is no point listening to machines because they simply make noise. To understand them we need to observe them, and they can be observed because they are built up from separate and distinct parts. Consequently, the symbolic universes of the societies of Western Europe were unfolded by a process of desymbolization that created a variety of fragments that were assembled into reality-like universes. Humanity's primary reliance on symbolization gradually shifted to a primary reliance on seeing and observing, which required that everything be divided and separated into its constituent elements. Moreover, the orders created by the sciences (including physics, chemistry, astronomy, botany, and zoology) have nothing in common with the order of a symbolic universe. The new orders were purely intellectual and technical. They could conveniently be catalogued in alphabetical order and entered into the encyclopaedias of that time, or they could be constituted into taxonomies. In addition, these orders could no longer be integrated into something equivalent to a symbolic universe because all this division and separation led to the constituent elements being studied as much as possible in laboratories, and thus away from the complex surroundings to which they were integral but which could not be dealt with by the emerging scientific approach to knowing and the technological approach to doing. It was these complex contexts that had made the entities under study what they were. Observing a plant in an artificially prepared bed of soil or a fish in an aquarium, all in a laboratory setting, severely restricts what can be learned about it. What was divided, separated, defined, measured, quantified, and occasionally mathematically represented were not the entities in relation to their contexts, without which they could not have come into being and without which they cannot adapt and evolve. Nevertheless, it was this emerging reality apprehended by seeing and observation that dealt with everything in its dividedness and separation as a distinct element. All enfolded relationships were eliminated, as were all the dialectically enfolded ones.

The above developments were coupled to a growing mistrust of symbolization and culture, which were seen as creating a subjective, religious, and superstitious interpretation of human life in the world. There was therefore no way back from all this division towards a new kind of liveable order. Human beings and societies had to be satisfied with orders of non-life that were increasingly created for, and in the image of, the kind of reality structures found in machines. Nevertheless, if secular religious attitudes had not blocked the recognition of new emerging limits to human knowing and doing, more synergistic relationships could have been created between the old and the new approaches, which would have transformed both of them.

Once set in motion, the above developments became self-reinforcing. The dialectical tensions that had characterized human conversations, social relationships, and groups in evolving traditional societies were based on people's enriching each other's lives by being both similar and different. These dialectical tensions were gradually banished to interpersonal relationships within a shrinking culture-based connectedness. This raises all kinds of questions regarding the relationship between individuals and their society, which were also raised in the previous chapter from a sociological perspective. What characterizes tradition-directed, inner-directed, and other-directed persons? If our metaconscious social selves are shaped by our relationships with others, what happens when our social bonds are progressively weakened and qualitatively transformed by desymbolization? What happens to these social selves when communities can no longer sustain their members by a fabric of social ties? In France, the process of desymbolization was met with a declaration of equality and freedom eventually leading to the French Revolution, but no solution was found to the community's dependence on social constraints and thus on hierarchical relationships. In contrast, the process of desymbolization in England was transformed by having to deal with entirely new kinds of constraints related to a growing dependence on the technical division of labour that transformed society's need for matter and energy. Since these constraints were not culturally mediated, they required an economic approach for organizing the technology-based connectedness, while an increasingly desymbolized cultural approach continued to organize the culture-based connectedness. The efforts to make this split liveable completely transformed symbolization and culture. The form desymbolization took in England was therefore somewhat different from the form it took in France.

What we are summarily describing is a complete derangement of the symbolic universes of the eighteenth century. This opened the door to further developments that previously were unknowable, unthinkable, and certainly undoable. Without this derangement it would have been very difficult to reorganize human work by means of the technical division of labour and thus prepare for subsequent mechanization and industrialization. It would also have been very difficult to adjust everything to accommodate a relatively distinct industry and economy from which the cultural approach had all but been displaced.

All these developments led to some aspects of human life becoming unconscious, and eventually it led to the recognition that an unconscious had been created.[5] It is not difficult to understand that Freud could overlook the role of a desymbolized culture in human life and thus had to attribute many repressed experiences to events from early childhood. However, desymbolization caused a great many elements in human experience to be repressed, which were internalized and linked within what today is referred to as the unconscious. For example, all who participated in the technical division of labour and the subsequent mechanization and industrialization as knowers, doers, or intermediaries had to repress something essential in their experiences and lives. We have noted that observers such as Adam Smith and Karl Marx recognized how destructive this would be. Anything that was repressed in human experience could no longer play its part in human consciousness. To some extent, a technical division of labour had occurred in earlier traditional societies, but these had highly integrated ways of life, which could be internalized as mental maps able to reintegrate divided social selves without the kind of unconscious elements that emerged in the eighteenth century. As a result, the members of these traditional societies were able to retain the unity and lucidity of their awareness of themselves in the world. In contrast, when a community becomes divided the way it was in the eighteenth century, the internalization of its way of life can no longer reintegrate multiple and divided social selves. Its members lose their awareness of some elements of their lives, bringing an unconscious into existence because a great deal of metaconscious knowledge is repressed and unable to truly participate in people's lives. They become divided against themselves and lose much of their peace of mind, gradually developing the kinds of anxieties that are so prevalent in our time.

There is a strong correlation between people's level of unconsciousness and the level of derangement of their community. If people can experience the entire way of life of their community to varying levels of depth, the corresponding metaconscious knowledge will be able to integrate any possible divided existence to some extent, and their consciousness will coincide with the life of their community. In contrast, if a society becomes so complex that this is no longer possible, then elements of its way of life can no longer be enfolded into the lives of its members, giving rise to many repressed unconscious elements that metaconsciously form an unconscious. Much of this chapter will deal with the relationships between this kind of human life and anti-societies.

There appears to be a strong correlation between desymbolization, repressed experiences, growing unconscious elements, and the eventual formation of an unconscious. Significant limitations on embodiment, participation, commitment, and freedom will restrict the grip people have on the corresponding situations, thus limiting their awareness of what is happening at particular moments in their lives. Approaching anything in life as if it were structured as a reality will repress an awareness of the dialectically enfolded aspects of human life in the world. The only time this does not happen is when we are dealing with classical and information machines and everything built up with them, because these are organized according to what we have referred to as a reality.

A distinction must therefore be made between metaconscious knowledge and unconscious metaconscious knowledge. Metaconscious knowledge arises from experiences with average or low levels of desymbolization, while unconscious metaconscious knowledge comes from experiences in which desymbolization represses significant aspects of our lives. In the case of average or low levels of desymbolization, neural and synaptic changes to the organization of the brain-mind that symbolize those experiences appear to be able to participate normally in a person's life. In contrast, in the case of highly desymbolized experiences with significant repressed elements, the neural and synaptic changes appear to be so weak or non-existent that they cannot adequately participate in the way a person's life works in the background. It is this unconscious metaconscious knowledge that pollutes people's lives on the symbolic level.

Along with the unfolding of the symbolic universes in the eighteenth century came an unfolding of people's lives. When people became wage earners in factories, their lives were divided into a variety of activities such as working, shopping for the necessities of life, spending time with friends and family, and relaxing. Each of these activities was carried out with different groups of people, including fellow workers, fellow shoppers, friends and family, and fellow members of clubs, parishes, and other organizations. The experiences with each group of activities required people to be present in somewhat different ways, with the result that along with the metaconscious knowledge of each group of people and their members there was also a distinct development of a metaconscious social self. At work, people may have done everything to improve their chances of promotion, or they may have attempted to escape the monotony and boredom of their work by "living" in their heads as much as possible by daydreaming. When shopping they were often in crowds and thus subject to crowd behaviour and psychology that were usually uncharacteristic of themselves. With family and friends, on the other hand, they were likely engaged in more intimate and lasting relationships that would have resulted in a great deal of metaconscious knowledge of these significant others as well as of themselves. When they relaxed with others they might share a passion for certain things; once again this would have brought out other aspects of who they were. In a sense, they were likely to develop multiple metaconscious social selves that may or may not have been well integrated depending on the kind of society in which these people lived. In other words, it would not be possible to understand people's psychic and social selves apart from the community in which they lived.

Personal Lives, Reality, and Myths

One of the most important things we have learned from physics is the extraordinary complexity of our relationships with matter, energy, and the physical surroundings that they help to constitute. Many prominent physicists have made important observations on this subject, including Niels Bohr, Albert Einstein, David Bohm, and Bernard d'Espagnat. Nevertheless, the social sciences have been little affected by these discoveries. I have always found this difficult to comprehend, since I was unable to develop a theory of symbolization, experience, and culture without struggling with this issue. It led me to the conclusion that human life is necessarily suspended in myths.[6] This is undoubtedly a disconcerting conclusion, given our supposedly secular age, and it may well lead to what Georges Devereux has called counter-transference reactions, which cause discoveries to be interpreted in such as way as to make them more liveable in terms of the spirit of our age.[7] The above possibility must be faced head-on if we are to understand what J.H. van den Berg has referred to as a divided existence in a complex society.[8] Without myths, all this division, separation, and desymbolization would have simply shattered individual and collective human life.

We have noted that it is possible for human communities to confidently proceed with individual and collective human life in an ultimately unknowable universe, eliminating the threat of the unknown by symbolizing it as more of what is already known and lived. We are thus in touch with the universe except for some missing "details" yet to be discovered and lived. Myths are the metaconscious creation of a kind of lived ontology. We have also noted that these myths perform two relatively distinct functions, of which only one appears to be affected by desymbolization.

By symbolizing the unknown as more of what we already know and live, we render all alternative ways of life unknowable and unliveable. I am not suggesting that it is impossible to understand the symbolic universes of food-gathering and hunting clans, for example. What I am suggesting is that it is the kind of knowledge that will always remain separated from our experience, our culture, and thus our lives. Even with the most vivid imagination, we can never completely enter into these symbolic universes. In other words, it is by means of our deepest metaconscious knowledge that the way of life into which we are born as our cultural inheritance from previous generations imposes certain limits on our lives beyond which we cannot know anything and certainly not live anything. These limits are undetectable because, through our limited experience of an ultimately unknowable universe, it is symbolically transformed in the image of our experiences, to constitute our symbolic universe. Ruling out anything that is radically different appears not to be affected by desymbolization.

The other function of our deepest metaconscious knowledge is to situate our symbolic universe in time, which requires guidance and direction. Despite a great deal of routine, individual and collective human life never repeats anything in quite the same way because everything is related to, and evolves in relation to, everything else. As a result, the symbolic universe in which the members of a community live constantly evolves in ways that are mostly cumulative by fitting new discoveries and events into it. To the extent that this cannot be entirely accomplished, a non-cumulative aspect may remain unconscious. These unconscious elements will gradually build into what, during a certain time, will be unconscious metaconscious knowledge, in the sense that it cannot directly participate in the historical journey of the community. However, as the unknown increasingly imposes itself on the way it is symbolized by the members of a culture, the strength of the influence of the metaconscious knowledge working in the background of people's lives will weaken relative to the unconscious metaconscious knowledge. The inevitable result is that at some point the latter will give rise to intuitions that people's grip on their lives and the world is not quite adequate. Such a historical situation occurs near the end of what (with hindsight) will be referred to as an epoch and is usually characterized by the door to relativism, nihilism, and anomie being cracked open, resulting in a variety of symptoms of a psychological, psychosocial, sociological, and cultural-anthropological character. Malfunctions such as anxiety, depression, anomie, and suicide may rise in these unique "cultural niches." It is this function of myths that appears to be deeply affected by desymbolization. It influences the dominant personality type, how this type is embedded in an individual life, how in their turn all these individual lives are embedded in a way of life as a unique mode of social and historical adaptation and evolution, and how all this is suspended in myths, which orient individual and collective human life within what is noble and liveable.

The transition from a tradition-directed person in a pre-industrial society to an inner-directed person in a transitional society, and the mutation into the other-directed person in a mass society, was integral to the transformations of technology, society, and the relations with the biosphere, as discussed previously. An ever-expanding part of the symbolic universes of the members of the industrializing societies was reorganized in the form of what we have referred to as reality. The dividing line between what was treated as living and what was dealt with as non-living ran through everyone's life. In order to make this situation liveable, myths had to play a somewhat different role in these societies from what they had in all earlier ones, and this will shed additional light on our secular myths.

We have noted that the secular myths of the first generation of industrial societies had to assure people that all previous cultures had been mistaken in exclusively relying on symbolization, experience, and culture to make sense of and live in the world. The good life would not be found through their cultural moralities and religions. Instead, the culture-based connectedness of human life and society had to be largely ignored in order to develop the technology-based connectedness through hard work; the resulting material progress would eventually translate into social and spiritual progress. Capital was the new lifeblood of a nation. Its expression would be the "value of values" capable of assigning a monetary value to everything by means of the Market. In this way, the first generation of secular myths, those of capital, work, progress, and happiness, did not anchor the daily lives of people as such. What they did anchor were only those fragments that were necessary to build the new economic order while trivializing or omitting all other elements. Alienation began to take on an entirely different form, now dominated by reification. In one context a person would be a *homo economicus*, while in others a hand, brain, or intermediary supervisor or manager. There was no place for the human subject except in what little remained of personal life, since individual moral, religious, and political convictions had been banished from public life. Public life could no longer be lived according to what little remained of the symbolic universes of these societies. It had to be conducted in terms of ideologies and world views that took certain fragments of human lives and society as primary, others as secondary, while neglecting still others altogether. All these developments added up to an incredible existential poverty: individual lives no longer making any sense and a collective life increasingly in the service of non-sense. A humanity that had been in the grip of false ultimate meanings was now beginning to be desymbolized, to become "human resources" for an entirely different order.

A second generation of secular myths was required once mass societies began to rely almost entirely on discipline-based approaches to knowing and doing, to build and evolve a technical order around and over what little remained of a highly desymbolized culture-based connectedness of human life and society. It must be emphasized that this added considerably to the division and separation of human lives and communities because the domains of disciplines, when mathematically represented, have nothing in common with the symbolic universes from which they are separated. That this is the case is evident from the development of logic, where at some point it became necessary to cut all ties with experience and culture.[9] The logical internal consistency of these domains has no equivalent in individual and collective human life. This is the previously noted difference between "reality" and a dialectically enfolded universe. We have also noted that domain-related experiences differ from those of daily life by a corresponding discontinuity in their foregrounds and backgrounds, thus contributing to the development of two kinds of metaconscious knowledge.[10] These developments therefore sharpened the historically earlier separation between knowers, doers, and supervisory or managerial intermediaries. They also extended the approaches originally developed in relation to the technology-based connectedness of human life and society to all aspects of a way of life, thus separating it from people's personal lives. In relation to the technical order that has replaced the cultural order, there can be no "experiences" of a human "subject" but only very different kinds of experiences: those of manual, brain, or intermediary functions. The result was that a person's organization of the brain-mind acts as a bifocal mental lens, with one part related to discipline-based mediation and the other part related to desymbolized cultural mediation.[11] It follows that the highly desymbolized experiences of daily life related to what remains of the culture-based connectedness cannot be used to make sense of the technical order and its influences on individual and collective human life. Thus deprived of our inner resources to effectively participate in our collective life, we have no choice but to rely on human techniques and integration propaganda.[12] As a result, mass societies require us to become other-directed persons, supplemented by the limited and highly desymbolized "culture" of our personal lives with significant others, especially when we are trying to hold onto moral, religious, and spiritual values and convictions.

The new secular myths required to make all this liveable were technique, the nation-state, science, and history.[13] The central myth of technique sustains our living as if there are no limits to our technical doing and, by implication, that culture-based doing has no merit other than subjective satisfaction. The sacred of the nation-state helps us to live as if everything is political and thus in the orbit of the state, and, by implication, as if the loss of the self-regulating character of human communities based on their culture is of no importance. The myth of science allows us to live as if scientific knowing has no limits, thereby ruling out the need for all other approaches to knowing. Finally, having delegated all knowing to science, all doing to technique, and all organizing to the nation-state, there remains but one historical path to the future, with the result that everything is within history: time can tell, and history will judge. History is now all in all, and nothing can challenge it. It would appear, therefore, that we continue to be utterly dependent on myths to make our divided lives and societies liveable. This raises the possibility that, when desymbolization weakens the role of myths, the door to relativism, nihilism, and anomie will crack open and a flood of symptoms including anxiety, depression, and aimlessness may engulf us.

As previously noted, technically mediated experiences limit embodiment, participation, commitment, and freedom, to the point that it is very difficult to "live" them. In other words, these technically mediated experiences have a considerable unconscious component, with the result that the organizations of people's brain-minds in contemporary mass societies have a minimal *meta*conscious knowledge because it is engulfed by *un*conscious metaconscious knowledge. It must not be concluded that we live in a kind of unconscious civilization, but that a symbolic species is busy turning itself into *homo informaticus*. Insofar as people have received a higher education in discipline-based approaches and are scientific or technical practitioners, the organizations of their brain-minds will be used as a bifocal mental lens, with one part directed towards what is lived as a reality and the other part towards what continues to be lived as a (fragmented and desymbolized) culture-based connectedness in their personal lives and those of their significant others. The portion of the brain-mind that is directed towards reality acts as a *homo informaticus*, while the other acts as what remains of a highly desymbolized symbolic species. We all start out with the latter, although primary socialization into a culture is now characterized by two streams of experiences: one that would have normally led to the creation of a symbolic universe, and another that permits us to participate in the reality of the technical order.[14]

For example, when we make decisions by means of discipline-based approaches, we may later have to admit to ourselves that we are not entirely sure how we feel about them and how we are going to live with them. In other words, having made decisions outside of our lives, so to speak, decisions that are largely separated from that life, we may have to admit that we have no idea whether we can actually live with them. We have made a decision but we have not lived it. It is symptomatic of the division between our relating to reality as a *homo informaticus* and as a member of a symbolic species. The latter may realize that "in a few days I may feel terrible about these decisions," or "I may be even more proud of them as I begin to live with them." There is a complete emotional disconnect, however, between discipline-based decisions and our lives as members of a symbolic species. These anxieties may well lead to a kind of secondary symbolization of the decisions we have made. However, in most cases the disconnect endures because our lives are suspended in myths, which do not encourage integration in this way. What integration is encouraged is between everything directly and indirectly related to our way of life that adapts and evolves a technical order by means of the cult of the fact, the cult of efficiency, and the cult of economic growth while discounting or obliterating everything that does not belong to this.

Many people were deeply disturbed by the Nuremburg Trials of Nazi war criminals, but we need to ask ourselves whether this was not an extreme case of dissociating the many simple decisions that had to be made: to round up Jewish people and other undesirables, to schedule the trains, to schedule the ovens, etc., on the one hand, and the meaning the Holocaust would have for humanity on the other hand. It is also in this way that, as a global community, we can acknowledge that our contemporary ways of life are unsustainable, while in all our decisions we continue business as usual except for the "secular magic" of sprinkling words such as "sustainability" and "green" over everything we do. Is this not another kind of Holocaust threatening future generations?[15] The emotional divide we have created between our *homo informaticus* side and our being a member of a symbolic species creates all kinds of psychological tensions and disturbances that endanger social relationships because we are making decisions based not on our lives and the lives of others but on domains radically separated from these lives. It creates a complete social disconnect from ourselves and from others. If I have difficulty living with my own decisions, how will others live with them? How will others see me as a consequence of these decisions, and how will I sleep with that? In the long term it may make all social relationships next to impossible, were it not for our being suspended in secular myths. If desymbolization continues to weaken our ability to live highly divided lives in a separated world, will it bring hell on earth, in the original sense of that term as a place where human relationships are impossible? It is in this division of life from non-life that public relations techniques play an important but utterly destructive role, using the language by which a community evolved life to legitimate and make non-life liveable. Public-relations-speak treats language the way we deal with the world by using our vision. It makes the reality we create into something it is not: a fake pseudo-liveable symbolic universe. We are beginning to see glimpses of the emerging anti-person who lives as if our being a symbolic species can be ignored most of the time, only to surrender ourselves to becoming *homo informaticus*: a being in a world ruled with discipline-based approaches.

At this point I need to make a confession. For a long time I have been deeply troubled by the fact that when we go to our doctors there is an almost complete disconnect between their diagnoses and the likely connections of a great many ailments to our nutrition, our work, our lifestyle, and the fact that many issues arising in our relationships are technically mediated in a technical order. For example, the many effects on people's health from assembly-line work are well known, yet I still have to meet a doctor who is aware of the literature on healthy work based on social epidemiology. This shows the essential difference between health care and end-of-pipe disease care. I can well understand that doctors have little or no time to engage in health care, since it would rapidly take them out of their areas of competence. On the other hand, if a much higher percentage of people working on assembly lines were diagnosed with a consideration of the ailments related to work characterized by high demand and low control, perhaps some pressure would build to redesign workplaces using preventive approaches. I find the same disconnect when I look at the psychiatric literature, on the one hand, and the kinds of observations I have been making in this book, on the other. The endless division and separation of our world and consequently our lives in that world surely must have many implications for our mental health, but the connections are mostly not being made. Once again, I can well understand that with their workload, most psychiatrists have little or no time to explore all this. In this way, we are all trapped into end-of-pipe approaches to our health and well-being. The equivalent of negative feedback loops have been all but severed in our reliance on discipline-based approaches in a technical order. To change this situation will require radical university reforms of the kind I have outlined elsewhere.[16] In the meantime, we will continue to treat whatever physical and mental symptoms derive from living highly divided lives in a divided world.

It would appear, therefore, that desymbolization continues to make anything other than the cult of the fact, the cult of efficiency, and the cult of economic growth largely unthinkable, undoable, and unliveable. Nothing radically other can find a place in our contemporary mass societies. Within the limits of what is knowable, doable, and liveable, desymbolization has all but eliminated the traditional ways in which people's lives were oriented and guided by means of cultural values. These are replaced by a rigid conformity to the technique-based connectedness of contemporary mass societies coupled to an other-directedness fed information from the outside by means of human techniques and integration propaganda; with that we must find our own way in what remains of the culture-based connectedness in our personal lives. Desymbolization thus creates a social self that is no longer rooted in the present by means of the past and an unanticipated future. Life is now dominated by public opinions, a statistical morality, and a personalized spirituality or religiosity of the day, which ebb and flow without any discernible patterns. To celebrate someone living this way as the new post-modern person or to acknowledge that perhaps we can do no better than a kind of protean approach[17] may constitute a tragic error that amounts to resigning ourselves to being alienated and reified. Surely it is preferable to understand this alienation and reification as best we can and struggle against them, since humanity generally accepts that slavery is not an acceptable form of life.

From Unfolded to Divided Lives

In the psychological literature there is a long tradition of recognizing a divided social self and its dependence on a community for maintaining a measure of unity. For example, William James acknowledged it, George Herbert Mead reported a pathological case, and Harry Stack Sullivan elaborated a kind of plural existence.[18] The phenomenon of a divided and thus a plural self appears to have intensified over the course of the twentieth century. Much of it had been in the making for a long time, being integral to the transformation of traditional societies into anti-societies and the tradition-directed person into the other-directed person. In terms of our model of the organization of the brain-mind building up a great deal of metaconscious knowledge, the phenomenon can be somewhat understood.

In a traditional society where people lived their lives in close contact with relatives, friends, and neighbours comprising a small and stable "world," a person shared a great many experiences with other people. It may therefore be expected that the metaconscious knowledge in each set of experiences corresponding to an ongoing relationship with a particular person would reflect who someone was with this other person. Obvious differences occurred in who a person was with each and every other person in his or her life, but such differences were greatly constrained by the social fabric to which these relationships belonged. Everyone's life was strongly guided by the same morality, religion, and way of life. Moreover, these lives overlapped and interpenetrated a great deal, with the result that the differences had more to do with shared interests, issues to be avoided with the other person, social obligations, and other such details. The social fabric transformed how the lives of two people were dialectically enfolded through social relationships and groups and how all of them maintained and evolved a dialectical tension between individual diversity and cultural unity. The threat to these communities was more likely to be the result of the weakening of individual differences than a growing diversity that could not be contained.

Much of this began to change in transitional societies where inner-directed persons had to steer their lives through the turbulence that came with an increasingly divided community and world. It became very difficult to be the same person with everyone else: fellow factory workers, bosses, strangers on the street, neighbours barely recognized, and family members with little energy left after long working days. In some cases the technical role of a "hand" was dealing with a "brain" or an intermediate supervisor or manager. In other cases the relationship was between "hands." Outside the factory, more and more strangers were encountered. Family life was worn down by fatigue, undermining its vitality. There were now many relationships in which people themselves were divided as they were obliged to suppress their hands or their brains and parts of their lives. Moreover, different activities were often conducted with different groups of people belonging to different social strata. Communities were divided differently as the traditional social division of labour was replaced by a technical division of labour. The symbolic quality of the experiences of life varied according to the levels of embodiment, participation, commitment, and freedom, and so did the metaconscious knowledge built up with them. As a result, the metaconscious knowledge of the social self related to the experiences with another person labelled A could be very different from the one with person B or person C. All this was further reinforced by the reciprocal character of these relationships, causing others to reflect back such differences. During the early phases of transition, what remained of a traditional society could push back these differences by means of the dialectical tensions around which social relationships and groups evolved. As these dialectical tensions weakened, entirely new kinds of social constraints emerged. These no longer respected the integrity of the human person and the integrality of a human life. For example, it was labour and not work that society demanded of its members. The remainder was regarded as mere "superstructure" or of little consequence, since what really mattered was self-interested anti-social behaviour. The new secular myths sustained these developments and simultaneously undermined all dialectical tensions between individual diversity and the cultural unity of a community. It became less important to know who a person's parents were than to know the kind of technical role a person played in strengthening the technology-based connectedness of the community. It was this technical role that dominated a person's life, the family, and the community. The inner-directed person remained capable of holding together a divided existence with plural "lives," participating in a new technical division of labour as well as in what remained of the social division of labour, and using the latter to make sense of the former. All this had a profound influence on symbolization, experience, and culture.

George Herbert Mead reported on a professor of education who one day left his work and disappeared and was later found in a logging camp, having completely repressed his social self and his life at the university.[19] It had apparently become impossible for this person, having grown up with a simple life in the bush, to maintain a dual existence. Since this pathological case was a rare exception, it may be concluded that for most people the divided existence was much less extreme and that society was able to hold together a plurality of social selves corresponding to lives lived through different technical, social, and cultural roles. There is no evidence that a "cultural niche" was created in which an extremely divided existence would be resolved by retaining a single social self and making all the others unconscious. In other words, if the way of life of a society is adequately experienced through the many social roles by which people participate in it, there is a basis for enfolding the various social selves associated with these roles through deeper layers of metaconscious knowledge. When this is not the case, a variety of mental disorders may be the result. The historical record is difficult to interpret because it is only recently that we have begun to recognize ourselves as a symbolic species whose brain plasticity is at the centre of how the organizations of our brain-minds function. The role of desymbolization during the last hundred years or so is even more difficult to interpret, since the observers at the time did not use the categories associated with symbolization or desymbolization in conjunction with culture or rationality and technique.

We have noted that the first generation of secular myths did not attempt to integrate individual lives as such into a way of life. These myths sought to integrate the technology-based connectedness with the culture-based connectedness, labour with life, material progress with happiness, and so on. It was an entirely different kind of integration, which permitted a much greater diversity of social selves associated with technical and social roles corresponding to a technology-based connectedness divided from a culture-based connectedness. Humanity had embarked on an entirely different kind of historical venture, in which the relationships with matter and energy and everything built up with them were far more important than those with the gods, which had bound individual lives to that of a community. The first generation of secular myths did not suspend human lives and a community in an ultimately unknowable universe, but instead suspended those aspects that were necessary for hard work, material progress, and a vague future happiness under the jurisdiction of money and capital. In other words, it was the desymbolized aspects of human life that were integrated, while the symbolized aspects were played down or neglected altogether.

Much of this continues to be the case for contemporary mass societies. The second generation of secular myths integrates all the desymbolized elements of human lives and societies by discounting or entirely ignoring whatever continues to depend on symbolization, experience, and culture. Technique accomplishes this by means of the cult of efficiency, science by the cult of the fact, and the nation-state by the political cult of economic growth; history eliminates any bridgehead for resymbolization. For all those who rely on scientific knowing and technical doing, the organizations of their brain-minds have constituted a kind of bifocal mental lens enabling them to focus either on disciplinary domains or on the daily-life world. In other words, one part of this lens deals with desymbolized approaches and the other part with what is left of symbolization. Because of the discontinuity in the metaconscious knowledge associated with each part, the structure of the metaconscious social self is similarly divided.

Participation in the technical order as a knower, doer, or intermediary supervisor and manager is not as a human subject but as someone playing a technical role. Similarly, on the receiving end of technique, the human subject is reified as well and generally relies on human techniques and integration propaganda to make some technical or political sense of it all. It is only in people's personal lives with significant others that a fragment of the desymbolized culture-based connectedness is actually lived. It would appear, therefore, that the metaconscious social self of the other-directed person in a mass society is composed of a very limited metaconscious psychic self as subject and a variety of "social selves" related to a variety of social and technical roles. All this is made somewhat liveable because the myths of our mass societies integrate all the desymbolized elements of the social selves while trivializing the subjective self lived through whatever relationships a person has with significant others. The situation appears to be even more complex than what is generally described in the psychological literature because so little attention is paid to desymbolization and its effects.

It would appear that, since the eighteenth century, there has been a gradual unfolding of dialectically enfolded metaconscious social selves into a diversity of social selves associated with people's involvement with other persons in social or technical roles that are loosely integrated through myths, with the psychic selves lived as subjects within the remaining culture-based connectedness. This unfolded complexity appears to be manifested by a proliferation of learning disabilities, anxiety, depression, sexual and gender issues, anomie, and much of our unconscious life.

The transition from traditional to anti-societies has reversed the hierarchy between the symbolic and the desymbolized aspects of human lives. In traditional societies, desymbolization was primarily related to their members being alienated as subjects by having their lives suspended in myths. In contemporary mass societies, people's lives are reified whenever they are on the receiving end of technically mediated relationships, which treat them as a limited set of facts to be determined by a survey, for example, or as a reified element to be made efficient in a particular activity, as necessitated by a technical order. It is possible, therefore, to hypothesize that the traditional person has been transformed into a kind of anti-person corresponding to societies being transformed into anti-societies.

To use the example of portrait painting, the other-directed person has the eyes of someone who is always watching everyone else and who knows that everyone else is doing the same. Their eyes express this and thus are very different from those of a tradition-directed or inner-directed person. Other-directed persons look at themselves with the eyes of everyone else, thus expressing a different self-awareness through their "radar" eyes.[20] Along with the eyes comes a face that is open to others (symbolized by a smile and a partially open mouth), but this openness comes with a risk of the portrait not being flattering to themselves and pleasing to the all-important other.

People whose lives work in the background via the organizations of their brain-minds as a mental map, a gyroscope, or a radar show decreasing levels of embodiment, participation, commitment, and freedom and thus a growing level of desymbolization. Mental maps embody the lives of people according to their participation in a way of life via which everything is related to everything else in their lives. Gyroscopes use the "deepest" vestiges of past experiences for guidance, thereby discounting the new emerging way of life. Consequently, people's lives can no longer fully work in the background except in a vague way that depends on deep metaconscious knowledge. Embodiment, participation, and commitment decline to the point that this is more an orientation of another kind of life rather than a person's own life. Finally, the "inner" symbolic resources for living a life become depleted to the point that "outer" signals must be detected for orientation. There is little embodiment other than the acquired skill of using the organization of the brain-mind to detect these outer signals and to use them to live a life, with the result that that life is lived according to the behaviour of others supplemented by human techniques and integration propaganda. People live by how others see them, and this ongoing experience is used to anticipate the reactions of others in order to be prepared for them.

During the transition, what it is to live a life changes. An experience lived with a life as a mental map in the background is integral to the totality of a life. An experience lived by a gyroscope is a kind of last-ditch attempt to have the past illuminate the present while discounting how the present has evolved away from the past. An experience lived by a radar is lived by the way others see the person and the necessities imposed by a technical order. It all shows a growing level of desymbolization – not the kind of desymbolization that existed in traditional societies but a desymbolization resulting from recreating human lives and society in the image of the machine and thus of non-life. It leads to reification taking precedence over alienation.

We can now reinterpret the events of the last two hundred years of industrialization as being primarily characterized by desymbolization and a shift from a primary reliance on a cultural approach to a primary reliance on discipline-based approaches separated from experience and culture and a secondary reliance on a highly desymbolized cultural approach. There is therefore no longer a clear boundary between our lives and the lives of others: we see ourselves through their eyes as they do through ours. Architecturally, this has manifested itself by opening up our living spaces to the outside by large expanses of glass and the limited use of curtains. On either side of that glass, we learn from others and the media every aspect of social behaviour, from smiling to frowning and everything in between. A mass society is a great equalizer, pressing everyone into the same mould, copied from the rich and famous as deeply as people can afford it. Almost everyone desires the same thing: to appropriate the goodness of technique as much as their resources permit. All the consumer goods in the world are aimed at the same kind of lifestyle. Individuality has become a threat to the uniformity being strengthened through our "radars" on a daily basis. We have difficulty tolerating genuine departures from what is determined to be "normal" because we have apparently intuited that this throws off our ability to make that determination. Desymbolization is almost certainly one of the factors that makes teenage life so terribly stressful. It all began in the eighteenth century when observation replaced understanding, thus implying a spiritual decision to trust our eyes more than our ears. Desymbolization implied the interpretation of life as non-life and a symbolic universe as reality. This has created a divided existence between an ongoing but limited engagement as a symbolic species and a "life" dominated by an indirect participation in the technical order of non-life. The former produces a stream of highly desymbolized experiences, while the latter produces a stream of what may be called anti-experiences. These are so lacking in embodiment, participation, commitment, and freedom that they can hardly be fitted into people's metaconscious lives because there is simply too little of a person or a person's life in them. In the extreme, such experiences produce an "unconscious noise" in people's metaconscious lives.

The third stream of experiences results from a direct participation in the technical order. For these "experiences" in the domains of disciplines, it is the non-symbolic elements that matter. These help to build the patterns of another kind of metaconscious knowledge characteristic of a kind of "insert" into the organization of a person's brain-mind. This "insert" turns the organization of the brain-mind into the equivalent of a bifocal mental lens for discipline-based knowers and doers. One part of this lens is aimed at domains separated from people's experience and culture, while the other part remains aimed at the remainder of their lives. The second generation of secular myths transformed these two parts into an integrated bifocal lens. We will explore this in the next subsection.

Growing Up Divided

Children growing up in contemporary mass societies encounter two of the above three streams of experiences. Elsewhere,[21] I have shown that television introduces children to the "real." Watching television is fundamentally different from being told stories because it reverses the hierarchy between the word and the image. The vantage point in the television show is the result of a long process involving camera people, editors, and producers, and the decontextualization and recontextualization of a great many fragments. The show has to "work" for an audience that the advertisers wish to target. The efficiency of doing so is measured by ratings. Unlike illustrated children's books, television is visual-audio, not to be confused with audio-visual. The latter has become virtually extinct even in the classroom, where the laptop is plugged in and the teacher or professor becomes a mere talking head in the background of the slides and video clips. Television teaches children to passively participate in a "world" that delivers so many perceptual "jolts" that their daily-life world appears boring in comparison.

For young children, the experiences of watching television are differentiated from the ones of learning a language and a culture in order to make sense of and live in the world of their significant others. In the former, the foregrounds of these experiences are characterized by the image dominating the word, while in the background the word continues to dominate the image. In the latter, their daily-life experiences are characterized by the word dominating the image in both the foreground and the background.

There are many other important differences between these two streams of experiences. To the extent that television can be mistakenly be regarded as a window on the world, this world is distorted into a fragmented and discontinuous one in contrast with the world of daily life.

A further distortion is introduced by the television image having clear and distinct edges, which do not occur in daily life. The manipulation of the boundaries of television images profoundly affects the context taken into account, thus creating the opportunity to skilfully manipulate the embodiment and participation of the viewer. What is excluded from a context affects the meaning and value of an image as much as what is included in it, and this creates the all-important equivalent of a kind of hidden curriculum. For example, by including the one plane that crashed and excluding the tens of thousands that did not crash, news programs are making the world appear to be much more dangerous than it really is, and this communicates profound messages about the risks we are willing and unwilling to take. The advertisements from a particular corporation will include what their advertising agency and public relations staff want us to know; by taking this out of the context of its global operations, they leave us with a highly distorted impression of what the company is doing. Because of our daily-life experiences, we "live" our television images as if beyond their edges there is more of the same, but in almost all cases the opposite is the case. Hence the tremendous power of public relations techniques and integration propaganda, because we tend to extrapolate these images to what lies beyond as more of the same. For adults, it is a remnant of our being a symbolic species, and for children it is a preparation for living in reality.

All these effects must be interpreted in the context in which they occur, namely, that of an other-directed person in a highly desymbolized anti-society. We know very well that two identical trucks will perform very differently on a highway and in a farmer's field. The effects of public relations techniques and integration propaganda cannot be understood apart from the contexts in which they occur, and countless studies have completely overlooked the unique context of an anti-society. It is no exaggeration to claim that the bath of images into which television immerses children is a technical substitute for what symbolic cultures used to accomplish by means of a tradition and way of life. As other-directed people, we are shown how to behave in every situation that matters for the maintenance and evolution of the technical order and the nation-state. By claiming the right to free speech as corporate legal persons, large corporations expose children to influences against which parents are largely powerless. It is another example of how we have two classes of citizens in a democracy.[22]

Moreover, the attitudes of adults with regard to watching television have a profound influence on children. Because of the second generation of secular myths, all cultural values have been desymbolized to the point that what is knowable, doable, and liveable is restricted to what most people do: what is considered normal becomes normative, which is completely incomprehensible in the symbolic universe of any traditional culture. In a mass society, what is normal as portrayed on television is far more significant than what appears to be normal in our so-called subjective daily lives.

The appearance of scientific and technical experts on television adds a further distortion to children's perceptions of our world. Discipline-based knowing and doing derive from domains that are entirely separated from experience and culture. As a result, there is no context to the "facts" presented by the experts. This context is added to in part by the beliefs, values, and employment of the specialists themselves and by everyone else working on the television show. Lawyers know all too well that the same "facts" put into different contexts can tell two different stories. Neither the prosecution nor the defence has difficulty finding specialists to support its case, which is rather strange, since the same "facts" cannot possibly be on both sides. To a limited extent, every specialist "symbolizes" what they say to make it accessible to non-specialists, but this interpretation is done on the basis of their daily-life experience and not on their expertise. It is not surprising, therefore, that television shows with experts can be made to argue almost any case. In my opinion, the influence of editorial policy and ownership has been greatly exaggerated because the daily-life interpretations are suspended in the second generation of secular myth, thus introducing children to the endless commonplaces projected by the media.

For children, television watching tends to establish and reinforce the "reality" of everything in their world as opposed to its meaning and value for their lives. What they observe becomes much more "real" than what they understand by means of language and culture. It prepares them for a growing dependence on human techniques and integration propaganda to complement their increasingly desymbolized culture and way of life.

Their experiences of television watching initiate children into desymbolizing experiences. In our previously developed model of culture[23] we have shown that the organizations of the brain-minds of babies and children create metaconscious knowledge, first of their physical selves, then their social selves, and eventually their spiritual selves. Moreover, this metaconscious knowledge acts as a metalanguage that launches them towards a symbolic mediation of their relationships with others and the world by means of a culture. Initially, the visual dimension of experience is fully integrated into these developments by contributing to an order of what is true in their lives, and this order evolves towards a cultural order in a traditional society. Such a cultural order cannot be entered into other than by means of language. This process of socialization begins to be undermined by children's experiences of television watching. It is likely their initial introduction into what is real and into their paying more attention to the visual than to the aural dimension of experience. This is accompanied by unprecedented levels of desymbolization as television watching restricts children's embodiment, participation, commitment, and freedom.

The other introduction of toddlers and children to what is real comes from their experiences of computers and the internet. These open the gates towards an entirely new playground, as different from all traditional playgrounds as reality is from what is true for their lives. Much of what we have already noted regarding their experiences of television watching also applies to the use of computers and the internet, with one important and frequently occurring difference. The foreground of any computer-generated and mediated experience corresponds exactly to what we have referred to as a domain. It is produced by rules and algorithms based on the principles of non-contradiction and separability instead of dialectical enfolding, as well as on closed definitions instead of dialectically enfolded meanings and values, in order to create a reality open to measurement, quantification, and mathematical representation.

The resulting programmed environments are thus domain-like. Within them any human function is simulated by technically dividing it into a sequence of independent and endlessly repeated steps that can only be changed by other steps. Whether any step is carried out by a human being or a device makes no difference. In other words, a person has to utilize a programmed device on its own terms. With our current computer architectures, no matter how much programs increase in their sophistication and speed of execution, the illusion that such quantitative changes will produce qualitative ones is reassuring for us and necessary to make all this liveable.

We do well to recall once more that machines thrive on repetition while human beings are destroyed by it, because our nervous systems have evolved to cope with a living world in which nothing ever repeats itself in quite the same way. Hence, any talk of computers being tools is ignoring the obvious. Computers will never be tools in the hands of a symbolic species because they will change the symbolic character of that species. There is a great deal of evidence of this. The corporations have finally learned that they cannot use computers other than on the terms laid down by these machines. Hence, to make effective and economic use of these devices and the associated information technology, the corporation had to be re-engineered, which involved its reorganization in the image of the machine.[24] At present, we are watching the re-engineering of the service sector, with organizations such as Walmart and Amazon leading the way.[25] Soon there will be very few jobs that have not been tailored to the image of the information machine. Worse, everything that is being turned into reality can be monitored in ways that are historically unprecedented. Even most slaves were not subjected to the equivalent of the new computer-based whip. This re-engineering will also allow us to be enclosed in our individually customized "reality bubble," with everything within it having been selected according to profiles established from previous interventions in reality.[26]

The point is that children are also being inserted into this reality as they learn to "play" in it. The new playgrounds of reality are decisively different from the traditional ones. It is possible to win in the simulated worlds of computer games, but this was not the case in a symbolic universe. Since the latter was dialectically enfolded, every human action had a great many consequences, and it was usually impossible to separate the positive, negative, and neutral ones. Since the domain of a computer program is entirely separated from experience and culture, it is populated by distinct and separate elements acting in causally limited ways that leave little room for anything except performance. In the case of games, this performance translates into decisively winning or losing. Unlike the games in traditional playgrounds, the games in the new playground of reality so lack embodiment, participation, commitment, and freedom that they result in a stream of highly desymbolized experiences. As adults we know how easy it is to become absorbed in a computer game or a computer-mediated task, but this must not be confused with commitment. Sooner or later we have to get back to our lives because we cannot live on a screen. This is equally the case for the computer-mediated playing of children. Moreover, children absorbed in the new playground of reality are drawn away from their daily-life contacts with others. Such face-to-face experiences are open-ended, uncertain, and ambiguous; to deal with this requires a commitment to face the risk of making yourself vulnerable to others. It takes a great deal of practice. By depriving children of this in the new playgrounds, those who are shy may prefer to be alone with the machine. Some children may intuit something of the differences between the new playground and face-to-face conventional play. They may prefer the latter because it appears more intriguing, complex, and rewarding as they enjoy the support of others and as they support them. Being absorbed by the new playground of reality appears to lead to addiction, while being immersed in living a life with others leads to participation, commitment, and friendship. It does not take long before children are attracted to the one or the other. Program domains allow them to escape their vulnerability with others and enjoy a definite solution and an unambiguous winning or losing against a machine. Building the culture-based connectedness of their symbolic lives allows them to enjoy the subtle complexity of a dialectically enfolded life with others and to be sustained by them. In anti-societies this possibility may well become increasingly rare.

Generally speaking, the schools appear to be doing everything to push children into the "life skills" related to the non-life of computer-mediated tasks. All this has a significant effect on their metaconscious knowledge of who they are, what it is to be alive, and how all is differentiated from the way computers as dead devices mimic their activities. In turn, this metaconscious knowledge deeply affects the way they express themselves and live their lives. The social constraints of culture-based communities are now being replaced by those of what is real. Children may learn to value in computers what they cannot find within themselves, and they may value themselves for what the computer lacks. This will further affect their perception of what it is to be human and what it is to be a thing. In any case, anything on a screen adds to the real, thus dominating what is true in human life. It also sharpens the distinctions between thinking and feeling, because the latter appears to children to be unique to human lives, while thinking may be thought of as something they share with computers. Hence, thinking may become associated with cold, detached, non-involved parts of the self, and feeling with a more basic and non-logical part of the self. It may even become difficult to understand how the two are expressions of one and the same person. This may divide their social selves as part machine and part emotion. Children learning to use cell phones, text messaging, twitter, facebook, and more are submerged into the life-milieu of technique and the technical order in which their relationships are technically mediated. The life-milieu of technique acts as a filter via which certain things can be experienced, while other things are distorted or filtered out entirely. There is thus a kind of dual socialization of toddlers and children, with one stream of experiences contributing towards their development as members of a symbolic species, and the other towards a highly desymbolized and reified life in a technical order.

Based on our model of symbolization, experience, and culture, the addition of a second stream of highly desymbolized experiences comes at a very critical time and could interfere with the development of language and everything built up with it. The transformation of vocal signs into symbols depends on a double referencing system.[27] One is that the differentiated experiences of children's lives act as a metalanguage in the context of which a variety of vocal signs have taken on a meaning differentiated from all other sounds. The other develops from experiences with a language foreground gradually becoming directly differentiated from each other and jointly from all other experiences, thus constituting what may be imagined as a cluster of language-experiences differentiated from all other clusters. The direct differentiation of vocal signs (and later phrases) from each other, and indirectly from all other experiences, begins the creation of a metaconscious "world" of language as the gateway into the symbolic universe of a community. Intuitions of these developments soon begin to affect the behaviour of children. When highly desymbolized experiences characterized by low embodiment, participation, commitment, and freedom become intermingled with the language development of children living in mass societies, the process of differentiation faces a much more complex task of sorting all this out; the desymbolized experiences may well act as the equivalent of noise, undermining the metaconscious formation of meanings and values. It could be a likely source of a variety of learning disabilities. The danger was well recognized a few decades ago when I knew several professors of computer engineering and computer science who refused to have computers in their homes until their children were well into their teenage years. Some of them spent a great deal of time convincing school boards to get computers out of the classrooms of young children. Of course, with the growing dependence on the internet, doing without them at home became increasingly difficult, and it was not long before many teachers became convinced that the new information technology was the saviour of education. Nevertheless, people who have carefully examined these kinds of issues continue to report disturbing findings.[28]

Because of the differences between computer-mediated experiences and culturally mediated ones, each associated with opposite hierarchies of the visual and aural dimensions of experience, it is highly probable that they involve different pathways to and within the brain-minds of children. It is tempting to cite some evidence based on magnetic resonance imaging (MRI) studies suggesting that already, after the first half-hour of computer-based internet use, detectable changes had occurred.[29] What the meaning and value of the changes are for human lives is almost impossible to know. First, we have no knowledge of how people's brain-minds deal with symbolic functions on the level of experience and culture. Second, we know nothing about how desymbolization has altered these functions and thus what kinds of brain-minds current MRI studies are observing. Third, we know nothing about how MRI-based observations emphasize or de-emphasize symbolic functions versus highly desymbolized ones. Fourth, we know absolutely nothing about how human brain-minds represent a dialectically enfolded human life in a symbolic universe and whether or not this can be exclusively captured by local brain activity. What we do know is that the development of Western logic required that this discipline be entirely separated from experience and culture.[30] We also know that a Turing machine can only deal with symbolization, experience, and culture via "facts" represented by context-independent zeros and ones as the ultimate building blocks of reality. Consequently, the application of computers to anything symbolic and cultural depends entirely on the cult of the fact and the cult of efficiency, thus transforming it into a domain of reality. Computers cannot deal with a dialectically enfolded symbolic universe of which "facts" are a specific manifestation, with the result that all these "facts" are suspended in myth. It is not surprising, therefore, that when mainframe computers began to be built, their applicability was limited to scientific and technical calculations; it was not until societies began to rely primarily on technique and only secondarily on culture that this limit essentially disappeared.

In a growing number of elementary classrooms, the playground of reality is extended into formal education. Of course, all this is done with the best will in the world as being creative and relevant, preparing life skills, and creating a workforce for the new economy, but rarely as a recognition of the necessities imposed by a technical order. Soon children are asked to make power-point presentations or write blogs. The former requires a search for relevant facts to be turned into bullets, and somehow these bullets are to be magically turned into patterns of understanding, but this is largely left to the audience. In other words, rather than emphasizing the meaning and value of each point in relation to another and the broader context of a topic, teachers are expected to play down or ignore the symbolic aspects of the task. From the outset, schools teach children to observe, collect distinct and relevant elements, package these elements into a power point presentation, and consider the job done. This amounts to an affirmation of reality at the expense of the symbolic element and is at best an uncritical acceptance of technique over culture and at worst the promotion of a secular religious attitude towards technology as the saviour of education. It furthermore emphasizes the need to become an extroverted person who has no difficulty speaking to others, preferably without notes. Introverted deeper thinkers must be socially adjusted to their classmates in a manner previously discussed in relation to human techniques and group dynamics.

Concerning the writing of blogs, children are also diverted as much as possible from everything symbolic. Blogs are destined for the internet and not for the world of magazines, books, and libraries.[31] Hyperlinks in the text are a good indicator of the vast differences between these two media. Originally, educators thought that putting hyperlinks into a text would facilitate learning because the student could then rapidly be acquainted with alternative perspectives and additional details, to mention only a few ways of enriching the context. Hyperlinks would encourage critical thinking rather than a regurgitation of the text. It turned out that the consequences of hypertexts are substantially negative.[32] They increase the cognitive load and thus interfere with the possibility of arriving at a deeper understanding of what is being read. With hindsight all this is terribly obvious. Clicking from link to link and from page to page makes it next to impossible to follow an argument.

Comprehension appears to be inversely proportional to the number of links in a text, and, generally speaking, readers following a "linear" conventional text instead of a hypertext tend to achieve a higher level of comprehension.[33] A hypertext externalizes much of what symbolization sought to accomplish by symbolizing what is being read in relation to everything in a person's life to reach a deeper understanding, because everything is differentiated in relation to everything else. In terms of our model of symbolization and culture, children do not learn the meaning of something by means of closed definitions. Instead, by differentiating everything in their lives from everything else, they refine the meaning of something in relation to the meanings of everything else and vice versa. Hyperlinks create no such synergy between what is being read and everything else previously understood by a child. Worse, they draw mental energy away from concentrating on a text as a result of cognitive overload. Any reading or writing mediated by a computer is subjected to a variety of distractions presented on the screen. Music may be played in the background, text alerts may sound, attention may be diverted away to check something that has suddenly come to mind, and so on. There is the ongoing temptation of becoming mindless consumers of "facts" and interesting sound bytes. The attention of the child is divided, and all kinds of extraneous issues interfere with the task at hand. It is not difficult to understand why all such interference is not very helpful in teaching children how to symbolize a text, grasp its deeper underlying thrust, and share what they learn with others. The end results are all too clear among university students, many of whom have great difficulty understanding what equations mean in terms of the kinds of relationships they establish among various entities in a domain, let alone what they may represent in our lives in the world. As a result, students cannot understand what the implications are for good design.

In sum, much of the effort of elementary school education is directed to socializing children into a reality as opposed to an equivalent of a symbolic universe for a mass society. It is a complete and uncritical surrender to the necessities imposed by the technique-based connectedness of a technical order. To argue that this prepares children for what lies ahead is a tragic mistake. It will simply multiply our current problems, because we cannot design the actions, devices, and organizations to achieve more of what we desire and less of what is undermining human life on our planet. It is a betrayal of our being a symbolic species.

Following the extension of the playground of reality by means of primary education, this reality takes on a secular mythological status when teenagers learn discipline-based sciences in high school. They are introduced to mythological detached and objective observers able to make neutral observations and discover facts by means of experimental designs that are universal. All this trivializes these observers as being people of a time, place, and culture, as if this would liberate them from all the superstitions of the past, as well as the myths within which we are all suspended in order to make our world a liveable one. That these observers are internally connected to what they observe, that these facts depend on a context, that the experimental designs depend on a discipline in a particular stage of development, and other such dependencies are taken to be trivial details that require no critical examination. As a result of this scientific mythology, the playground of reality is moved closer to reality itself in the lives of teenagers. Their domain-based experiences become far more objective, factual, value-free, and trustworthy to them than the other stream of experiences related to symbolization, experience, and culture. Once again, the symbolic aspects of human beings involved in discipline-based sciences are played down or entirely ignored. When these limitations of discipline-based approaches to science are pointed out, it is usually interpreted as "science-bashing."

When teenagers learn physics in high school, they encounter a domain that is populated with entities such as forces, masses, flat planes, inclines, pulleys, and more. None of these have anything in common with the teenagers' daily-life world: forces are pure by not being embodied in anything; masses have no volume and can occupy a single point; planes or inclines are non-material because they have no friction; and the surroundings are also non-material because they produce no drag. As students advance in the subject they will learn to compensate for all of this, but two important issues are almost always glossed over. First, it is assumed that, by taking physics, teenagers learn to scientifically examine the world in which they live. As noted, physics does no such thing. In their physics classes, students are asked to imagine a virtual domain populated exclusively with physical phenomena that are observed and not understood in the symbolic sense. Perhaps many teenagers would have less difficulty learning discipline-based sciences if all this was explained a little more carefully. For example, the teacher might point out that they are going to be simulating physical phenomena in an imagined space from which all other kinds of phenomena are excluded, but that nevertheless this is a useful exercise because the simulations may be used to understand physical phenomena when they are not significantly affected by others.

The second issue is closely related to the first. What they are learning about these physical phenomena rarely applies to their daily lives. In other words, none of the daily-life intuitive physics implied in their abilities to walk, run, play ball, bicycle, climb trees, and more (which is therefore symbolic and cultural) can be simulated, for a variety of reasons: these physical phenomena, along with others, are enfolded into these activities and determine the boundary conditions for any equations; these activities are governed by the intentions of the people carrying them out; and, even if some of the teenagers eventually earn doctorates in physics, none of this will change. The time is long overdue that these teenagers learn something about the differences between intuitive physics and school physics, and that the applicability of modern physics is limited to situations in which all other kinds of phenomena may be neglected, such as the big bang or subatomic situations, but mostly in technical systems structured as a reality. If this were to be an integral part of the learning of physics in high school, a great many confusions and misunderstandings could be avoided. Moreover, it would lay a foundation for the students' learning when to use intuitive physics and when to use school physics, and thus something about the limits of discipline-based approaches to knowing in relation to those based on symbolization and culture. In turn, this would keep the door open to learning that the domains of disciplines have the "nature" of reality, while the students' world is symbolized as a symbolic universe. In the meantime, the most important contribution that the learning of discipline-based sciences makes to the lives of teenagers is their acceptance of reality and thus the myth of science. We have now come full circle and are back to the kinds of issues discussed in the opening chapters.

The "experiences" students gain when solving problems in the domain of physics will be differentiated from all other experiences, and thus, in the terms of our model, may be regarded as a unique cluster among all others in a person's life. Over time such a cluster will be subdivided into some clusters essentially corresponding to the different categories of problems encountered in the domain of physics. If students continue to study physics in university, the domain may be "filled" with a continuum or a fluid to examine the physics of solid objects, liquids, or gases. The more these clusters become differentiated from those related to daily-life experiences, the more they are on the way to becoming a sub-part in a person's organization of the brain-mind, and the more the two will together act as a kind of bifocal mental lens. However, this is not likely to occur until many years of study followed by a great deal of experience in applying this discipline-based knowledge to further differentiate it from all other daily-life approaches.

The daily-life experiences of children and teenagers are permeated by physical phenomena. These are encountered as being intermingled with many others, and it is precisely this intermingling that makes each situation unique. In differentiating these experiences from many others in their lives, children acquire a great deal of metaconscious knowledge of intuitive physics, but there is no separate knowledge of the physical phenomena themselves. They are part of the dialectically enfolded character of what in traditional societies would have become a symbolic universe but which in a mass society is a reality. Discipline-based physics is separated from experience and culture, which is why high school physics does not begin with a kind of Socratic dialogue to uncover and build on students' metaconscious knowledge of physics. It cannot be done. Nevertheless, our civilization still does not clearly distinguish between knowing and doing embedded in experience and culture and knowing and doing separated from experience and culture, with all the confusion this entails for high school education. Learning science will enlarge the stream of highly desymbolized experiences related to reality and by implication will discount the other stream, which all too soon will become largely restricted to the students' personal lives.

Access to the domain of physics relies exclusively on the visual dimension of experience and on graphical representation in the form of free-body diagrams. The latter remove all the details experienced in daily life because physical phenomena are examined in a mathematical domain. Nothing can enter it without first having been mathematically defined, and all relationships with the previously introduced entities must be represented by means of mathematical equations. This remains the case no matter how far students advance in the subject and learn to account for additional "details" such as friction, inertia, and air resistance. Even on the frontiers of physics, its mathematical domain has nothing in common with the world in which we live because it remains exclusively populated by physical phenomena. All other phenomena are externalized and relegated to other disciplines.

Despite all these limitations, the mythology of science that accompanies the learning of discipline-based science in high school ensures that the playground of reality is turned into reality itself. It is much more "real" than what little remains of the culture-based connectedness in people's lives in a contemporary anti-society. High school education thus plays a fundamental role in socializing teenagers into the secular myths of anti-societies. If the universe and human life within it have the "nature" of reality, then there are no more limits to science, and our so-called secular education contributes to science becoming one of our important myths. The so-called detached, objective scientific observers are the people whose organizations of their brain-minds function as a bifocal mental lens by which they take on this mythical activity, which human beings can only hope to approach with a great deal of discipline. If the observer influences what is observed, can the discipline of physics be separated from all other disciplines? If the observer must be included in the experiment, would it not compel the consideration of many other disciplines? In other words, if we desire to be truly scientific, the discipline of physics would have to be substantially modified because physicists cannot entirely detach themselves from symbolization, experience, and culture. The level of desymbolization would have to be decreased, and that would shake the very secular myths on which our civilization is built and into which the next generation is being socialized.

Reality and the Desymbolization of Personal Lives

In surveying the vast literature attempting to make sense of what is happening to our children, our schools, and our societies, we see a relatively clear pattern. One part of the literature stems from people having a deep sense that something has fundamentally changed, that this is causing a variety of significant issues, and that we must therefore step back and attempt to make some sense of it all if we are to intervene in an appropriate and responsible manner. This literature includes glimpses of what we have referred to as desymbolization, attempts at resymbolization, and an occasional implicit questioning of the secular myths that guide our lives and communities. The other part of the literature appears to be more firmly guided by these secular myths in its attempts to make sense of our situation and in responding to it. This literature assumes that our latest technologies are opening up all kinds of new opportunities, and though the unavoidable dislocations that accompany them may at times be painful and disturbing, everything generally continues to move in the right direction. As always, we would be better off if each part of the literature was used to assess the plausibility of the other, and fortunately, there are elements of this throughout. We will select a few works as entry points into our attempt to symbolize what is happening in the vast and powerful currents of desymbolization.

The title of the first work, *Right-Brained Children in a Left-Brained World*, by Jeffrey Freed and Laurie Parsons, gets right to the heart of the matter.[34] Humanity may always have had some so-called right-brained members, but under the current circumstances their numbers may well be rising rapidly. What is commonly diagnosed as attention deficit disorder (ADD) or attention deficit hyperactivity disorder (ADHD) appears to coincide with cognitive approaches to making sense of and living in the world that are dominated by brain function commonly associated with the right hemisphere. It appears that in these people the visual dimension of experience is dominant to the point of inhibiting the development of the symbolic functions associated with the left hemisphere, which rely primarily on the aural dimension of experience. A page full of text or a page full of images may well be cognitively dealt with in much the same way. The letters are treated like spatial objects as opposed to symbols. For example, the letters "b" and "d" would be seen as the same, just as a teacup remains itself whether the handle faces to the right or the left. This condition also appears to coincide with a hypersensitivity of the other senses: hearing, touching, tasting, and smelling. When such children are touched or hugged, it may feel bone-crushing. They may be picky eaters because the odours and tastes of certain foods may be exaggerated. Sounds are often experienced as highly invasive. Consequently, almost everything that happens in classrooms will distract these children. In the extreme, they may experience a kind of sensory bombardment of stimuli that can become so unbearable that they "shut down."

It is difficult to imagine how, as toddlers, these children learned to deal with symbolic functions: differentiating vocal signs from other sounds, learning to turn these vocal signs into symbols in any language, and entering into a symbolic universe. Moreover, encountering and dealing with other people almost entirely in terms of mental images must make it very difficult for such children to join in the community of a symbolic species. The usual sequential integration of phonemes into words, words into sentences, sentences into ideas, and so on may require their translation into a sequence of mental images, but it is unlike watching a kind of mental movie because each image is incomplete in itself. The difficulties these children have in entering into a language and a culture must be very great, to the point that it may account for a tragically common low self-esteem because they appear to be "dumb" compared to other children. To avoid endless humiliation, many children adopt a strategy of attempting to be perfectionists. They watch other children do things until they are able to figure things out to a level at which they believe they have a chance of successfully doing them. The alternative strategy of learning by trial and error may simply add to further humiliation and disapproval by others. When an adjustment is made in the teaching style to accommodate the gifts these children do have in excelling at right-brain activities, learning to read can be greatly facilitated, but it is almost certainly done in a very different manner. There is also an overlap between this disorder and autism, in the sense that these children deal with the world almost exclusively in terms of pictures. Words may invoke images, and conversations may turn into a kind of lived movie.

It is tempting to speculate that this disorder may involve an inability of the brain-mind to coordinate the various dimensions of experience associated with the senses. It may explain why these children are often hypersensitive in several dimensions of experience in order to compensate what they cannot accomplish in the oral one. This compensating would correspond to what we know about low-vision or blind children. Each of the senses may potentially contribute to experiencing a situation, but these contributions are often unnecessary for those who have normal vision. For example, when a blind person crosses a street, the angle of the feet provides sensations indicative of where the person is spatially. Since the road is rounded for drainage, the feet will be angled slightly upward until the middle of the road is reached and slightly downward when crossing to the other side. A sighted person does not need to pay attention to this because it is much easier to learn from the visual dimension of experience. In other words, the sensitivity of the tactile dimension is heightened to provide the kind of contribution to an experience that for most people is entirely unnecessary. Again, all this is little more than informed speculation, but learning a language and a culture with a visual approach requires that all the other senses contribute what they can if a child with ADD or ADHD is to have a chance at joining a community. It must be very difficult to turn pictures into words and words into ideas.

There is an obvious convergence between the situations in which children with ADD or ADHD find themselves and those all of us confront. Nevertheless, I do not believe that the former constitute a kind of cognitive asymptote towards which all of humanity in the life-milieu of technique is moving. All of us face a setting characterized by sensory overload, social overload, crowding, and noise as we keep up with a fast-paced way of life. It is impossible to focus our attention on anything for too long because this may lead to missing other "bits." Everything must be done as instantly as possible in text messages, sound bytes, and newsflashes. There is no time to dwell on anything because there is so much more stimulation than we can possibly deal with. The technical mediation by television, computers, the internet, cell phones (now turned into mobile computers), and dealing with the world via discipline-based approaches helps constitute a medium whose message is the following: a unique stream of highly desymbolized experiences easily differentiated from what little remains of culturally mediated ones; the need for short attention spans in order to keep up with the sensory bombardment and not miss anything; the elimination of "deeper" understanding and behaviour requiring symbolic functions; the inability of lives to fully work in the background; and a lack of time to integrate everything into a life lived in a meaningful world. Educators have added their infatuation with one educational technology after another: educational television, computers on every desk, the information highway connecting these computers to the world, the internet as a library, and every imaginable screen-based device.

To all this must be added the techniques of group dynamics, pedagogy, and human relations. They extend the stream of highly desymbolized experiences generated by television watching, computer and internet use, and the like at home, as well as by our involvement in discipline-based approaches. Nearly half a century of building a technical order by means of discipline-based approaches to knowing and doing has made reality more primary than what we perceive our lives in the world to be via symbolization, experience, and culture. There is an ever-greater reliance on the approaches associated with getting a grip on reality, which tend to be right-brained at the expense of those related to language and culture associated with left-brained functions. As a result, it is becoming more and more difficult for children to integrate the visual dimension of experience into the aural dimension on the latter's terms. It undermines their development as members of a symbolic species, and our long-term future may be hanging in the balance. A great many studies justify our being mesmerized by all the new technologies and techniques that we create because of the many unique opportunities they will offer, without looking further ahead to examine how these will be offset by what we have referred to as "technique changing people." For example, the social media are mostly intensifying the trends we have described. All this would not be significant if we were not a symbolic species and could live full and responsible lives in what we have referred to as reality instead of a symbolic universe. Without any exception, these technologies and techniques significantly diminish our embodiment, participation, commitment, and freedom in our lives, to produce a stream of highly desymbolized experiences that involve more right-brained than left-brained activities. One would expect, therefore, that children with ADD or ADHD would thrive in this "cultural niche." Unfortunately, this is certainly not the case, and the pleas for our schools to accommodate their unique learning styles must be heard if these children are to have a chance. Nevertheless, this will not remove the threat desymbolization poses to a symbolic species. Our schools simply are not coping with the difficulties of desymbolization. They continue to carry on as if we live in a traditional society now equipped with all kinds of new technologies and techniques; that is, as if technique does not change people. Unless we embark on a mega-project of resymbolizing our lives and societies, beginning in the universities, we are facing a losing battle.[35]

This is particularly evident when we pause to reflect on the differences between living in a reality and living in time and history. For people to participate in a historical journey, their lives must be inserted into time. The overlap between ADD or ADHD with autism is instructive on this point. Living in something that resembles what we have referred to as reality is like living in a "solid object-world" that more easily shatters than it evolves. Doing so implies corresponding changes to the development of the metaconscious social self. It too must become a kind of solid object instead of a historical person. Hence, relationships with others become excruciatingly difficult. In order for a metaconscious social self capable of joining the historical journey of a community to be developed, the reality in which a person lives must be inserted into time and history; this implies that the symbolic function must dominate the spatial functions. By implication, the rest of humanity faces a similar problem. If babies and children face a stream of highly desymbolized experiences that undermine their ability to enter what little remains of the culture-based connectedness as a kind of childhood symbolic universe, a growing dominance of the visual dimension of experience over the aural dimension of experience may well have similar consequences for the development of their metaconscious social selves and eventually their metaconscious spiritual selves. This development may well be particularly tragic in Western cultures which have been profoundly influenced by Judaism, possibly the first culture to create human life in a linear time and thus history. Living this way is a sharp contrast with living within natural cycles without history, sometimes symbolized as a cycle of reincarnations.

It would appear that autistic people live in a spatial reality that takes different forms on different days of the week, but within any of these forms changes are almost unbearable. Each day has its own characteristic reality that exists outside of time. Nothing can be made sense of as constantly evolving and adapting. In this sense, autistic persons have no life the way we know it. There is a set of routines that must be adhered to, to the point where their meanings and values cannot be established by differentiating everything from everything else in the living of a life. For example, there is no difference between reading a phone book and a book with stories. Both must be converted into mental images, and the memorization of half a phone book is little different from revisiting a habitual story in a book. The meaning and value of these activities appear to be more or less the same. In other words, autistic people appear to live in a timeless sequence of realities that are like solid worlds filling their appropriate days in a rigid sequence. All this is experienced as a set of objective facts, much like the elements in a spatial landscape of life.

It may well be that children diagnosed with ADD or ADHD face similar issues. For example, one child was able to make sense of what a person was writing from the sounds the pen made on paper.[36] Another child had learned to sense the mood of a parent from incredibly meticulous observation of body language and facial expressions rather than by engaging in conversation. It is not entirely clear how these children make sense of time – presumably via mental images. We are slowly learning to better help these children, but the disorderly and disturbing symptoms of their condition have labelled them, much as little or no vision turns a person into a member of "the blind" or little or no hearing turns a person into a member of "the deaf." What this implies is that all one needs to know about the lives of such people is that their visual or aural dimension of experience does not function and that this permeates all of their persons and their lives. Everything else is secondary. Such labelling deals a crushing blow to these people's metaconscious social selves, which helps to explain why the overwhelming majority of blind persons are unemployed and without much of a place in what is left of our societies. A segment of the deaf community has argued that sign language is the entry into a deaf culture distinct from the culture of non-deaf people, but this involves a confusion as to the role a culture plays in the lives of the members of a symbolic species.

What we may learn from all this is that contemporary anti-societies are unintentionally doing almost everything to ensure that as many experiences as possible are dominated by the visual dimension at the expense of the aural dimension. Doing so is inseparable from a significant restructuring of people's brain-minds. The short attention span which became necessary to cope with the sensory overload of texts, tweets, sound bytes, news flashes, and everything else "instant" undermines symbolization and thus leads to an unfolding of the dialectically enfolded brain-mind of a member of a symbolic species. After all, if we pay attention to something for too long a time we may miss out on the next "bit" of desymbolized instantaneity. This ensures that we are unable to replace our public opinions with our own; we slavishly follow what others do in order to be normal with our own freely made ethical decisions; and we faithfully attend the secular "Google church" to participate in the cult of the fact and the cult of efficiency. There is no point bemoaning our children's short attention spans. Their children may have even shorter ones unless as an entire civilization we commit to resymbolization. The effects of desymbolization find their expressions in our art and music. For example, much of our popular music is dominated by some kind of beat that is accompanied by often-repeated short phrases that have no beginning, middle, or end. These override what little remains of a musical narrative, traditionally referred to as a song. The same is true for the visual arts, but it is offensive to most people to suggest that artists faithfully express their intuitions of a deep metaconscious knowledge of technique, producing an art of non-sense as a reflection of or escape from technique.[37]

All this points towards the possibility that our admiration for machines in general and for the computer in particular may become a self-fulfilling prophecy, but not in the way we had imagined. Computers will not become more and more like us, as was envisaged in the fifth-generation computer project of some time ago, but we are becoming more and more like our computers. These are machines that can deal with reality but not with symbolic life. As we are pushed more and more towards the former, we become more and more like machines.

The differences between a reality and a symbolic universe are amply confirmed by the way we relate them to discipline-based approaches to knowing and doing. Benson Snyder found that students at MIT could be divided into four groups based on the relative dominance of two approaches to knowing and doing, of which one is predominantly associated with the applied and natural sciences and the other with the social sciences and humanities.[38] He called the former approach numeracy and the latter literacy. One of the four groups started out by predominantly using the numeracy approach, but after graduation they experienced its limitations in the their professional and private lives and were compelled to compensate by increasing their use of the literacy approach. In a second group this greater complementarity of the two approaches did not develop. Students in the third group successfully used both approaches and kept on doing so after graduation. Surprisingly, the fourth group was strongest in the literacy approach and strengthened their numeracy approach.

The differences between the numeracy and literacy approaches may be further illustrated by the fascinating research undertaken by Sheila Tobias.[39] She explored the possibilities of full and equal participation of women in mathematics and the physical sciences, to refute the traditional prejudice that women are weak in these areas of study. What this research implicitly shows is that a comparison of the learning of a subject dealing with something organized in terms of what we have referred to as reality with the learning of a subject more closely related to a symbolic universe confirms their organizational differences, and that these differences help to explain traditional gender preferences.

Diaries were kept by a group of graduate students and faculty in the social sciences who had been persuaded to take a course in university physics or chemistry, and also by a group of graduate students and faculty in the physical sciences who had signed up for a course in English literature. Some members of the first group found the lack of an overview at the beginning of the course a significant barrier to learning the subject matter. The lack of such an overview would be inconceivable in their areas of expertise because it formed the basis for integrating what was being presented in the lectures and placing each and every topic in the context of what they already knew. They found the relatively linear and cumulative development of the course, during which the bigger picture emerged only towards the end, to be a significant obstacle to learning the material. The assurances of the professor that they did not need to know such a bigger picture to learn the subject were baffling to them because the lack of an overview prevented them from putting what they were learning in their own terms, that is, in the context of everything they knew. They felt that they could not fully engage themselves in the material. They were lost because they had no idea where the course was going, where they were in its trajectory, and how they could relate to it. Others felt that they were being treated like children by being spoon-fed with small amounts of material and having to trust that somehow it would all come together in the end. Some people found it very difficult to do the assigned problems because they could not grasp their significance for the overall subject matter. The professor was rather baffled by these kinds of reactions; proceeding step by step appeared to him to be the most pedagogically sound approach, since it was not shrouded in a host of unnecessary details. The students were more than a little frustrated at being limited to imitating demonstrated problems in order to master a set of techniques as opposed to struggling with fundamental concepts, including their broader implications. The feedback they received on their problem assignments confirmed this because it gave the students no clue as to how well they were mastering the subject matter. That the professor appeared to be baffled by their reactions was taken to be yet another indicator that their engagement in the subject was of little concern to him. Reciprocally, the professor was unable to understand the reactions of the students, because he regarded the broader context they were requesting as extraneous material that would stand in the way of learning the subject. In other words, the hidden curriculum appeared to suggest that the intellectual context, as well as the history of the subject, was of no importance. From this the students deduced that physics or chemistry had no general structure or context, since only a few things were connected in each chapter and no larger connections appeared to be required. Physics or chemistry amounted to a portfolio of techniques for solving different categories of problems. No personal opinions or intellectual pushing the limits of the subject was welcome or necessary. One student summed it all up when she said that if she could not put things in her own words, she felt excluded from the subject.

In the other experiment, in which graduate students and faculty with expertise in the physical sciences took a course in English literature, much the mirror image situation occurred. Since this kind of subject is much more contextual and cannot be broken up into relatively self-contained building blocks, a certain tolerance for ambiguity was constantly required. The subject matter could not be reduced to mathematical equations without extreme reductionisms. Worse, no question had a single right answer. Consequently, some members of the group complained that there were too many words, which was incomprehensible to the professor. After all, it is by means of words that concepts and an understanding are built up. The members of the group felt that the advancement of the subject was not sufficiently linear and meandered all over the place. There was too much context, since concepts could not be stripped of their historical and intellectual baggage.

The two experiments illustrate some of the differences between the numeracy and literacy approaches. Numeracy corresponds to learning something that is assumed to have the structure of what we have referred to as reality, while literacy examines those subjects whose structure belongs to a symbolic universe of a people and a culture. Numeracy is associated with the highly desymbolized domain of a discipline separated from experience and culture. Literacy may appear to imitate the discipline-based approaches of the physical sciences by sticking to one category of phenomena, but the level of desymbolization cannot be matched. Clearly, people with a highly developed numeracy approach find the literacy approach more than a little confusing, and the reverse is equally the case: Teaching someone the mastery of a domain that has the structure of reality can be done one element at a time. Each element can then be connected into the previously introduced ones by means of clear-cut relationships that can be cast into equations. The complexity of a domain is devoid of any enfoldedness: each element can be separated from all the others; it can then be defined on its own terms; next, it can be measured, quantified, and mathematically expressed; and it can be added into a complexity that can be built up one element at a time. The opposite is the case for subjects whose structure is related to a symbolic universe that nevertheless attempt to imitate the disciplines of mathematics and the physical sciences by limiting their field of study to one category of phenomena. They can never be successful because any such category belongs to a dialectically enfolded symbolic universe as opposed to a mathematical reality. Hence, each "element" enfolds something of the whole, and something of the whole expresses itself in every element. As noted, the brain-minds of the members of a traditional society enfold its way of life and culture as experienced and lived by each person. Technological artefacts enfold something of that culture, which is why archaeology is possible. All this starts to change when societies begin to build a technical order by means of discipline-based approaches to knowing and doing that are separated from experience and culture. Nevertheless, to regard these societies as megamachines constitutes a level of reductionism that is absolutely untenable, as we have seen. Hence, what little is left of qualitative approaches in the social sciences remains quite distinct from the numeracy approach in the form of quantitative methods.

There is no question that if the universities embark on a project of resymbolization,[40] the numeracy and literacy approaches will begin to converge somewhat. For example, a physics problem on an exam may require an answer that solves the problem, explains its significance for understanding physical phenomena, and comments on the limited applicability of the solution to a daily-life situation. Similarly, a question on a social science examination may require an answer that explains a particular theory, assesses its applicability to society given its exclusive focus on one category of phenomena, and explains what may be filtered out if quantitative methods are used.

All these issues have vast implications well beyond the university. It is for this reason that Snyder's longitudinal study of the lives of members of the four groups is so important.[41] He showed that after graduation, in both their private and professional lives, people who failed to develop their literacy approach had difficulties with interpersonal relationships and with context-rich issues, both of which grew in importance as they advanced in their careers and were promoted to supervisory and managerial positions. These difficulties also affected their marriages and friendships in a negative manner. This once again confirms the problems that we create when a way of life is exclusively surrendered to discipline-based approaches to knowing and doing, which implies that the numeracy approach has no limits and that, by implication, the literacy approach has little or no scientific merit.

Escape into Play

In an article entitled "Care of the Dying in America," Laurie Zoloth examines what may be happening to those on the receiving end of the technical order in suburban America.[42] It is a moving commentary on how American culture deals with the "double anguish of life and death" from a perspective based on the Jewish tradition. As the situation of Jewish people continued to deteriorate in Western Europe prior to the Second World War, and especially when their extermination began, the few who managed to flee to America and were permitted to enter found it impossible to join its culture of escape while thinking about the terror of those they left behind. They provided a unique vantage point and perspective.

No greater sign of our alienation is provided than by life in the suburbs. As a physical habitat, the suburbs constitute a seemingly endless sprawl of mass-produced houses (interchangeable parts) that are occupied by people who are not much different from their neighbours, as if they were popped out of similar moulds, living a lifestyle that defines who they are by what they have. Life in the suburbs leaves an impression on its inhabitants of being so ordinary that there is little to tell. It is a sign of reification that may help us understand the prevalent high levels of anxiety and depression as a portent of what is to come as the levels of desymbolization continue to rise. It is very difficult to find some meaning and direction in this habitat and social milieu – an anti-society reified by a technical order. Life has become so unbearable for so many people that they refuse to assume the responsibilities of adulthood and instead revert to acting as children. Children live in a happy place where sorrow can be fixed with a hug and a treat. Why can adults not go there by making the suburb a gigantic Disneyland, where people play instead of struggling with their difficult lives? Maybe we can all achieve this with a little Prozac, adult toys, soft food, playful dress, and a life lived via a computer screen as a virtual door into games and a "life" that is a little less scary because of the lack of embodiment, participation, commitment, and freedom.

All this is made liveable by advertising that shows us how to be a better, more spontaneous, more authentic, and freer person by playing as opposed to living life. There is a confusion between childhood and adult life when, for example, a parent calls a child their buddy and plays with electronic gadgets as their toys. There is also a kind of nostalgia for the past, which somehow was more authentic; by playing at life it may be possible to get a little closer to the past. Of course, if a great many experiences so lack embodiment, participation, commitment, and freedom that they can hardly be lived, they cannot be lived as an adult, leaving only play-acting as an alternative. Every possible escape is planned on weekends and holidays, until people can make it to the longest escape of them all: retirement as an endless playtime and a flight from intergenerational responsibilities. In the meantime, all signs of aging must be carefully hidden as a denial of the death to come.

From a cultural perspective, childhood is a time of limited embodiment, participation, commitment, and freedom. It is also the time when life is not yet suspended in myths, permitting life and the world to be played with, yielding endless surprises and a few sorrows that can be readily be remedied.

All this escapism led to a crisis when "technique changed people." Many people had to deal with their jobs being outsourced, their debts mounting, their relationships floundering, their children encountering difficulties in school, their friends becoming fewer, and their moral and religious compasses failing. The pressures of desymbolization rose, and people's metaconscious social selves became more and more divided. It should have come as no surprise that more people struggled with anxiety and depression. This became one of the most profitable opportunities for the pharmaceutical industry, which provided end-of-pipe remedies to the situation. It is true that drugs can prevent daily-life problems from putting people into a rage against something they cannot understand. It is also true that when people take these medications, many of them are themselves again, and some even feel better than normal.

Zoloth's article describes the Talmudic rituals in which people remember their great sorrows of the past in order to prepare themselves to face life and celebrate it in preparation for another year. If our diagnosis of the threat of desymbolization is correct, then we all must learn to live with our metaconscious knowledge of separation and death. As adults we have been exiled from childhood and have become integral to a fabric of relationships between our lives and others, and thus share a historical journey anchored in, and oriented by, secular myths.

Escape into Screens

We have all seen it and probably done it as well. As parents taking our kids to the park, we cannot prevent ourselves from spending a good deal of that time on our cell phones. At our business lunches, we are keeping one eye on our cell phones next to our plates. Family life at home is constantly interrupted by whatever arrives on one screen or another. In theatres, many people divide their attention between the movie and their own screens. Others carefully check their phones as soon as the credits roll to make sure they did not miss something important. Two close friends may be strolling together in a park, but the one or the other is constantly on the phone. Even when we are engaged in activities that ought to receive our full attention (such as driving the car), we cannot help endangering others and ourselves by sending and receiving text messages. It is not difficult to multiply these kinds of examples. It is obvious that our behaviour implies either that the social relationships in traditional societies were not essential for sustaining human life, or that relationships can be maintained just as well or even better through a technical mediation. We simply ignore the many studies that show that this mediation is not neutral, to the point that it encourages and exaggerates undesirable personality traits, with significant negative consequences for our lives together. Some of our technologies appear to correlate with certain mental disorders, referred to as idisorders.[43] A study by Larry D. Rosen and others shows that five factors contribute towards these idisorders: the user-friendliness of the devices; their appeal to our senses through colourful screens and clear sounds; our ability to hide behind our devices and remain more anonymous than in face-to-face relationships; our resulting inability to detect the effects we have on others (mainly communicated through eye etiquette, facial expressions, and body language); and the instant availability of others' attention, almost regardless of what they are involved in. It appears that most of us are willing to do things through our devices that would not come up in our heads when we are face to face with one another. So-called idisorders include narcissistic personality disorder, obsessive compulsive disorder, attention deficit hyperactivity disorder, depression, anti-social personality disorder, social phobia, and others. It must not be concluded that these disorders are symptoms of our other-directed personalities having been amplified by these technologies to the point of enslaving our persons and our lives. Our technologies are fundamentally transforming the other-directed personality, since so many of our relationships are now technically mediated.

For example, social media like Facebook do not facilitate our relationships with others – they completely transform them. Many people become totally absorbed by a strategy to project a certain image of themselves, to be seen on all the right pages, to have all the right friends, and to multiply these desirable contacts as much as possible because you can never have enough. It leads to a cult of self-importance and can become a full-time occupation, not only to have others recognize this image but to exploit them to gain popularity and respect. The living of our lives can thus be taken over by a cult of self-importance acted out through a domain-like Facebook page, which has little to do with who people really are and what their lives are like. It is all about how people want others to see them and the art of acting this out in a few actual face-to-face encounters. For most of the time, others are "screened out," with the result that people's own lives are "screened in." We have already encountered the ways this is accomplished. Technical mediation limits embodiment, making it much more difficult to sense how others are being affected; it limits participation to less than full and complete involvement, since so much is stripped off by this mediation; it reduces the need for commitment because it is possible to "back out" with greater anonymity and occasionally with full anonymity; and it diminishes one's freedom because the manoeuvring room to be yourself, to be sensitive to others, and to take responsibility for what is said and done is significantly curtailed. For example, meeting a close friend face to face may reveal that something is fundamentally wrong, and this usually entirely unrelated to what is being discussed. All this is but one symptom of what we have referred to as the anti-society, characterized by its inability to adequately sustain its members. There is little room for any sustaining care for others in an all-consuming need for self-promotion, attention seeking, and narcissism. Again, we must remember that our relationships with our technologies are reciprocal in character, with the result that people with narcissistic tendencies may well be attracted to the social media as an ideal platform, while in other cases becoming more narcissistic amounts to the necessity to survive on social media.

Escape into Medical Drugs and Enhancement Technologies

The inability of contemporary anti-societies to sustain their members may lead to situations in which drugs like Prozac become necessary for survival, and certainly for success. If the technical order makes human life more stressful, makes human relationships more instantaneous, and thus speeds up the pace of life, then anything that helps us cope will give us an advantage over those who refuse to use such coping mechanisms. In other words, anything that helps us cope feeds dangerously into an obsession with success and prestige. How long will it be before the refusal to take a particular drug to enhance a person's performance during an interview will simply make that person uncompetitive? How long will it be before the stress levels become unbearable without the Prozacs of this world? To treat these kinds of issues as belonging to mental health is to confuse our inner physic states with the conditions under which we live our lives. I may be deeply distressed over the death of a close friend and mourn the loss for a long time, but is this what human beings normally do under such circumstances or is it a mental disorder?[44] Is it possible to speak of a mental disorder apart from the conditions that produce it, unless the disorder results from a biophysical malfunction of the organ of the brain itself? Consider alienation as an example. It must always be considered in relation to the conditions that enslave us. Our secular myths hide from us our deepest enslavement in order to make life liveable during a historical epoch. It is often possible then to proceed with hope and confidence. Under conditions of desymbolization such hope and confidence may be eroded, thus creating a cultural niche ready to generate widespread anxiety and depression. In this case, we ought to be talking about a cultural malfunction related to an inability to sustain human life in an ultimately unknowable universe. Treating the various disorders this alienation may produce is better than nothing, but such an end-of-pipe approach quickly becomes enormously expensive and thus less and less practical. Under these circumstances, what is required is a project of resymbolization in parallel with treating the accompanying disorders, in the expectation that the former will steadily decrease the need for the latter. It is here that we are currently trapped by our secular myths. When our technical civilization is faced with serious issues, we spontaneously look towards appropriate discipline-based approaches, since the question of their limits never arises. However, doing so will aggravate the problem, and resymbolization will remain unthinkable and undoable.

Carl Elliott has struggled with many of the same issues related to life and death without linking them to technique.[45] Each and every human generation is born into a certain way of making sense of and living in the world, although this may draw on many influences in so-called multicultural societies where we build our lives by adapting and evolving our cultural inheritances. Some of us may sense that all of this is built on thin air, and worse, that therefore we are built of the same. When this happens we may well ask: Why do I bother with this job? Why live in this house? Why belong to the church, synagogue, or temple? Why do I continue doing what I do when I get up? In other words, some of us may experience glimpses of relativism, nihilism, and anomie. It does not matter what form our lives have taken, they appear to lack whatever is required to live, and no other form is expected to do any better. For Elliott, this attitude is not merely a need to face our mortality. We must respond to what any particular age and culture have made of our lives as mortal beings. The desymbolization of our secular myths is leading into a sense that somehow the bottom has dropped out of our lives. People turn to doctors, therapists, psychiatrists, and spiritual advisors, very few of whom have any idea why this is happening. They are often struggling with these issues in their own lives. Many of them do what they can, but it is very difficult for all of us. The alienated meanings and values of traditional cultures were probably more liveable in general than our alienated and highly desymbolized meanings and values.

In the same article,[46] Elliott points out that there are retirement homes that provide people with everything that they could possibly need and want, and yet there will be people who sense that there is something missing in their lives. Sending in the experts on our inner psychic being can do very little good because they are all members of disciplines (psychiatry, psychology, social work, spiritual counselling, etc.) who have little or no idea of the broader context beyond their own fields. Consequently, they will attempt to remove sadness, anxiety, and lack of fulfilment from this inner being as best they can. Hence, all manifestations such as anxiety, depression, phobias, obsessions, and compulsions must be treated and eliminated. A concept of alienation cannot have a place, let alone any significance in this work. But what if alienation is integral to human life? What if we do anything to avoid the possibility of having to face this alienation by being the most extrovert we can be, consuming what we can afford, taking whatever love we can get, and spending as much time as possible in reality of one kind or another? The most tragic symptom of our desymbolization is that it permits our civilization to apply techniques to our inner psychic being by unfolding it from our belonging to a time, place, and culture, and (for some of us) from our being a creature (i.e., a created being) above all. It may be argued that there is nothing new here. Has every human civilization not enclosed itself within its myths to exclude anything radically different? Is this triumph of discipline-based science and technique over our inner psychic being not a temporary situation until our secular myths are shattered, as were all previous ones? Will our grandchildren, great-grandchildren, or their descendants not be compelled to symbolize the unknown in very different ways than we do? All this is the case if our interpretation of the historical record has any merit.

Nevertheless, there appears to be one all-important difference that may make our situation unique. The desacralization of our secular myths will depend on civilization and the continuation of our being a symbolic species.[47] This is now in doubt for the first time in our human journey. If the interpretation that we have presented is correct, we are currently caught between two functions of our secular myths that appear to have been separated by desymbolization. These myths continue to do what they have always done: to exclude anything radically other than our present way of life. What these myths do increasingly less well is to make human life liveable within these limits, because they sustain the kinds of activities that undermine our ability to symbolize the meanings and values of our creations by exposing their limits and thereby contributing to their eventual desacralization as something else is sacralized at the same time. What are commonly interpreted as widespread anxiety and depression are thus symptoms of the symbolic bottom falling out of our lives as well as of our inability to restore some meaning and direction to our lives because of a loss of symbolic capacity. What many people need is what our civilization is undermining with all its might. It is turning health care into end-of-pipe disease care, which is a kind of therapeutic interpretation of life that makes us think we can make ourselves whole by curing all our diseases and disorders.[48] As Elliott points out, at one point Valium was regarded as a symbol of the soul-deadening character of American suburban life.[49] We have forgotten this at our peril as we now use the Prozacs of this world as the "secular opium of the people."

Can we drug the members of a technical civilization out of their sadness, or must we face what our lives are by symbolizing them in order to transcend the limits of science, technique, and the nation-state, in the recognition that history cannot tell anything and will certainly not judge anything? It would launch us on a way of life and a mutation of our civilization that hopefully will be more liveable and sustainable. The alternative is to create other ways of sustaining human lives by means of the next generation of Prozacs: pills that combine the manufacturing of happiness with the manufacturing of a greater endurance to make us into the efficient beings the technical order requires. If we choose this course of events, we will guarantee our destruction as a symbolic species, and we will most likely take the planet down with us.

Escape into Technique as the Measure of All Things

By translating everything into reality the way earlier societies mapped everything into a symbolic universe, discipline-based approaches to knowing and doing have made efficiency the measure of all things. With the gradual appearance of anti-societies whose members are less and less sustained by webs of social relationships, human lives have become the measure of themselves. In other words, all of us need to decide for ourselves what it is to be human; once we decide this, "my way" is what defines it. We can no longer rely on others, and via them, on previous generations by means of symbolization, a way of life, and a culture. There are two apparently contradictory approaches to this: one is a kind of self-justification and the other is an attempt at resymbolization, which is rare. In the case of the former, with the emergence of anti-societies and the other-directed personality, there was the gradual acceptance of a discourse of authenticity.[50] I "name" what it is to be human for myself because desymbolization makes it impossible to do this through relationships with others. Once I have named how it is that my life must be human, I need to judge everything I intend to do and the results of doing it in relation to that. If there is conformity between the two, I am authentic and true to myself. In other words, I need to do what people used to do together by means of a culture: decide what is right and wrong and live accordingly. A person needs to be a superhuman being to do the work that began with the cultural inheritance from previous generations and the adaptation and evolution of this inheritance through the collective effort of a community. As a result, I cannot help but look at others as I make all these decisions. I cannot help but measure my face, body, and sexuality against those of others; my career success against those of others; my material satisfaction and consumption against those of others; my friends and popularity against those boasted of by others; my ability to use the media to stay in touch and ahead of most others against their success; my authenticity and worthiness against those of others; and so on. I become keenly aware that my chosen form of life competes with those others have chosen for themselves. Hence, I become keenly aware of competing against everyone in every possible measure of my life. I am left with no choice but to adopt every possible enhancement technology and supercharging drug available. Of course, my initial reaction is that this or that new development goes too far and that I cannot possibly fit it in. I appear to share this opinion with many others. However, jobs are scarce, life is lonely, and the world is tough; soon more and more of my friends and acquaintances fit it into their authentic lives, and in order not to be left behind, I have no choice but to do the same. After all, I can do with my body and life as I see fit. We know the pattern all too well. We have adopted one enhancement technology and drug after the other: Valium, Prozac, Ritalin, Viagra, growth hormones, sex reassignment surgery, gene therapy, liposuction, implants, Botox, all manner of plastic surgery, and the many more soon to follow. It fits a well-established pattern of enhancing our senses, our bodies, our intelligence, our immune systems, and our lives.

However, these extensions are means as well as mediation, and in the latter capacity there is a non-neutral filtering with its own message. All this cannot be interpreted one technology and technique at a time, with the result that technique has become a life-milieu and system.[51] In the same vein, we celebrate the miracles of biotechnology, nanotechnology, stem cell research, implants, and the limitless potential of processing ever-larger quantities of data. I must confess that as a blind person I cringe every time I hear of the miracles in store for the "disabled." Once again, all of this may be perfectly summed up in what Jacques Ellul warned against: that technique feeds on its own problems in a kind of self-augmenting progress that has very little to do with genuine human values established independently from technique through symbolization and culture.[52] It would appear that the creation of an anti-society, an other-directed person, life in the fast lane, everything I am worth and my authenticity is all a slavish accommodation to technique.

The standard reply to people making remarks like this is that they must be pessimists and thus against science and technology. Such reactions simply reveal the enslavement of the people making the pronouncements. The real question is: where are we now and where are we going? In order to answer these questions, we need to symbolize our situation, including our historical journey, establish the meanings and values of all our creations, and redesign them accordingly. We are back to what we have already argued in the previous four chapters. The fact that we even have to discuss what it is to have an authentic life is a direct consequence of our having desymbolized our cultures as a symbolic species. We act as though an authentic life is one that is as fulfilled as possible in the terms of our secular myths. Even our most intimate sexual relationships have been turned into transgressions of technique.[53] As in earlier civilizations, our secular sacred is organized in relation to a sacred transgression which orders human life into a sacred and a profane.[54] Within the broadest possible context, it would appear that our search for fulfilment is a secular replacement for the search for a religious good created by a culture. Acknowledging that science and technique have limits is to de-sacralize them and to return them to their status as human creations.

In traditional societies, the search for meanings and values was done through relationships with others, and the joint evolving of a culture was guided by its myths. Self-improvement was directed towards others.[55] It may have included a resolve to pay more attention to others and not to talk too much about yourself and your own feelings, to think carefully about what you said to others, to take seriously everything you did with others, and to maintain a certain self-restraint and dignity. In an anti-society, improving yourself is to do everything you can with the time and financial resources you have to look and dress the best you can, to lose weight, to get a more stylish haircut, to drive something as impressive as you can afford, and so on. In the first case, it is all about improving yourself through social ties and reciprocal responsibilities to one another. In the latter case, it is all about improving yourself by using whatever technique has to offer. Life must be sacrificed to technique so that technique can be all in all: it is therefore simply the secular replacement of the religious good created by traditional societies. I do not believe that this overstates our situation. Carl Elliott points out that when we decide to get our hair tinted, it is often the result of a great deal of advertising complemented by what we see others doing to their hair.[56] The manufacturer has done exhaustive technical studies of how hair colour affects our appearance and how this in turn affects the impression we make on others. We buy not so much the tint but the results of that kind of technical research. The same is true for plastic surgery. What we are really consuming are the results of a great deal of research on how to make a "perfect" body. The same is true for tooth whitening and straightening. All this and more communicates the message that you are authentically the best you can be and that you can afford to take good care of yourself. In every way today we are tempted to consume the latest technical research in our quest of being true to ourselves.[57] It is difficult to ignore a significant level of despair in all of this. Our lives have been so emptied of meaning as a result of their desymbolization by technique that now there is nothing to hang on to except the very technique that caused the problem in the first place. There no longer is anything else beyond the desymbolized person and community. There is only technique as our human destiny of self-fulfilment to replace what has been lost.

All of this is self-reinforcing. As we see ourselves in the social mirror of others, it is rare that we do not become aware of something that could be improved in order to be even more true to ourselves in our quest to be authentic. We are never our true selves and always in the process of becoming more authentic, in this way making ourselves into exactly what technique requires us to be in an endless pursuit of performance and efficiency. We are always ready for the next generation of technical improvements. We must constantly be reborn in the image of technique. In coming to these conclusions, it is important not to interpret them as a moral judgment. Our search for authenticity is a question of necessity in the context of a symbolic species suffering under desymbolization. It heightens the anxiety of our lives as we recognize that most of the time we fall short of being as true and authentic as we believe we need to be, and yet we may be powerless to do much about it. It is clear from the epidemic use of antidepressants that most of us are not having a good time. Only a few people have the abilities and stamina to remain endlessly competitive in the pursuit of being the best one can be, as a result of having our lives become the measure of themselves. Out of this are born new sins and new means for salvation, but no new morality will emerge other than the pursuit of performance in the name of authenticity.

We appear to have declared ourselves to be completely open to technique and ready to be remoulded as required. Is a greater sign of alienation possible? If we recognize it for what it is, can a bridgehead for resymbolization be established in our lives? If we do not intervene, we will continue to increase our anxiety levels because we no longer know who we are, but we do know who we need to be in relation to technique. Getting there, however, is as impossible as a megamachine society. Everything is being desymbolized into reified fragments of the reality of our technical order, with what little remains of our lives, communities, ecosystems, and the biosphere pushed to its fringes. None of these have their identity or integrality intact. I am not sure that in the long term we will find the post-modern rhetoric about performances, masks, role playing, and narratives very reassuring. If we no longer live lives but follow the scripts of the technical approaches all around us as knowers or doers, if we abandon the conviction that we have identities, if we are now forced to assume the "identity of the day," then freedom and love will become concepts of the past. Without them we will have to give up all hope of liberation and being genuinely loved for who we are.

Speaking of love and freedom presupposes that there continues to be a self. We have shown that in an anti-society our metaconscious social selves are enfolded into a variety of selves turned towards the people with whom we share particular activities in our lives. The characteristics of our personality related to the people with whom we work, maintain a family, relax, or participate in any other activity will form a criss-crossing and overlapping network, and it is this network that is constantly evolving in the context of our lives.[58] It is the key to understanding what binds together an unfolded but core identity. We have become who we are with all the different people who accompany us in all our daily-life activities.

We have not given up on living a life, including the self who does this. What we have done is to become terribly anxious to the point of depression. The anxiety and depression would be inconceivable if we had simply become without selves and without lives as cogs in a megamachine. No matter how reified the elements of our persons and lives may be, we continue to reach out for love and thus accept our capacity for being loved, as many songs on the media endlessly proclaim. It is a call of despair arising out of living in anti-societies among strangers demanding that we must make ourselves stand out as special before that love can be found. It is also the despair of our inability to suppress what others disapprove of and to enhance what makes us lovable. Such despair comes from our inner lives' continuing dependence on some level of symbolization and culture, while our outer lives must conform more and more to technique and the authenticity it necessitates.

Here we encounter another aspect of the role of integration propaganda in our lives. As a result of pharmaceutical and medical advertising, we are made to feel defective in some way, and since this can be fixed by means of a drug or procedure, we have an obligation to "repair" ourselves. It is yet another manifestation of the medicalization of human life: either we are worth everything technique can offer, or we have no choice but to be the best we can be in order to survive the ever-heightening competition between strangers brought on by technique. In other words, our concept of identity becomes self-reinforcing when it is used to justify our enslavement to the gigantic enterprise of medical and pharmaceutical research. We do it for ourselves, with the result that we need technique as much as we need this integration propaganda to complement our highly desymbolized culture. This also helps to explain why people in the most affluent societies never have enough. We are all placed on a treadmill to perform and stand out among strangers in the crowds that surround us in almost every activity. Consumerism has thus undergone a qualitative change by being related to the authenticity of our inner selves via the behaviour of our outer selves. We consume the products of our technical research to continue to fulfil our (frequently induced) desires, but now we also need help to overcome our lack of authenticity.[59]

All this is further reinforced by the people we admire: the movie stars, the sports heroes, the rich and famous, who all embody what we could be and should be if we stick to our journey of self-fulfilment by every possible technical means. As a result, the true self is increasingly becoming defined by science and technique, thus deepening our social, psychic, and spiritual dependence on them.

To the extent that "technique changing people" permeates our persons and lives, to the same extent will we find it more and more difficult to engage in non-technically mediated relationships of face-to-face encounters with partners, children, parents, relatives, and close friends. Especially when difficulties in such relationships arise, we no longer have the experiences and social support more common in traditional societies. Fewer of us have the kinds of close friends or trusted wise persons we can turn to for advice. Technically mediated relationships with their limited embodiment, participation, commitment, and freedom screen out the kinds of subtle clues that something is going amiss in a relationship. These often go unnoticed because the party making the somewhat larger contribution to the issue may not have the experience to pick up and interpret these subtle clues. It is also becoming more difficult to express such clues because they are no longer learned, given the scarcity of non-technically mediated experiences. In turn, all this is further aggravated because of an impoverishment of the metaconscious knowledge built up from these experiences. This knowledge is thus lacking the deeper dimensions, and this can affect our metaconscious social selves. Moreover, the lack of significant face-to-face experiences and the difficulties this creates for our lives to fully work in the background will almost certainly get worse from generation to generation, as our electronic gadgets and the media to which they are connected multiply to occupy an expanding proportion of the daily lives of children. Some children may begin to prefer technically mediated relationships because they are generally less risky than the corresponding face-to-face ones. It may even make computer-based counselling programs acceptable.

Joseph Weizenbaum was the first to write such a program for amusement and was shocked when many people took it utterly seriously.[60] He might have been prepared if he had known that the rules for conducting interviews with the workers at the Hawthorne Plant in Chicago led to a kind of psychotherapeutic management of the workers.[61] These rules caught the attention of Carl Rogers and inspired an entirely new approach to counselling.[62] It proceeds much as Weizenbaum's computer program did. The counsellor must not provide any opinions, advice, praise, or criticism, or agree with anything. The counsellor must act as a neutral sounding board by being the perfect stranger listening to another stranger who is deeply upset by something. Frequently, the counsellor takes the words of the other person and repeats them as a question, which is essentially what the computer program was designed to do. In an anti-society with a highly desymbolized culture, the person-to-person contact that characterized earlier counselling has become too risky. In this sense as well, it would appear that we are becoming more like computers. We are less and less embodied and thus have no vantage point on others and the world, making it very difficult to advise someone. We participate less and less in technically mediated relationships, reducing them to little more than exchanges of words. No commitment is necessary other than to the rules first pioneered in the Hawthorne experiment. Once again, it would appear that we are being re-engineered in the image of technique. The result is two kinds of people: those who can keep up with the fast pace of technique-based connectedness by intense concentration or by using technical means of enhancement, and those who are unable to do so and who therefore must have some kind of disorder that should be diagnosed and treated.

A great deal of what we have been discussing in this subsection is related to living as if we ourselves, our relationships, our lives, and our world have the structure of a reality in which context plays a trivial role. If everything is perceived as an arrangement of separate elements, there is no context to restrict anything. An element is no longer integral to a context, and thus only has a significance of its own or in relation to similar elements. For example, we no longer see faces in the following ways: a person may have an appealing warm smile, with the result that her somewhat large nose is of little consequence. Another person may have eyes that are full of intelligence and humour, and this overshadows his big chin. Instead, we see a variety of features that are judged on their own terms, with the consequence that any imperfections stand out and must be "fixed." Among teenagers this can lead into the extreme reaction of bullying someone over something rather trivial in the context of overall appearance, personality, and life. Without a context, everything in life can be distorted out of proportion and thus opened up for technical "improvement."

Hope Is Possible

Must we conclude that the person of the past has been turned into the anti-person of our time? In order to answer this question, the findings of this and the previous chapter need to be integrated. The more mass societies turned into anti-societies by taking on the characteristics diametrically opposite to traditional societies in terms of their ability to socially sustain their members, the more people sought to escape the effects of these societies by "living" on the net. Such "anti-living" represents an escape from, and a compensation for, the effects of technique on these societies. People were in awe of all the new possibilities. Imagine being liberated from much of our embodiment of society, with all its ethnic, religious, gender, and cultural trappings. Next, imagine that on the net we can hide from the other person anything we do not wish them to know, such as having a disability or being unattractive in some way. Add to this the possibilities of a participation that can be switched on or off at a whim. Top all this off with no longer having to commit your person and your life, thus permitting endless experimentation without risks. Surely the crown of all these possibilities will be an unprecedented freedom. There is but one small problem: Do these possibilities not converge towards what traditionally has been referred to as anomie? Does this not amount to moving towards creating what in the Jewish and Christian traditions was referred to as hell? In the Jewish tradition, hell referred to the garbage dump outside of Jerusalem, where people threw anything that was no longer of any use to life. In the Christian tradition, hell referred to a condition of the complete impossibility of any relationships and also of freedom and love, and thus also to non-life. Because of the minimal constraints resulting from embodiment, participation, commitment, and freedom, anything can be imagined except what humanity used to experience and live. Are we in danger of becoming so reified by technique that such an absence of experience and life is no longer our concern? Will the possibilities become almost limitless if we can shed what little remains of our embodiment, participation, commitment, and freedom that subsists despite the technical order? What will happen to humanity if it sheds much of being a symbolic species, which accounts for our historical journey thus far? If we look around in our lives and communities, it would appear that more and more people are paying a very high price for all the new technical possibilities. All this adds another important dimension to the self-augmenting character of technique, which continues to feed on the very problems it creates. Is there any other possible explanation for the explosive global need for the Googles and Facebooks? Is this not responding to deep and universal needs as global as technique itself: a lack of being socially sustained as experienced in the anti-societies and the need to re-create experience and life into their opposites on the net? Is humanity becoming united through its attempt to re-create all of life in the image of non-life through discipline-based approaches and the substitution of reality for symbolic universes? Does the desymbolization of humanity thus represent a kind of collective death wish anchored in our myths?

Much of what has been happening during the second half of the twentieth and the beginning of the twenty-first centuries may be summarized as a kind of desire to no longer live as a symbolic species by handing over all knowing, doing, and organizing to discipline-based approaches that have little validity outside of our technical systems. It is making us appendages of such systems. It is worth remembering that Western civilization gave birth to these developments by betraying what it had achieved in terms of a unique dialectical tension between the individual and society, as a consequence of Greek philosophy, Roman law, and the Jewish and Christian approaches.[63] This dialectical tension has now largely collapsed, at best functioning between the individual and what little remains of the culture-based connectedness of human life. This remnant of our symbolic past also contains everything that has been externalized by all our discipline-based approaches. Nevertheless, what remains of this culture-based connectedness could constitute a beachhead for the resymbolization projects humanity will require in order to continue to make use of such approaches where they are applicable but not to become enslaved by them. We must create a civilization that includes science and technique on our terms established through their resymbolization. It will help transform the anti-persons we have become into persons living lives enriched by what discipline-based approaches can contribute within their limits, but not enslaved by the cults of the fact, efficiency, growth, and disembodied communal and individual life. Resymbolization will bring a halt to living with our creations as limitless secular gods.

Given our current secular myths, this conclusion will undoubtedly result in accusations of pessimism or of science and technology bashing, since our gods can have neither limits nor flaws. Resymbolization will de-sacralize these gods, restore them to the status of human creations, control their applications (by design or other culture-based equivalences), and ensure that such applications do not transcend their limitations. However, there must be enough of a remnant of our being a symbolic species to make this possible. Consequently, we are in a race against time before our desymbolization becomes irreversible, and human history as we have known it comes to an end. If this were to happen, our discipline-based approaches to knowing, doing, and political organizing will continue to slice through the fabric of living relationships that make up human life, society, and the biosphere, until their final collapse.

In other words, this work is a call to awaken us from our secular religious illusions in order to challenge the empire of non-sense with sense. Without such an effort, our future will become steadily less liveable and sustainable. The continuation of our being a symbolic species, as well as our historical journey, depends on it. What stands in our way is not the impossibility of the task we face. We could begin the resymbolization of our universities any day, as proposed in this work as well as the previous ones.[64] The only genuine obstacle that stands in our way is our secular idolatry of science, technique, and the nation-state. We have moved from traditional idolatry to secular idolatry, but the former never led to desymbolization and the barring of resymbolization. We must live in the knowledge that science, like any other human creation, has real limits to what it can know, that technique has substantial limits to what it can accomplish, and that politics is incapable of bringing about the necessary changes. This liberation will result in a withdrawal of our admiration for, and commitment to, the "system" we have built for ourselves. We can withdraw our allegiance from what makes no sense. Our secular cults can achieve no more than those of the past in ensuring a liveable and sustainable future for ourselves, our children, and our grandchildren. This call for change therefore begins with a conversion of the intelligence with which we regard others and the world, which will translate into a preoccupation with what makes sense for our lives as well as for those of others and a complete indifference to the empire of non-sense.

Epilogue

If the preceding analysis has merit, our civilization faces a decisive choice in this twenty-first century: to continue reorganizing life in the image of non-life, or to choose life. Since we have no idea how far we can travel down our present road, it may well be the wager of our time. It is not a matter of humanity not having faced such fundamental choices before, but that our means and their consequences are so much more far-reaching. It may not be much of an exaggeration to suggest that ultimately it is a choice of love: who or what do we really love – the system our civilization has built, or our lives? Our analysis has led us to the conclusion that we cannot have both. We do well to pause and look at one another and especially our children and grandchildren before deciding what it will be.

If we choose life, will we be able to translate this choice into something concrete? I recognize that none of us is capable of influencing the system that poses the threat. None of us can change its course, but all of us can say no and withdraw our implicit allegiance and support. Some may well scoff at this as a trivial act, like all the talk about love we daily hear on the media, and say that we are better off putting our faith in the secular political religion of democracy. Our analysis also shows that this is a political illusion.

Choosing to love life can indeed change the very system that our civilization has built during the last two centuries. Two conditions are required for it to continue to function as a system in relation to our civilization. First, it must have its own internal mechanism of development, acquired through the global flow of scientific and technical information that connects all disciplines and, via them, every area of specialization. It is a mechanism precisely because it excludes human beings from intervening as such, because our participation in disciplines precludes the choice of selecting alternatives that will improve human life, society, and the biosphere and reduce our alienation and reification. Instead, everything will be improved on its own terms, as if there was no such life on this planet. The second condition is that for this internal mechanism to function, it must be shielded from any outside interference. Therefore, individually and collectively, we must adore this system as the most important, decisive, and valuable element in our experience. If we say no, then we ourselves will come under scrutiny as we attempt to make sense of this system of non-sense by means of resymbolization. We will then discover how unscientific our science is in relation to all life, how inhuman our technique is, how uneconomic our unlimited growth is in producing debt rather than wealth, how the nation-state is a major source of disorganization and chaos, how we are "spaced out" in our societies thanks to our disembodied participation in many activities and on the internet, and how deeply anxious and distressed we have become in our reification. Let us be honest with ourselves: the dreams we once shared, the hopes we once invested in the current system, and the confidence we once had in our future have all vanished. The evidence of brokenness is all around us.

There is nothing new in all of this. Throughout history, humanity has always managed to adore its own creations by implicitly withdrawing all critical attention from them. Doing so makes these creations into powers in themselves, and this always turns out very badly for us. In this sense, we have essentially replaced the cults of the past with our secular cults of the fact, efficiency, growth, and disembodied life. With the hindsight of two centuries, it could be argued that our cults of scientific and technical production are little different from the fertility cults of agricultural societies.

In choosing life, we will begin the slow but all-important process of changing the intelligence by which we apprehend our lives and our world by resymbolizing them. Such an intellectual conversion will produce results that may inspire people that things can and must be done differently. When this idea spreads, the second prerequisite for our system to function as such will disappear, and in the process trigger a mutation towards what can only be a more liveable and sustainable way of life because it will be based first on symbolization, and only secondarily on technique insofar as it serves our goals and aspirations. Love can bring us a different future when it becomes the driving force of change. There is no point in declaring our love for our partners, children, and close friends without embedding it in what we are doing to ourselves and how we take responsibility for it. It is with this observation that I will end my five-volume study of the relationship between technique and culture as the unleashing of the forces of desymbolization on our ongoing dependence on symbolization.

Notes

Introduction: Our New Spiritual Masters

1 Hubert Dreyfus, *What Computers Still Can't Do: A Critique of Artificial Reason* (Cambridge, MA: MIT Press, 1992).
2 Willem H. Vanderburg, *Our War on Ourselves* (Toronto: University of Toronto Press, 2011).
3 Ibid.
4 Ibid.
5 Max Weber, *The Theory of Social and Economic Organization*, ed. Talcott Parsons (New York: Oxford University Press, 1947).
6 Jacques Ellul, *The Technological Society*, trans. John Wilkinson (New York: Vintage Books, 1964).
7 Vanderburg, *Our War on Ourselves*, 95–6.
8 T.W. Deacon, *The Symbolic Species: The Co-evolution of Language and the Brain* (New York: W.W. Norton, 1998).
9 Willem H. Vanderburg, *The Growth of Minds and Cultures: A Unified Theory of the Structure of Human Experience* (Toronto: University of Toronto Press, 1985). For an updated summary, see Vanderburg, *Our War on Ourselves,* chapter 1.
10 Ibid.
11 Lucien Malson, *Les enfants sauvages* (Paris: Union Générale d'Editions, 1964). A good account in English can be found in Roger Shattuck's *The Forbidden Experiment* (New York: Washington Square, 1981).
12 Helen Keller, *The Story of My Life* (New York: Doubleday, 1902). This and other cases are discussed by Ernst Cassirer, *An Essay on Man* (New Haven: Yale University Press, 1944).
13 Vanderburg, *The Growth of Minds and Cultures*.

14 Ibid.
15 Willem H. Vanderburg, ed., *Perspectives on Our Age: Jacques Ellul Speaks on His Life and Work* (Toronto: House of Anansi Press, 2004).
16 Deacon, *The Symbolic Species.*
17 Jacques Ellul, *The New Demons*, trans. C. Edward Hopkin (New York: Seabury, 1975); Vanderburg, *The Growth of Minds and Cultures.*
18 Willem H. Vanderburg, *Living in the Labyrinth of Technology* (Toronto: University of Toronto Press, 2005); Vanderburg, *The Growth of Minds and Cultures.*
19 P.A. Colinvaux, *Why Big Fierce Animals Are Rare: An Ecologist's Perspective* (Princeton, NJ: Princeton University Press, 1978).
20 Andrew Bard Schmookler, *The Parable of the Tribes: The Problem of Power in Social Evolution* (Berkeley: University of California Press, 1984).
21 Jacques Ellul, *Histoire des institutions: l'antiquité* (Paris: Presses Universitaires de France, 1961).
22 Eric Voegelin, *Order and History*, vol. 2, *The World of the Polis* (Baton Rouge: Louisiana State University Press, 1987); Eric Voegelin, *Order and History*, vol. 3, *Plato and Aristotle* (Baton Rouge: Louisiana State University Press, 1987).
23 Ellul, *Histoire des institutions.*
24 Jacques Ellul, *On Freedom, Love and Power*, comp., ed., and trans. Willem H. Vanderburg (Toronto: University of Toronto Press, 2010).
25 Ibid.
26 Ibid.
27 Jacques Ellul, *The Meaning of the City*, trans. Dennis Pardee (Grand Rapids, MI: William B. Eerdmans, 1970).
28 Ellul, *On Freedom, Love and Power.*
29 Vanderburg, ed., *Perspectives on Our Age*, chapter 4.
30 Jacques Ellul, *La pensée marxiste*, comp. and ed. Michel Hourcade, Jean-Pierre Jézéquel, and Gérard Paul (Paris: La Table Ronde, 2003).
31 Jacques Ellul, *The Humiliation of the Word*, trans. Joyce Main Hanks (Grand Rapids, MI: Eerdmans, 1985).
32 Vanderburg, *Our War on Ourselves.*
33 Ibid.
34 Adam Smith, *An Inquiry into the Nature and Causes of the Wealth of Nations*, 2 vols. (Chicago: University of Chicago Press, 1976).
35 Vanderburg, *Our War on Ourselves.*
36 Smith, *An Inquiry into the Nature and Causes of the Wealth of Nations.*
37 Vanderburg, *Our War on Ourselves.*
38 Ellul, *La pensée marxiste.*

39 John Kenneth Galbraith, *The New Industrial State* (New York: New American Library, 1985).
40 Vanderburg, *Living in the Labyrinth of Technology*.
41 Karl Polanyi, *The Great Transformation* (New York: Farrar and Rinehart, 1944).
42 Vanderburg, *Living in the Labyrinth of Technology*.
43 Ellul, *La pensée marxiste*.
44 Ibid.
45 Ellul, *The New Demons*.
46 Max Weber, *The Protestant Ethic and the Spirit of Capitalism*, trans. Talcott Parsons (New York: Charles Scribner, 1985).
47 Vanderburg, *Living in the Labyrinth of Technology*; Vanderburg, *Our War on Ourselves*.
48 Ibid.
49 Ibid.
50 S. Giedion, *Mechanization Takes Command* (New York: Oxford University Press, 1948); Jacques Ellul, "Remarks on Technology and Art," *Bulletin of Science, Technology and Society* 21, no. 1 (February 2001); Jacques Ellul, *L'empire du non-sens: l'art et la société technicienne* (Paris: Presses universitaires de France, 1980); Willem H. Vanderburg, "Comments on the Empire of Non-Sense: Art in a Technique-Dominated Society," *Bulletin of Science, Technology and Society* 21 (February 2001): 38–54.
51 Jacques Ellul, *Les successeurs de Marx*, comp. and ed. Michel Hourcade, Jean-Pierre Jezequel, and Gérard Paul (Paris: La Table Ronde, 2007).
52 Vanderburg, *Living in the Labyrinth of Technology*.
53 Ellul, *La pensée marxiste*.
54 Stuart Dreyfus and Hubert Dreyfus, with Tom Anathasiou, *Mind over Machine: The Power of Human Intuition and Expertise in the Era of the Computer* (New York: Free Press, 1986). See also Stuart E. Dreyfus, "System 0: The Overlooked Explanation of Expert Intuition," chapter 2 in *Handbook of Research Methods on Intuition*, ed. M. Sinclair (Cheltenham: Edward Elgar Publishers, 2014).
55 David Riesman, Nathan Glazer, and Reuel Denney, *The Lonely Crowd: A Study of the Changing American Character* (Garden City, NY: Doubleday Anchor, 1950), 507.
56 Vanderburg, *Living in the Labyrinth of Technology*. See especially chapters 4–6.
57 Vanderburg, *Our War on Ourselves*, 289–304.
58 Willem H. Vanderburg, "Placing Engineering and Other Professions under Public Oversight," *Bulletin of Science, Technology & Society* 32, no. 2

(March 2012): 171–80; Vanderburg, *The Labyrinth of Technology* (Toronto: University of Toronto Press, 2000).

59 Vanderburg, *Our War on Ourselves*, chapter 2.

60 H.H. Gerth and C. Wright Mills, eds., *From Max Weber: Essays in Sociology* (New York: Oxford University Press, 1963); see also Rogers Brubaker, *The Limits of Rationality: An Essay on the Social and Moral Thought of Max Weber* (London: Allen and Unwin, 1984).

61 Vanderburg, *Our War on Ourselves.*

62 Ibid.

63 Gerth and Mills, *From Max Weber*.

64 Ellul, *The Technological Society*; Jacques Ellul, *The Technological System*, trans. Joachim Neugroschel (New York: Continuum, 1980); Vanderburg, *Living in the Labyrinth of Technology*, chapter 10.

65 Ellul, *The Technological Society*.

66 Ellul, *The New Demons*.

67 Georges Devereux, *From Anxiety to Method in the Behavioral Sciences* (New York: Humanities Press, 1967).

68 Vanderburg, *Living in the Labyrinth of Technology*.

69 Ellul, *The Technological Society*.

70 Vanderburg, *Our War on Ourselves*, chapter 4..

71 Ibid.

1. The Cult of the Fact: What Discipline-Based Science Will Never Know

1 Benson Snyder, "Literacy and Numeracy: Two Ways of Knowing," *Daedalus* 119 (1990): 233–56.

2 Willem H. Vanderburg, *The Labyrinth of Technology* (Toronto: University of Toronto Press, 2000), chapter 4.

3 Stuart Dreyfus and Hubert Dreyfus, with Tom Athanasiou, *Mind over Machine: The Power of Human Intuition and Expertise in the Era of the Computer* (New York: Free Press, 1986).

4 Schulamit Reinharz, *On Becoming a Social Scientist: From Survey Research and Participant Observation to Experimental Analysis* (New Brunswick, NJ: Transaction Books, 1984).

5 Jacques Ellul, *The Technological System*, trans. Joachim Neugroschel (New York: Continuum, 1980), 16; Willem H. Vanderburg, *Living in the Labyrinth of Technology* (Toronto: University of Toronto Press, 2005), chapter 9.

6 Willem H. Vanderburg, *Our War on Ourselves* (Toronto: University of Toronto Press, 2011). See chapter 1.

7 Ludwig Wittgenstein, *Philosophical Investigations*, trans. G.E.M. Anscombe (Oxford: Basil Blackwell, 1967).
8 Vanderburg, *Living in the Labyrinth of Technology*.
9 John Kenneth Galbraith, *The New Industrial State* (New York: New American Library, 1985).
10 Jacques Ellul, *The Technological Society*, trans. John Wilkinson (New York: Vintage Books, 1964); Radovan Richta, *Civilization at the Crossroads: Social and Human Implications of the Scientific and Technological Revolution* (Prague: International Arts and Sciences Press, 1969).
11 Herman Daly, *Steady-State Economics* (Washington, DC: Island Press, 1991). See especially chapters 12 and 13.
12 B. Lietaer, *The Future of Money: A New Way to Create Wealth, Work and a Wiser World* (New York: Random House, 2001); see appendix.
13 Vanderburg, *Our War on Ourselves*; Philip Coggan, *Paper Promises: Money, Debt and the New World Order* (New York: Penguin Books, 2011).
14 Adam Smith, *An Inquiry into the Nature and Causes of the Wealth of Nations*, 2 vols. (Chicago: University of Chicago Press, 1976).
15 David C. Korten, *When Corporations Rule the World* (West Hartford, CT: Kumarian Press, 1995).
16 Galbraith, *The New Industrial State*.
17 Vanderburg, *Living in the Labyrinth of Technology*.
18 Galbraith, *The New Industrial State*.
19 Jacques Ellul, *Propaganda: The Formation of Men's Attitudes*, trans. Konrad Kellen and Jean Lerner (New York: Vintage Books, 1965).
20 Herman Daly and John B. Cobb, Jr, *For the Common Good: Redirecting the Economy toward Community, the Environment, and a Sustainable Future* (Boston: Beacon, 1989).
21 Ellul, *Propaganda*, chapters 2–4.
22 Vanderburg, *Our War on Ourselves*.
23 Ibid.
24 Ellul, *Propaganda*.
25 Vanderburg, *Our War on Ourselves*.
26 José Goldemburg, Thomas Johansson, Amulya Reddy, and Robert Williams, *Energy for a Sustainable World* (New York: Wiley, 1988).
27 Vanderburg, *The Labyrinth of Technology*.
28 Ellul, *The Technological System*.
29 Vanderburg, *The Labyrinth of Technology*.
30 Jacques Ellul, *The Technological Bluff*, trans. Geoffrey W. Bromiley (Grand Rapids, MI: Eerdmans, 1990).
31 Jacques Ellul, *Métamorphose du bourgeois* (Paris: Calmann-Lévy, 1967).

32 Raymond Williams, *Key Words: A Vocabulary of Culture and Society* (London: Fontana, 1983).
33 Ernst Cassirer, *An Essay on Man* (New Haven: Yale University Press, 1944); T.W. Deacon, *The Symbolic Species: The Co-evolution of Language and the Brain* (New York: W.W. Norton, 1998); Willem H. Vanderburg, *The Growth of Minds and Cultures: A Unified Theory of the Structure of Human Experience* (Toronto: University of Toronto Press, 1985); Norman Doidge, *The Brain That Changes Itself: Stories of Personal Triumph from the Frontiers of Brain Science* (London: Penguin Books, 2007).
34 Vanderburg, *The Growth of Minds and Cultures*. For updates, see Vanderburg, *Living in the Labyrinth of Technology*, chapter 2, and Vanderburg, *Our War on Ourselves*, chapter 1.
35 Vanderburg, *Our War on Ourselves*.
36 Joyce Main Hanks, *Jacques Ellul: An Annotated Bibliography of Primary Works* (Stamford, CT: Jai Press, 2000).
37 Vanderburg, *Living in the Labyrinth of Technology*; Vanderburg, *Our War on Ourselves*.
38 Georges Devereux, *From Anxiety to Method in the Behavioral Sciences* (New York: Humanities Press, 1967).
39 Vanderburg, *Living in the Labyrinth of Technology*.
40 Ibid.
41 Vanderburg, *Our War on Ourselves*, chapter 5.
42 Jacques Ellul, *La pensée marxiste*, comp. and ed. Michel Hourcade, Jean-Pierre Jézéquel, and Gérard Paul (Paris: La Table Ronde, 2003).
43 Jacques Ellul, *On Freedom, Love and Power*, comp., ed., and trans. Willem H. Vanderburg (Toronto: University of Toronto Press, 2010); Jacques Ellul, *On Being Rich and Poor: Christianity in an Age of Globalization* (Toronto: University of Toronto Press, 2014).
44 Ellul, *Propaganda*; Daniel J. Boorstin, "Statistical Morality," in *The Americans: The Democratic Experience* (New York: Vintage Books, 1974), 238–44; Daniel J. Boorstin, "How Opinion Went Public," in *Democracy and Its Discontents: Reflections on Everyday America* (New York: Random, 1974), 12–21; Edward S. Herman and Noam Chomsky, *Manufacturing Consent: The Political Economy of the Mass Media* (New York: Pantheon, 1988); Noam Chomsky, *Media Control: The Spectacular Achievements of Propaganda* (New York: Seven Stories, 1997).
45 Jacques Ellul, *The Ethics of Freedom*, trans. Geoffrey W. Bromiley (Grand Rapids, MI: Eerdmans, 1976).
46 Jacques Ellul, *Hope in Time of Abandonment*, trans. C. Edward Hopkin (New York: Seabury, 1973).

47 Lewis Mumford, *The Myth of the Machine: The Pentagon of Power* (New York: Harcourt, Brace, Jovanovich, 1970).
48 Vanderburg, *The Labyrinth of Technology*, chapter 10.
49 Vanderburg, *Our War on Ourselves.*
50 Robert J. Lifton, *History and Human Survival* (New York: Random House, 1961).
51 Devereux, *From Anxiety to Method in the Behavioral Sciences.*
52 Thomas S. Kuhn, *The Structure of Scientific Revolutions*, 2nd ed. (Chicago: University of Chicago Press, 1970).
53 Vanderburg, *The Growth of Minds and Cultures.*
54 Vanderburg, *Living in the Labyrinth of Technology*, part 2; Vanderburg, *Our War on Ourselves*, chapter 4.
55 Vanderburg, *Living in the Labyrinth of Technology*, part 2.
56 Ibid. See chapter 10; Vanderburg, *Our War on Ourselves*, chapter 4.
57 Vanderburg, *Our War on Ourselves.*

2. The Cult of Efficiency: What Technical Means Cannot Accomplish

1 Willem H. Vanderburg, *The Labyrinth of Technology* (Toronto: University of Toronto Press, 2000); see section 3.3.
2 Ibid.
3 Willem H. Vanderburg, "Placing Engineering and Other Professions under Public Oversight," *Bulletin of Science, Technology & Society* 32, no. 2 (March 2012): 171–80.
4 Willem H. Vanderburg, "The Hydrogen Economy as a Technological Bluff," *Bulletin of Science, Technology and Society* 26, no. 4 (August 2006): 299–302.
5 Vanderburg, *The Labyrinth of Technology*, 271–6.
6 Ibid., 289–90.
7 Willem H. Vanderburg, *Our War on Ourselves* (Toronto: University of Toronto Press, 2011).
8 Ricardo Semler, *Maverick: The Success Story behind the World's Most Unusual Workplace* (New York: Warner Books, 1993).
9 Donald Schön, *The Reflective Practitioner: How Professionals Think in Action* (New York: Basic, 1983).
10 Jacques Ellul, *The Technological Society*, trans. John Wilkinson (New York: Vintage Books, 1964); *The Technological System*, trans. Joachim Neugroschel (New York: Continuum, 1980).
11 Vanderburg, *Our War on Ourselves.*
12 Jane Jacobs, *The Death and Life of Great American Cities* (London: Pelican, 1961).

13 Ibid.
14 Ibid.
15 Vanderburg, *The Labyrinth of Technology*, 374–7.
16 Vanderburg, *Our War on Ourselves*, 289–304.
17 Vanderburg, *The Labyrinth of Technology*, 284–285.
18 Adam Smith, *An Inquiry into the Nature and Causes of the Wealth of Nations*, 2 vols. (Chicago: University of Chicago Press, 1976).
19 Willem H. Vanderburg, *Living in the Labyrinth of Technology* (Toronto: University of Toronto Press, 2005), 400–7.
20 Ibid.
21 Ibid.
22 Vanderburg, *Our War on Ourselves*.
23 Ibid.
24 Jacques Ellul, *The Humiliation of the Word*, trans. Joyce Main Hanks (Grand Rapids, MI: Eerdmans, 1985).
25 Georges Devereux, *From Anxiety to Method in the Behavioral Sciences* (New York: Humanities Press, 1967).
26 Ibid.
27 Vanderburg, *The Labyrinth of Technology*.
28 Stephen E. Ambrose, Robert W. Rand, and Carmen M.E. Krogh, "Wind Turbine Acoustic Investigation: Infrasound and Low-Frequency Noise – A Case Study," *Bulletin of Science, Technology & Society* 32, no. 2 (March 2012): 128–41. See also vol. 31, nos. 4 and 5, which deal with the same topic.
29 Russell L. Ackoff, "The Future of Operational Research Is Past," *Journal of the Operational Research Society* 30, no. 2 (1976).
30 Stuart Dreyfus and Hubert Dreyfus, with Tom Athanasiou, *Mind over Machine: The Power of Human Intuition and Expertise in the Era of the Computer* (New York: Free Press, 1986).
31 Vanderburg, *Our War on Ourselves*.
32 Ibid.
33 Ibid.
34 Schön, *The Reflective Practitioner*; Christopher Alexander, *The Nature of Order, Book 3: A Vision of a Living World* (New York: Basic Books, 1983).
35 E.P. Thompson, *The Making of the English Working Class* (New York: Random House, 1963).
36 Martin Shain, "Stress and Satisfaction," *Occupational Health and Safety Canada* (April–May 1999): 33–47; "The Fairness Connection," *Occupational Health and Safety Canada* (June 2000): 22–8.
37 N. Khan and W.H. Vanderburg, *Healthy Work: An Annotated Bibliography* (Lanham, MD: Scarecrow Press, 2004).

38 Thomas Engstrom, Jan A. Johansson, Dan Jonsson, and Lars Medbo, "Empirical Evaluation of the Reformed Assembly Work at the Volvo Uddevalla Plant: Psychosocial Effects and Production Responsibility," *International Journal of Industrial Ergonomics* 16, no. 5 (1995): 293–308; Thomas Engstrom and Lars Medbo, "Production Systems Design – A Brief Summary of Some Swedish Design Efforts," in *Enriching Production: Perspectives on Volvo's Uddevalla Plant as an Alternative to Lean Production*, ed. Ake Sandberg (Aldershot: Avebury, 1995), 65; Thomas Engstrom, Lars Medbo, and Dan Jonsson, "An Assembly-Oriented Product Description as a Precondition for Efficient Manufacturing with Long Cycle Time Work," in *Production and Quality Management Frontiers – IV*, ed. David J. Sumanth, Johnson A. Edosomwan, Robert Poupart, and D. Scott Sink (Norcross, GA: Industrial Engineering and Management Press, 1993), 460–1; Christian Berggren, "The Fate of Branch-Plants – Performance versus Power," in *Enriching Production*, ed. Sandberg, 107–8.

39 Robert Karasek and Tores Theorell, *Healthy Work: Stress, Productivity and the Reconstruction of Working Life* (New York: Basic, 1990).

40 Vanderburg, *Our War on Ourselves*, 256–60.

41 Semler, *Maverick*.

42 Based on the studies of the curricula of business administration and management schools, it is highly unlikely that these CEOs would have fully comprehended what was happening, given the strong likelihood of the previously mentioned counter-transference reactions.

43 Semler, *Maverick*.

44 James P. Womack, Daniel T. Jones, and Daniel Roos, *The Machine That Changed the World* (New York: Macmillan, 1990).

45 Ibid.

46 Tim Jackson, *Material Concerns: Pollution, Profit and Quality of Life* (New York: Routledge, 1996); Nina Nakajima and Willem H. Vanderburg, "A Failing Grade for the German End-of-Life Vehicles Take-Back System," *Bulletin of Science, Technology and Society* 25, no. 2 (April 2005): 170–86.

47 Vanderburg, *The Labyrinth of Technology*, 252–92.

48 Jackson, *Material Concerns*.

49 Nina Nakajima, "Exploring the Potential of Preventive Approaches to Converge Business Interests with Environmental Sustainability," doctoral thesis, Department of Mechanical Engineering, University of Toronto, 2003.

50 Jackson, *Material Concerns*.

51 Vanderburg, *The Labyrinth of Technology*.

52 Vanderburg, *Our War on Ourselves*.

53 Jeff Rubin, *Why Your World Is About to Get a Whole Lot Smaller: Oil and the End of Globalization* (New York: Random House, 2009).
54 Vanderburg, *The Labyrinth of Technology*, chapters 8–11.
55 Dreyfus et al., *Mind over Machine*.
56 Karasek and Theorell, *Healthy Work*.
57 Shain, "Stress and Satisfaction" and "The Fairness Connection."
58 Vanderburg, *The Labyrinth of Technology*, chapter 10.
59 John Kenneth Galbraith, *The New Industrial State* (New York: New American Library, 1985).
60 Ellul, *The Technological System*, chapter 2.
61 Ellul, *The Humiliation of the Word*.
62 Peter Brödner, *The Shape of Future Technology: The Anthropocentric Alternative* (New York: Springer-Verlag, 1990); Jacques Ellul, *Changer de révolution: l'inéluctable proletariat* (Paris: Editions de Seuil, 1982).
63 Vanderburg, *Our War on Ourselves*.
64 Ellul, *The Technological Society*.
65 Vanderburg, *Our War on Ourselves*.

3. The Cult of Growth: The Anti-Economy

1 Jacques Ellul, *The Technological Society*, trans. John Wilkinson (New York: Vintage Books, 1964). See also David S. Landers, *The Unbound Prometheus: Technological Change and Industrial Development in Western Europe from 1750 to the Present* (Cambridge: Cambridge University Press, 1969).
2 Willem H. Vanderburg, Living in the *Labyrinth of Technology* (Toronto: University of Toronto Press, 2005).
3 Sherry Turkle, *The Second Self: Computers and the Human Spirit* (New York: Simon and Schuster, 1984); Turkle, *Life on the Screen: Identity in the Age of the Internet* (New York and Toronto: Touchstone/Simon and Schuster, 1995); Turkle, *Alone Together: Why We Expect More from Technology and Less from Each Other* (New York: Basic Books, 2011).
4 Joseph Weizenbaum, *Computer Power and Human Reason* (San Francisco: H.H. Freeman, 1976).
5 Hubert Dreyfus, *On the Internet* (New York: Routledge, 2001). Although the second edition brings the book up to date regarding advances made in the design of search engines, the first edition has a more comprehensive discussion of the relevance problem, which is fundamental for understanding this work.
6 Hubert Dreyfus, *What Computers Still Can't Do: A Critique of Artificial Reason* (Cambridge, MA: MIT Press, 1992).

7 Craig Brod, *Technostress: The Human Cost of the Computer Revolution* (New York: Addison-Wesley, 1984).
8 Peter Brödner, *The Shape of Future Technology: The Anthropocentric Alternative* (Berlin: Springer-Verlag, 1990).
9 Shoshana Zuboff, *In the Age of the Smart Machine: The Future of Work and Power* (New York: Basic Books, 1988).
10 Stuart Dreyfus and Hubert Dreyfus, with Tom Anathasiou, *Mind over Machine: The Power of Human Intuition and Expertise in the Era of the Computer* (New York: Free Press, 1986). See also Stuart E. Dreyfus, "System 0: The Overlooked Explanation of Expert Intuition," chapter 2 in *Handbook of Research Methods on Intuition*, ed. M. Sinclair (Cheltenham: Edward Elgar Publishers, 2014).
11 Vanderburg, *Living in the Labyrinth of Technology*.
12 John Kenneth Galbraith, *The New Industrial State* (New York: New American Library, 1985).
13 Vanderburg, *Living in the Labyrinth of Technology*.
14 Ibid.
15 Jacques Ellul, *La pensée marxiste*, comp. and ed. Michel Hourcade, Jean-Pierre Jézéquel, and Gérard Paul (Paris: La Table Ronde, 2003).
16 Ibid.
17 Willem H. Vanderburg, *Our War on Ourselves* (Toronto: University of Toronto Press, 2011).
18 Willem H. Vanderburg, *The Labyrinth of Technology* (Toronto: University of Toronto Press, 2000).
19 Jacques Ellul, *Pour qui, pour quoi travaillons-nous?* comp. and ed. Michel Hourcade, Jean-Pierre Jézéquel, and Gérard Paul (Paris: La Table Ronde, 2013).
20 Ibid.
21 Jacques Ellul, *On Freedom, Love and Power*, comp., ed., and trans. Willem H. Vanderburg (Toronto: University of Toronto Press, 2010).
22 Ellul, *La pensée marxiste*.
23 Jacques Ellul, *Métamorphose du bourgeois* (Paris: Calmann-Lévy, 1967).
24 Ellul, *La pensée marxiste*.
25 Jacques Ellul, *Changer de révolution: l'inéluctable proletariat* (Paris: Editions du Seuil, 1982).
26 Max Weber, *The Protestant Ethic and the Spirit of Capitalism*, trans. Talcott Parsons (New York: Charles Scribner, 1985).
27 Ellul, *On Freedom, Love and Power*.
28 Ibid.
29 Vanderburg, *Our War on Ourselves*.

30 Karl Polanyi, *The Great Transformation* (New York: Farrar and Rinehart, 1944).
31 Vanderburg, *Living in the Labyrinth of Technology*.
32 Polanyi, *The Great Transformation*.
33 Vanderburg, *Living in the Labyrinth of Technology*.
34 Vanderburg, *Our War on Ourselves*.
35 Ibid.
36 Vanderburg, *Living in the Labyrinth of Technology*.
37 Jacques Ellul, *The New Demons*, trans. C. Edward Hopkin (New York: Seabury, 1975).
38 Ellul, *The Technological Society*.
39 Willem H. Vanderburg, *The Labyrinth of Technology* (Toronto: University of Toronto Press, 2000).
40 Robert Theobald, *The Economics of Abundance: A Non-Inflationary Future* (New York: New York Pitman, 1970).
41 Vanderburg, *Living in the Labyrinth of Technology; Our War on Ourselves*.
42 Ibid.
43 Ibid.
44 Ibid.
45 Herman Daly and John B. Cobb, Jr, *For the Common Good: Redirecting the Economy toward Community, the Environment, and a Sustainable Future* (Boston: Beacon, 1989).
46 John Maynard Keynes, *The General Theory of Employment, Interest and Money* (London: Macmillan and Co., 1936).
47 Theobald, *The Economics of Abundance*.
48 Vanderburg, *Our War on Ourselves*.
49 H.H. Gerth and C. Wright Mills, eds., *From Max Weber: Essays in Sociology* (New York: Oxford University Press, 1963); see also Rogers Brubaker, *The Limits of Rationality: An Essay on the Social and Moral Thought of Max Weber* (London: Allen and Unwin, 1984).
50 Jacques Ellul, *The Technological Society*, trans. John Wilkinson (New York: Vintage Books, 1964).
51 Vanderburg, *Living in the Labyrinth of Technology; Our War on Ourselves*.
52 Ibid.
53 Galbraith, *The New Industrial State*.
54 Radovan Richta, *Civilization at the Crossroads: Social and Human Implications of the Scientific and Technological Revolution* (Prague: International Arts and Sciences Press, 1969).
55 Vanderburg, *The Labyrinth of Technology*, chapter 7.
56 Daly and Cobb, *For the Common Good*.
57 Vanderburg, *Our War on Ourselves*.

58 Herman Daly, *Beyond Growth: The Economics of Sustainable Development* (Boston: Beacon, 1996).
59 B. Lietaer, *The Future of Money: A New Way to Create Wealth, Work and a Wiser World* (New York: Random House, 2001).
60 Vanderburg, *Living in the Labyrinth of Technology*.
61 Philip Coggan, *Paper Promises: Money, Debt and the New World Order* (New York: Penguin Books, 2011).
62 Lietaer, *The Future of Money*.
63 The equivalent of a Tobin tax (named after the Nobel laureate who proposed it) would be a simple means for vastly reducing speculative bubbles by levying a modest tax, with the income being used to benefit the public good.
64 Jacques Ellul, *On Being Rich and Poor: Christianity in a Time of Economic Globalization*, ed. and trans. Willem H. Vanderburg (Toronto: Toronto University Press, 2014).
65 Ibid.
66 Ibid.
67 Galbraith, *The New Industrial State*.
68 Vanderburg, *Living in the Labyrinth of Technology*.
69 Ibid.
70 Ibid.
71 Ibid.
72 Ibid.
73 Carolyn Marvin, *When Old Technologies Were New: Thinking about Communication in the Late Nineteenth Century* (New York: Oxford University Press, 1988).
74 Jacques Ellul, *Propaganda: The Formation of Men's Attitudes*, trans. Konrad Kellen and Jean Lerner (New York: Vintage Books, 1965).
75 Galbraith, *The New Industrial State*; Vanderburg, *Living in the Labyrinth of Technology*.
76 See note 63.
77 Dwight D. Eisenhower, *The Military-Industrial Complex*, Public Papers of the Presidents, Dwight D. Eisenhower, 1960, 1035–40. http://coursesa.matrix.msu.edu/~hst306/documents/indust.html.
78 Galbraith, *The New Industrial State*.
79 Nick Turse, *The Complex: How the Military Invades Our Everyday Lives* (New York: Metropolitan Books, 2008); Joseph E. Stiglity, *The Three Trillion Dollar War: The True Cost of the Iraq Conflict* (New York: W.W. Norton, 2008); Wendy Kaminer, *It's All the Rage: Crime and Culture* (Reading, MA: Addison-Wesley, 1995).

80 Braden R. Allenby and Deanna J. Richards, eds., *The Screening of Industrial Ecosystems* (Washington, DC: National Academy Press, 1994), Introduction.
81 Vanderburg, *Our War on Ourselves*.
82 Vanderburg, *The Labyrinth of Technology*, chapter 4.
83 For an overview of this research, see Vanderburg, *Living in the Labyrinth of Technology*, chapter 9.
84 Robert Karasek and Töres Theorell, *Healthy Work: Stress, Productivity and the Reconstruction of Working Life* (New York: Basic Books, 1990).
85 Vanderburg, *The Labyrinth of Technology*, chapter 10.
86 Vanderburg, *Our War on Ourselves*.
87 Ibid.
88 Vanderburg, *The Labyrinth of Technology*. See also A. Jürgensen, N. Khan, and W.H. Vanderburg, *Sustainable Production: An Annotated Bibliography* (Lanham, MD: Scarecrow Press, 2001); N. Khan and W.H. Vanderburg, *Sustainable Energy: An Annotated Bibliography* (Lanham, MD: Scarecrow Press, 2001); N. Khan and W.H. Vanderburg, *Healthy Cities: An Annotated Bibliography* (Lanham, MD: Scarecrow Press, 2001); N. Khan and W.H. Vanderburg, *Healthy Work: An Annotated Bibliography* (Lanham, MD: Scarecrow Press, 2004).
89 Vanderburg, *Our War on Ourselves*.
90 Didier Nordon, *Les mathématiques pures n'existent pas* (Paris: Éditions Actes Sud, 1981).
91 Vanderburg, *The Labyrinth of Technology*, chapter 3.
92 Vanderburg, *Our War on Ourselves*, chapters 4 and 5.
93 Karasek and Theorell, *Healthy Work*.
94 Vanderburg, *The Labyrinth of Technology*, chapters 8–11.
95 Ricardo Semler, *Maverick: The Success Story behind the World's Most Unusual Workplace* (New York: Warner, 1993).
96 For a discussion of the Uddevalla plant, see chapter 2.
97 Jeff Rubin, *Why Your World Is About to Get a Whole Lot Smaller: Oil and the End of Globalization* (New York: Random House, 2009).
98 Joel Bakan, *The Corporation: The Pathological Pursuit of Profit and Power* (Toronto: Penguin Canada, 2004).
99 T.E. Graedel and B.R. Allenby, *Industrial Ecology* (Englewood Cliffs, NJ: Prentice-Hall, 1995).
100 Nina Nakajima and Willem H. Vanderburg, "A Description and Analysis of the German Packaging Take-Back System," *Bulletin of Science, Technology and Society* 26, no. 4 (October 2006): 438–46; "A Failing Grade for Our Efforts to Make Our Civilization More Environmentally

Sustainable," *Bulletin of Science, Technology and Society* 25, no. 2 (April 2005): 129–44; "A Failing Grade for the German End-of-Life Vehicles Take-Back System," *Bulletin of Science, Technology and Society* 25, no. 2 (April 2005): 170–86; "A Failing Grade for WEEE Take-Back Programs for Information Technology Equipment," *Bulletin of Science, Technology and Society* 25, no. 6 (December 2005): 507–17.

101 Vanderburg, *The Labyrinth of Technology*, chapter 8.

102 Graedel and Allenby, *Industrial Ecology*.

103 Vanderburg, *The Labyrinth of Technology*, chapter 8.

104 Ibid., chapters 8–11.

105 Jacques Ellul, *The Political Illusion*, trans. Konrad Kellen (New York: Knopf, 1967).

106 Vanderburg, *The Labyrinth of Technology*, chapter 9.

107 Willem H. Vanderburg, "The Hydrogen Economy as a Technological Bluff," *Bulletin of Science, Technology and Society* 26, no. 4 (August 2006): 299–302.

108 José Goldemberg, Thomas Johansson, Amulya Reddy, and Robert Williams, *Energy for a Sustainable World* (New York: Wiley, 1988).

109 Ibid.

110 Vanderburg, *Living in the Labyrinth of Technology*, 357–61.

111 Ibid.

112 Vanderburg, *The Labyrinth of Technology*, chapter 11.

4. The Cult of Disembodied Communal Life: The Anti-Society

1 Karen Horney, *The Neurotic Personality of Our Time* (New York: W.W. Norton, 1937).

2 Robert J. Lifton, *The Broken Connection* (New York: Basic Books, 1983).

3 Willem H. Vanderburg, *The Growth of Minds and Cultures: A Unified Theory of the Structure of Human Experience* (Toronto: University of Toronto Press, 1985).

4 Ernst Cassirer, *An Essay on Man* (New Haven: Yale University Press, 1944).

5 Lifton, *The Broken Connection*.

6 Howard Gardner, *The Mind's New Science: A History of the Cognitive Revolution* (New York: Basic Books, 1985).

7 T.W. Deacon, *The Symbolic Species: The Co-evolution of Language and the Brain* (New York: W.W. Norton, 1998); Vanderburg, *The Growth of Minds and Cultures*.

8 David Bohm, *Wholeness and the Implicate Order* (London: Routledge and Kegan Paul, 1980); Bernard d'Espagnat, *In Search of Reality* (New York: Springer-Verlag, 1983).

9 Vanderburg, *The Growth of Minds and Cultures*; Willem H. Vanderburg, *Our War on Ourselves* (Toronto: University of Toronto Press, 2011). See chapter 1.
10 Jacques Ellul, *The Humiliation of the Word*, trans. Joyce Main Hanks (Grand Rapids, MI: Eerdmans, 1985).
11 Stuart Dreyfus and Hubert Dreyfus, with Tom Anathasiou, *Mind over Machine: The Power of Human Intuition and Expertise in the Era of the Computer* (New York: Free Press, 1986).
12 Michael Polanyi, *Personal Knowledge* (Chicago: University of Chicago Press, 1962).
13 Carl G. Jung, *Psychological Types*, trans H.G. Baynes (New York: Pantheon, 1964).
14 Horney, *The Neurotic Personality of Our Time*.
15 Ibid.
16 Ibid.
17 Vanderburg, *Our War on Ourselves*, chapter 2.
18 Ibid.
19 Ibid.
20 Ibid.
21 Ibid.
22 Ibid. See chapters 3 and 4.
23 Ibid. See chapter 3.
24 Ibid. See chapters 3 and 4.
25 Willem H. Vanderburg, *Living in the Labyrinth of Technology* (Toronto: University of Toronto Press, 2005), chapter 2.
26 Willem H. Vanderburg, *The Labyrinth of Technology* (Toronto: University of Toronto Press, 2000).
27 Jacques Ellul, *La pensée marxiste*, comp. and ed. Michel Hourcade, Jean-Pierre Jézéquel, and Gérard Paul (Paris: La Table Ronde, 2003).
28 Vanderburg, *Living in the Labyrinth of Technology*.
29 Ellul, *The Humiliation of the Word*.
30 Max Weber, *The Theory of Social and Economic Organization*, ed. Talcott Parsons (New York: Oxford University Press, 1947); Rogers Brubaker, *The Limits of Rationality: An Essay on the Social and Moral Thought of Max Weber* (London: Allen and Unwin, 1984).
31 Vanderburg, *Our War on Ourselves*, chapter 4.
32 H.H. Gerth and C. Wright Mills, eds., *From Max Weber: Essays in Sociology* (New York: Oxford University Press, 1963).
33 David S. Landers, *The Unbound Prometheus: Technological Change and Industrial Development in Western Europe from 1750 to the Present* (Cambridge: Cambridge University Press, 1969).

34 Jacques Ellul, *The Technological Society*, trans. John Wilkinson (New York: Vintage Books, 1964).

35 Jacques Ellul, *The Technological System*, trans. Joachim Neugroschel (New York: Continuum, 1980); Vanderburg, *Living in the Labyrinth of Technology*.

36 Vanderburg, *Our War on Ourselves*.

37 Ibid.

38 Ibid. See also Jacques Ellul, *The Empire of Non-sense*, trans. Michael Johnson and David Lovekin (Winterbourne, Berkshire: Papadakis Publisher, 2014).

39 Jacques Ellul, *The Political Illusion*, trans. Konrad Kellen (New York: Alfred E. Knopf, 1967).

40 Jacques Ellul, *Propaganda: The Formation of Men's Attitudes*, trans. Konrad Kellen and Jean Lerner (New York: Vintage Books, 1965); *The New Demons*, trans. C. Edward Hopkin (New York: Seabury, 1975).

41 David Hounshell, *From the American System to Mass Production, 1800-1932* (Baltimore: Johns Hopkins University Press, 1984).

42 Ellul, *The Technological Society*, chapter 5.

43 Tim Blackmore, *War X: Human Extensions in Battlespace* (Toronto: University of Toronto Press, 2005).

44 John Kenneth Galbraith, *The New Industrial State* (New York: New American Library, 1985).

45 Raymond Aron, *Progress and Disillusion: The Dialectics of Modern Society* (New York: Praeger, 1986).

46 Paul Goodman, *Growing Up Absurd: Problems of Youth in the Organized System* (New York: Random House, 1956).

47 Ellul, *Propaganda*.

48 Ibid.

49 Bohm, *Wholeness and the Implicate Order*; d'Espagnat, *In Search of Reality*.

50 Ellul, *Propaganda*.

51 Vanderburg, *Living in the Labyrinth of Technology*; *Our War on Ourselves*.

52 Vanderburg, *Our War on Ourselves*, chapter 2.

53 Ellul, *The New Demons*; Vanderburg, *Living in the Labyrinth of Technology*; Vanderburg, *Our War on Ourselves*.

54 Vanderburg, *Living in the Labyrinth of Technology*.

55 Arnold Toynbee, *A Study of History* (New York: Oxford University Press, 1933–54). For an abridged version, the reader is referred to D.C. Somervell, ed., *A Study of History* (New York: Dell, 1978). See also Joseph E. Tainter, *The Collapse of Complex Societies* (Cambridge: Cambridge University Press, 1988).

56 Ellul, *Propaganda*.
57 Ellul, *The Humiliation of the Word*.
58 Ellul, *The Empire of Non-sense*.
59 Stephen Bayley, *Sex, Drink and Fast Cars* (New York: Pantheon, 1986); Stuart and Elizabeth Ewen, *Channels of Desire: Mass Images and the Shaping of American Consciousness* (Minneapolis: University of Minnesota Press, 1992); Stuart Ewen, *Captains of Consciousness: Advertising and the Social Roots of the Consumer Culture* (New York: McGraw-Hill, 1976).
60 Ellul, *Propaganda*.
61 Ellul, *The Humiliation of the Word*; Uwe Poerksen, *Plastic Words: The Tyranny of a Modular Language* (University Park: Pennsyvania State University Press, 1995).
62 Vanderburg, *Our War on Ourselves*, chapter 2.
63 Joel Bakan, *The Corporation: The Pathological Pursuit of Profit and Power* (Toronto: Penguin Canada, 2004).
64 The following analysis is extensively based on Jacques Ellul, "De la signification des relations publiques dans une société technicienne," *L'année sociologique* 13 (1963): 69–152.
65 Richard Stivers, *The Illusion of Freedom and Equality* (Albany: State University of New York Press, 2008).
66 Ellul, *Propaganda*.
67 Vanderburg, *Our War on Ourselves*, chapter 5.
68 This analysis continues to be based on Ellul's "De la signification des relations publiques dans une société technicienne."
69 Ibid.
70 Ellul, *The Political Illusion*; *Propaganda*.
71 Ellul, *Propaganda*.
72 Vanderburg, *Living in the Labyrinth of Technology*.
73 Ellul, *The Political Illusion*.
74 Richard Stivers, *Shades of Loneliness: Pathologies of a Technological Society* (New York: Rowman and Littlefield, 2004).
75 Ibid.
76 Herman E. Daly and John B. Cobb, Jr, *For the Common Good: Redirecting the Economy toward Community, the Environment, and a Sustainable Future* (Boston: Beacon Press, 1989).
77 Eric Schlosser, *Fast Food Nation* (New York: Houghton Mifflin, 2002).
78 Jean M. Twenge and W. Keith Campbell, *The Narcissism Epidemic: Living in the Age of Entitlement* (New York: Free Press, 2009).
79 Stivers, *Shades of Loneliness*.

80 Richard Wilkinson and Kate Pickett, *The Spirit Level: Why More Equal Societies Almost Always Do Better* (London: Penguin Books, 2009)
81 Lifton, *The Broken Connection.*
82 Ibid.
83 Wilkinson and Pickett, *The Spirit Level.*
84 Ibid.
85 Ibid.
86 Jacques Ellul, *On Being Rich and Poor: Christianity in a Time of Economic Globalization*, comp., ed., and trans. Willem H. Vanderburg (Toronto: University of Toronto Press, 2014).
87 Wilkinson and Pickett, *The Spirit Level.*
88 Ibid.
89 Ellul, *On Being Rich and Poor.*
90 Bernard Lietaer, *The Future of Money: A New Way to Create Wealth, Work, and a Wiser World* (New York: Random House, 2001).
91 Tim Jackson, *Material Concerns: Pollution, Profit and Quality of Life* (New York: Routledge, 1996).
92 Herman Daly, *Beyond Growth: The Economics of Sustainable Development* (Boston: Beacon Press, 1996). See part 6.
93 Lietaer, *The Future of Money.*
94 Ibid.
95 Ibid.
96 David K. Johnston, *Free Lunch: How the Wealthiest Americans Enrich Themselves at Government Expense and Stick You with the Bill* (New York: Penguin Group, 2007).
97 Ellul, *On Being Rich and Poor.*
98 Lietaer, *The Future of Money.*
99 Ibid.
100 Ibid.
101 Ibid.
102 Ibid.
103 Bernard Lietaer and Jacqui Dunne, *Rethinking Money: How New Currencies Turn Scarcity into Prosperity* (San Francisco: Berrett-Koehler, 2013).
104 Ibid.
105 Paul Hawken, *Blessed Unrest* (New York: Penguin Group, 2007).
106 Lietaer and Dunne, *Rethinking Money.*
107 Ibid.
108 Ibid.
109 Ibid.
110 Ibid.
111 Ibid.

5. The Cult of Disembodied Personal Life: Epidemics of Anxiety and Depression

1 J.H. Van den Berg, *Divided Existence and Complex Society: An Historical Approach* (Pittsburgh: Duquesne University Press; distributed New York: Humanities Press, 1974).
2 Ibid.
3 J.H. Van den Berg, *The Changing Nature of Man* (New York: Delta, 1961).
4 Jacques Ellul, *The Humiliation of the Word*, trans. Joyce Main Hanks (Grand Rapids, MI: Eerdmans, 1985).
5 The following is based on J.H. Van den Berg, *Divided Existence and Complex Society*.
6 Willem H. Vanderburg, *The Growth of Minds and Cultures: A Unified Theory of the Structure of Human Experience* (Toronto: University of Toronto Press, 1985).
7 Georges Devereux, *From Anxiety to Method in the Behavioral Sciences* (New York: Humanities Press, 1967).
8 Van den Berg, *Divided Existence and Complex Society*.
9 Willem H. Vanderburg, *Living in the Labyrinth of Technology* (Toronto: University of Toronto Press, 2005). See part 2.
10 Ibid.
11 Ibid.
12 Ibid.
13 Jacques Ellul, *The New Demons*, trans. C. Edward Hopkin (New York: Seabury, 1975); Vanderburg, *Living in the Labyrinth of Technology*.
14 Willem H. Vanderburg, *Our War on Ourselves* (Toronto: University of Toronto Press, 2011).
15 William R. Arney, *Experts in the Age of Systems* (Albuquerque: University of New Mexico Press, 1991).
16 Vanderburg, *Our War on Ourselves*.
17 Robert J. Lifton, *The Protean Self: Human Resiliency in an Age of Fragmentation* (New York: Basic Books, 1993).
18 Van den Berg, *Divided Existence and Complex Society*.
19 George Herbert Mead, *Mind, Self and Society* (Chicago: University of Chicago Press, 1934).
20 Van den Berg, *Divided Existence and Complex Society*.
21 Vanderburg, *Our War on Ourselves*, chapter 2.
22 Jacques Ellul, *Propaganda: The Formation of Men's Attitudes*, trans. Konrad Kellen and Jean Lerner (New York: Vintage Books, 1965). Ellul's interpretation of human life in a mass society is at the heart of my

disagreements with post-modernist views regarding the way audiences relate to television as well as to other technologies. These include the *reception theory* of Wolfgang Iser, the *reader response theory* of Stanley Fish, and what Michel de Certeau and John Fiske referred to as *resistant activity*. Although people continue to form their own meanings, they now do so within the confines of a highly desymbolized mass society, and this changes everything.

23 Vanderburg, *The Growth of Minds and Cultures; Our War on Ourselves*, chapter 1.
24 Vanderburg, *Living in the Labyrinth of Technology*, chapter 9.
25 Simon Head, *Mindless: Why Smarter Machines Are Making Dumber Humans* (New York: Basic Books, 2014).
26 Eli Pariser, *In the Filter Bubble: What the Internet Is Hiding from You* (New York: Penguin Press, 2011).
27 Vanderburg, *Our War on Ourselves*, chapter 1.
28 It is important to go back to a time when people could still differentiate between life with and without computers. See, for example, Hubert Dreyfus, *On the Internet* (New York: Routledge, 2001); Sherry Turkle, *The Second Self: Computers and the Human Spirit* (New York: Simon and Schuster, 1984); and Craig Brod, *Technostress: The Human Cost of the Computer Revolution* (New York: Addison-Wesley, 1984). These kinds of studies are now difficult if not impossible to carry out.
29 Nicholas Carr, *The Shallows* (New York: W.W. Norton, 2010).
30 Vanderburg, *Living in the Labyrinth of Technology*, part 2.
31 Dreyfus, *On the Internet*.
32 Carr, *The Shallows*.
33 Ibid.
34 Jeffrey Freed and Laurie Parsons, *Right-Brained Children in a Left-Brained World: Unlocking the Potential of Your ADD Child* (New York: Simon and Schuster, 1997).
35 Vanderburg, *Our War on Ourselves*.
36 Freed and Parsons, *Right-Brained Children in a Left-Brained World*.
37 Jacques Ellul, *L'empire du non-sens: l'art et la société technicienne* (Paris: Presses Universitaires de France, 1980).
38 Benson Snyder, "Literacy and Numeracy: Two Ways of Knowing," *Daedalus* 119 (1990): 233–56.
39 Sheila Tobias, *They're Not Dumb, They're Different: Stalking the Second Tier* (Tucson: Research Corporation Foundation for the Advancement of Science, 1990).
40 Vanderburg, *Our War on Ourselves*.

41 Snyder, "Literacy and Numeracy."
42 Laurie Zoloth, "Care of the Dying in America," in *Prozac as a Way of Life*, ed. Carl Elliott and Todd Chambers (Chapel Hill: University of North Carolina Press, 2004).
43 Larry D. Rosen et al., *Idisorder: Understanding Our Obsession with Technology and Overcoming Its Hold on Us* (New York: St Martins Press, 2012).
44 Allan Francis, *Saving Normal: An Insider's Revolt against Out-of-Control Psychiatric Diagnosis DSM-5, Big Pharma, and the Medicalization of Ordinary Life* (New York: William Morrow, 2013); Gary Greenberg, *The Book of Woe: The DSM and the Unmaking of Psychiatry* (London: Penguin Books, 2013); Robert Whitaker, *Anatomy of an Epidemic: Magic Bullets, Psychiatric Drugs, and the Astonishing Rise of Mental Illness in America* (New York: Crown Publishers, 2010).
45 Carl Elliott, "Pursued by Happiness and Beaten Senseless," in *Prozac as a Way of Life*, ed. Carl Elliott and Todd Chambers (Chapel Hill: University of North Carolina Press, 2004).
46 Ibid.
47 Ellul, *The New Demons*; Vanderburg, *The Growth of Minds and Cultures*.
48 Elliott, "Pursued by Happiness and Beaten Senseless."
49 Carl Elliott, *Better Than Well* (New York: W.W. Norton, 2003).
50 Ibid.
51 Jacques Ellul, *The Technological System*, trans. Joachim Neugroschel (New York: Continuum, 1980); Vanderburg, *Living in the Labyrinth of Technology*.
52 Ellul, *The Technological System*.
53 Ellul, *The New Demons*; Richard Stivers, *Evil in Modern Myth and Ritual* (Athens: University of Georgia Press, 1982).
54 Ibid.
55 Elliott, *Better Than Well*.
56 Ibid.
57 Ibid.
58 Ibid.
59 Ibid.
60 Joseph Weizenbaum, *Computer Power and Human Reason* (San Francisco: H.H. Freeman, 1976).
61 Van den Berg, *Divided Existence and Complex Society*.
62 Ibid.
63 Jacques Ellul, *Histoire des institutions: l'antiquité* (Paris: Press Universitaires de France, 1961).

64 I began my study of the relationship between technique and culture in *The Growth of Minds and Cultures*. It was followed by an examination of the reliance of engineering on discipline-based approaches and how the current approaches could be resymbolized (*The Labyrinth of Technology*). This work was followed by an examination of human life in contemporary mass societies (*Living in the Labyrinth of Technology*), and then by the first study of the possibilities of resymbolizing our universities (*Our War on Ourselves*). The present work completes my series on the relationship between technique and culture.

Index